Statistics

with

STATA

Updated for Version 10

Statistics
with
STATA

Updated for Version 10

Lawrence C. Hamilton
University of New Hampshire

BROOKS/COLE
CENGAGE Learning™

Australia · Brazil · Japan · Korea · Mexico · Singapore · Spain · United Kingdom · United States

BROOKS/COLE
CENGAGE Learning™

Statistics with Stata: Updated for Version 10
Lawrence C. Hamilton

Acquisitions Editor: Molly Taylor
Assistant Editor: Catie Ronquillo
Editorial Assistant: Rebecca Dashiell
Media Editor: Catie Ronquillo
Marketing Manager: Greta Kleinert
Marketing Communications Manager: Mary Anne Payumo
Project Manager, Editorial Production: Jennifer Risden
Creative Director: Rob Hugel
Art Director: Vernon Boes
Print Buyer: Paula Vang
Permissions Editor: Roberta Broyer
Cover Designer: Roger Knox
Cover Image: Chad Baker/Getty Images

For product information and technology assistance, contact us at
**Cengage Learning Customer & Sales Support,
1-800-354-9706**

For permission to use material from this text or product, submit all requests online at
www.cengage.com/permissions
Further permissions questions can be emailed to
permissionrequest@cengage.com

Library of Congress Control Number: 2008933832
ISBN-13: 978-0-495-55786-9
ISBN-10: 0-495-55786-2

Brooks/Cole
10 Davis Drive
Belmont, CA 94002-3098
USA

Cengage Learning is a leading provider of customized learning solutions with office locations around the globe, including Singapore, the United Kingdom, Australia, Mexico, Brazil, and Japan. Locate your local office at:
international.cengage.com/region

Cengage Learning products are represented in Canada by Nelson Education, Ltd.

For your course and learning solutions, visit
www.cengage.com

Purchase any of our products at your local college store or at our preferred online store **www.ichapters.com**

Printed in Canada
1 2 3 4 5 6 7 12 11 10 09 08

Contents

Preface

Statistics with Stata is intended for students and practicing researchers, to bridge the gap between statistical textbooks and Stata's own documentation. In this intermediate role, it does not provide the detailed expositions of a proper textbook, nor does it come close to describing all of Stata's features. Instead, it demonstrates how to use Stata to accomplish a wide variety of statistical tasks. Chapter topics follow conceptual themes rather than focusing on particular Stata commands, which gives *Statistics with Stata* a different structure from the Stata reference manuals. The chapter on Data Management, for example, covers a variety of procedures for creating, updating, and restructuring data files. Chapters on Summary Statistics and Tables, ANOVA and Other Comparison Methods, and Fitting Curves, among others, have similarly broad themes that encompass a number of separate techniques.

The general topics of the first six chapters (through ordinary least squares regression) roughly parallel an introductory course in applied statistics, but with additional depth to cover practical issues often encountered by analysts — how to aggregate data, create dummy variables, draw publication-quality graphs, test for normality, or translate ANOVA into regression, for instance. In Chapter 7 (Regression Diagnostics) and beyond, we move into the territory of advanced courses or original research. Here, readers can find basic information and illustrations of how to obtain and interpret diagnostic statistics and graphs; perform robust, quantile, nonlinear, logit, ordered logit, multinomial logit, or Poisson regression; fit survival-time and event-count models; construct composite variables through factor analysis and principal components; divide observations into empirical types or clusters; analyze time series and complex survey data; or fit multilevel and mixed models. Stata has worked hard in recent years to advance its state-of-the-art standing, and this effort is particularly apparent in the wide range of regression and modeling commands it now offers.

Finally, we conclude with a look at programming in Stata. Many readers will find that Stata does everything they need already, so they have no reason to write original programs. For an active minority, however, programmability is one of Stata's principal attractions, and it underlies Stata's currency and rapid advancement. This chapter opens the door for new users to explore Stata programming, whether for specialized data management tasks, to establish a new statistical capability, for Monte Carlo experiments, or for teaching.

Generally similar versions ("flavors") of Stata run on Windows, Macintosh, and Unix computers. Across all platforms, Stata uses the same commands and produces the same output. Datasets, graphs, and programs created on one platform can be used by Stata running on any other platform. The versions differ in some details of screen appearance, menus, and file handling, where Stata follows the conventions native to each platform — such as \directory\filename file specifications under Windows, in contrast with the /directory/filename specifications under Unix. Rather than display all three, I employ Windows conventions, but users with other systems should find that only minor translations are needed.

Notes on the Seventh Edition

I began using Stata in 1985, the first year of its release. (Stata's 20th anniversary in 2005 was marked by a special issue of the *Stata Journal*, filled with historical articles and interviews including a brief history of *Statistics with Stata*.) Initially, Stata ran only on MS-DOS personal computers, but its desktop orientation made it distinctly more modern than its main competitors — most of which had originated before the desktop revolution, in the 80-column punched-card Fortran environment of mainframes. Unlike mainframe statistical packages that believed each user was a stack of cards, Stata viewed the user as a conversation. Its interactive nature and integration of statistical procedures with data management and graphics supported the natural flow of analytical thought in ways that other programs did not. **graph** and **predict** soon became favorite commands. I was impressed enough by how it all fit together to start writing the original *Statistics with Stata*, published in 1989 for Stata version 2.

A great deal about Stata has changed since that book, in which I observed that "Stata is not a do-everything program . . . . The things it does, however, it does very well." The expansion of Stata's capabilities has been striking. This is very noticeable in the proliferation, and later in the steady rationalization, of model fitting procedures. William Gould's architecture for Stata, with its programming tools and unified syntax, has aged well and smoothly incorporated new statistical methods as these were developed. The broad range of graphs in this book's Chapter 3, the formidable list of modeling commands that begins Chapter 10, or the newest time series, survey, and mixed-modeling capabilities introduced in Chapters 13, 14, and 15 illustrate some of the ways that Stata became richer over the years. Suites of new techniques such as those for panel (**xt**), survey (**svy**), time series (**ts**), or survival time (**st**) data open worlds of possibility, as do programmable commands for nonlinear regression (**nl**) and generalized linear modeling (**glm**), or general procedures for maximum-likelihood estimation. Other major extensions include the development of a matrix programming capability, and the wealth of new data-management features. Data management, with good reason, has been promoted from an incidental topic in the first *Statistics with Stata* to the second-longest chapter in this seventh edition.

Stata version 8 marked a radical upgrade in Stata's history, led by the new menu system or GUI (graphical user interface), and completely redesigned graphing capabilities. A limited menu system, evolved from the student program StataQuest, had been available as an option since version 4, but Stata 8 for the first time incorporated an integrated menu interface offering a full range of alternatives to typed commands. These menus are more easily learned through exploration than by reading a book, so *Statistics with Stata* provides only general suggestions about menus at the beginning of each chapter. For the most part, this book employs commands to show what Stata can do; those commands' menu counterparts should be easy to discover. Conversely, if you start out working mainly through menus, Stata provides informal training by showing each corresponding command in the Results window. The menu system works by translating user clicks and choices into Stata commands, which it then feeds to Stata for execution.

The redesigned graphing capabilities of Stata called for similarly sweeping changes in Chapter 3, turning it into the longest chapter in recent editions. The topic itself is complex, as the thick *Graphics Reference Manual* (and other material scattered through the documentation) attests. Rather than try to condense the syntax-based reference manuals, I have taken a different and complementary approach based on examples. Chapter 3 thus provides an organized gallery of 58 graphs, from basic to creative, each with instructions for how it was drawn. Further examples appear throughout the book; even graphs in Chapter 16 demonstrate new variations. To an unexpected degree, *Statistics with Stata* has became a showcase for using graphs as an integral part of research.

Statistics with Stata version 10 differs from its version 9 predecessor in some notable respects. Most obvious are the two new chapters: Chapter 14 on Survey Data Analysis, and Chapter 15 on Multilevel and Mixed-Effects Modeling. Both chapters reflect Stata's developing strength in these areas. Within other chapters there are several new sections, including one on extended missing value codes in Chapter 2, sections about the Graph Editor and Creative Graphing in Chapter 3, and a section demonstrating ARMAX time series regression models in Chapter 13. The final chapter on programming contains two new sections: an example program that draws multiple graphs for reporting survey results; and a first look at matrix programming with Mata. Less visible changes occur throughout the book, reflecting updates in Stata commands, output, or features. New or new-ish methods for drawing time plots, fitting nonlinear models, estimating standard errors, and obtaining predictions or diagnostic statistics make their appearance in this edition, as do some further options for long-established commands. In a few instances, I gave up on the efforts of earlier editions to provide complete lists of functions or options for certain commands. Those lists had grown to a point where it became more practical to refer users to Stata's online help: "Type **help math functions** for a complete list with details."

Users coming from version 8 or older editions of *Statistics with Stata* might notice other changes as well, including new sections on robust standard errors (Chapter 9) and cluster analysis (Chapter 12). Much more has been added, of course, over the years since the 171-page first edition. Because Stata now does so much, far beyond the scope of an introductory book, *Statistics with Stata* presents many features telegraphically in the "Example Commands" sections that begin most chapters, or in partial lists of options followed by reference to the relevant **help** keywords.

As described in Chapter 1, Stata's help and search features have advanced to keep pace with the program. Behind the interactive documentation available through help files stand Stata's website, Internet and documentation search capabilities, user-community listserver, NetCourses, the *Stata Journal*, and printed documentation — presently over 7,000 pages and growing. *Statistics with Stata* provides an accessible gateway to Stata; these other resources can help you go further.

Acknowledgments

Stata's architect, William Gould, deserves credit for originating the elegant program that *Statistics with Stata* describes. Pat Branton pointed out the need for this newest edition of the book. She then helped greatly by sharing my initial chapter drafts with other smart people at StataCorp, as well as reading them closely herself. Bill Rising, Alan Riley, and Jeff Pitblado deserve special thanks for countless invaluable suggestions and troubleshooting assistance.

James Hamilton contributed key advice about time series, for Chapter 13. Leslie Hamilton carefully read and edited the final manuscript. At Cengage, Catie Ronquillo took over the job of editor for this latest edition. Jennifer Risden capably managed the final production.

The book is built around data. To avoid endlessly recycling examples, I drew on fresh sources, sometimes from my own research but also including the Granite State Poll, General Social Survey (Davis et al., 2005), and compilations of U.S. data on migration (Voss et al., 2005) and voting (Robinson, 2005). Data from agencies including the U.S. Census Bureau, Statistics Iceland, Statistics Greenland, Statistics Canada, the Northwest Atlantic Fisheries Organization, Greenland's Natural Resources Institute, and the Department of Fisheries and Oceans Canada can be found among the examples. A presentation given by Brenda Topliss inspired the "gossip" program example in Chapter 16. My work on the Community and Environment in Rural America (CERA) surveys, with Mil Duncan and Chris Colocousis, inspired the "multicat" program example in that chapter. Other people whose data or ideas contributed to this book are Amy Bassett, Igor Belkin, Robert Bell, Cliff Brown, Erich Buch, Anna Kerttula de Echave, Greg Goddard, Richard Haedrich, Dave Hamilton, Barry Keim, Paul Mayewski, Loren D. Meeker, David Moore, James Morison, Per Lyster Pedersen, Rasmus Ole Rasmussen, Jane Rusbjerg, Steve Selvin, Carole Seyfrit, Andrew Smith, Heather Turner, and Sally Ward.

Dedication

To Leslie, Sarah, and Dave.

Stata and Stata Resources

Stata is a full-featured statistical program for Windows, Macintosh, and Unix computers. It combines ease of use with speed, a library of pre-programmed analytical and data-management capabilities, and programmability that allows users to invent and add further capabilities as needed. Most operations can be accomplished either via the pull-down menu system, or more directly via typed commands. Menus help newcomers to learn Stata, and help anyone to apply an unfamiliar procedure. The consistent, intuitive syntax of Stata commands frees experienced users to work more efficiently, and also makes it straightforward to develop programs for complex or repetitious tasks. Menu and command instructions can be mixed as needed during a Stata session. Extensive help, search, and link features make it easy to look up command syntax and other information instantly, on the fly.

After introductory information, we'll begin with an example Stata session to give you a sense of the "flow" of data analysis, and of how analytical results might be used. Later chapters explain in more detail. Even without explanations, however, you can see how straightforward the commands are — **use** *filename* to retrieve dataset *filename*, **summarize** when you want summary statistics, **correlate** to get a correlation matrix, and so forth. Alternatively, the same results can be obtained by making choices from the Data or Statistics menus.

Stata users have available a variety of resources to help them learn about Stata and solve problems at any level of difficulty. These resources come not just from StataCorp, but also from an active community of users. Sections of this chapter introduce some key resources — Stata's online help and printed documentation; where to phone, fax, write, or e-mail for technical help; Stata's website (www.stata.com), which provides many services including updates and answers to frequently asked questions; the Statalist Internet forum; and the refereed *Stata Journal*.

A Typographical Note

This book employs several typographical conventions as a visual cue to how words are used:

- Commands typed by the user appear in a bold fixed-width font. When the whole command line is given, it starts with a period, as seen in a Stata Results window or log (output) file:

  `. list year boats men penalty`

- *Variable* or *file* names within these commands appear in italics to emphasize the fact that they are arbitrary and not a fixed part of the command.

- Names of *variables* or *files* also appear in italics within the main text to distinguish them from ordinary words.

- Items from Stata's menus are shown in the Arial font , with successive options separated by a dash. For example, we can open an existing dataset by selecting File > Open , and then finding and clicking on the name of the particular dataset. Note that some common menu actions can be accomplished either with text choices from Stata's top menu bar,

  <table>
  <tr><td>File</td><td>Edit</td><td>Data</td><td>Graphics</td><td>Statistics</td><td>User</td><td>Window</td><td>Help</td></tr>
  </table>

  or with the row of icons below these. For example, selecting File > Open is equivalent to clicking the leftmost icon, a tiny picture of an opening file folder . One could also accomplish the same thing by typing a direct command of the form

  ```
  . use filename
  ```

Thus, we show the calculation of summary statistics for a variable named *penalty* as follows:

```
. summarize penalty
```

Variable	Obs	Mean	Std. Dev.	Min	Max
penalty	10	63	59.59493	11	183

These typographic conventions exist only in this book, and not within the Stata program itself. Stata can display a variety of onscreen fonts, but it does not use italics in commands. Once Stata log files have been imported into a word processor, or a results table has been copied and pasted, you might want to format them in a Courier font, 10 point or smaller, so that columns will line up correctly.

In its commands and variable names, Stata is case sensitive. Thus, **summarize** is a command, but Summarize and SUMMARIZE are not. *Penalty* and *penalty* would be two different variables.

An Example Stata Session

As a preview showing Stata at work, this section retrieves and analyzes a previously-created dataset named *lofoten.dta*. Jentoft and Kristofferson (1989) originally published these data in an article about self-management among fishermen on Norway's arctic Lofoten Islands. There are 10 observations (years) and 5 variables, including *penalty*, a count of how many fishermen were cited each year for violating fisheries regulations.

If we might eventually want a record of our session, the best way to prepare for this is by opening a "log file" at the start. Log files contain commands and results tables, but not graphs. To begin a log file, choose File > Log > Begin ... from the top menu bar, and specify a name and folder for the resulting log file. Alternatively, a log file could be started by choosing File > Log > Begin from the top menu bar, or by typing a direct command such as

```
. log using monday1
```

Multiple ways of doing such things are common in Stata. Each way has its own advantages, and each suits different situations or user tastes.

Log files can be created either in a special Stata format (.smcl), or in ordinary text or ASCII format (.log). A .smcl ("Stata markup and control language") file will be nicely formatted for viewing or printing within Stata. It could also contain hyperlinks that help to understand commands or error messages. .log (text) files lack such formatting, but are simpler to use if you plan later to insert or edit the output in a word processor. After selecting which type of log file you want, click Save . For this session, we will create a .smcl log file named *monday1.smcl*.

An existing Stata-format dataset named *lofoten.dta* will be analyzed here. To open or retrieve this dataset, we again have several options:

> select File > Open > *lofoten.dta* using the top menu bar;
>
> click on 🖼 > *lofoten.dta*; or
>
> type the command **use *lofoten*** .

Under its default Windows configuration, Stata looks for data files in folder C:\data. If the file we want is in a different folder, we could specify its location in the **use** command,

```
. use c:\books\sws_10\chapter_01\lofoten
```

or change the session's default folder by issuing a **cd** (change directory) command,

```
. cd c:\books\sws_10\chapter01\
. use lofoten
```

or select File > Change Working Directory ... from the menus. Often, the simplest way to retrieve a file will be to choose File > Open and browse through folders in the usual way.

To see a brief description of the dataset now in memory, type

```
. describe
```

```
Contains data from C:\data\lofoten.dta
  obs:           10                          Jentoft & Kristoffersen '89
 vars:            5                          12 Jan 2008 15:48
 size:          170 (99.9% of memory free)
```

variable name	storage type	display format	value label	variable label
year	int	%9.0g		Year
boats	int	%9.0g		Number of fishing boats
men	int	%9.0g		Number of fishermen
penalty	int	%9.0g		Number of penalties
decade	byte	%9.0g	decade	Early 1970s or early 1980s

```
Sorted by:  decade  year
```

Many Stata commands can be abbreviated to their first few letters. For example, we could shorten **describe** to just the letter **d**. Using menus, the same table could be obtained by choosing

Data > Describe data > Describe data in memory > (OK).

This dataset has only 10 observations and 5 variables, so we can easily list its contents by typing the command **list** (or the letter **l**; or Data > Describe data > List data > (OK)):

. **list**

	year	boats	men	penalty	decade
1.	1971	1809	5281	71	1970s
2.	1972	2017	6304	152	1970s
3.	1973	2068	6794	183	1970s
4.	1974	1693	5227	39	1970s
5.	1975	1441	4077	36	1970s
6.	1981	1540	4033	11	1980s
7.	1982	1689	4267	15	1980s
8.	1983	1842	4430	34	1980s
9.	1984	1847	4622	74	1980s
10.	1985	1365	3514	15	1980s

Analysis could begin with a table of means, standard deviations, minimum values, and maximum values. Type **summarize** or **su**; or select from the drop-down menus, Statistics > Summaries, tables, & tests > Summary and descriptive statistics > Summary statistics > (OK)

. **summarize**

Variable	Obs	Mean	Std. Dev.	Min	Max
year	10	1978	5.477226	1971	1985
boats	10	1731.1	232.1328	1365	2068
men	10	4854.9	1045.577	3514	6794
penalty	10	63	59.59493	11	183
decade	10	.5	.5270463	0	1

To print results from the session so far, click on the Results window and then 🖶, or from the menus choose File > Print > Results .

To copy a table, commands, or other information from the Results window into a word processor, drag the mouse to select the results you want, right-click the mouse, and then choose Copy Text from the mouse's menu. Finally, switch to your word processor and, at the desired insertion point either right-click and Paste or click the word processor's "paste" icon.

Did the number of penalties for fishing violations change over the two decades covered by these data? A table containing summary statistics for *penalty* at each value of *decade* shows that there were more penalties in the 1970s:

```
. tabulate decade, sum(penalty)
```

Early 1970s or early 1980s	Summary of Number of penalties		
	Mean	Std. Dev.	Freq.
1970s	96.2	67.41439	5
1980s	29.8	26.281172	5
Total	63	59.594929	10

The same table could be obtained through menus:

Statistics > Summaries, tables, & tests > Tables > One/two-way table of summary statistics

Then fill in *decade* as variable 1, and *penalty* as the variable to be summarized. Although menu choices often are straightforward to use, you can see that they tend to be more complicated to describe than the simple text commands. From this point on, we will focus primarily on the commands, mentioning menu alternatives only occasionally. Fully exploring the menus, and working out how to use them to accomplish the same tasks, will be left to the reader. For similar reasons, the Stata reference manuals likewise take a command-based approach.

Perhaps the number of penalties declined because fewer people were fishing in the 1980s. The number of penalties correlates strongly ($r > .8$) with the number of boats and fishermen:

```
. correlate boats men penalty
(obs=10)
```

	boats	men	penalty
boats	1.0000		
men	0.8748	1.0000	
penalty	0.8259	0.9312	1.0000

A graph might help clarify these interrelationships. Figure 1.1 plots *men* and *penalty* against *year*, produced by the **graph twoway connected** command. In this example, we first ask for a twoway (two-variable) connected-line plot of *men* against *year*, using the left-hand y axis, **yaxis(1)**. After the separator || , we next ask for a connected-line plot of *penalty* against *year*, this time using the right-hand y axis, **yaxis(2)**. The resulting graph visualizes the correspondence between the number of fishermen and the number of penalties over time.

```
. graph twoway connected men year, yaxis(1)
       || connected penalty year, yaxis(2)
```

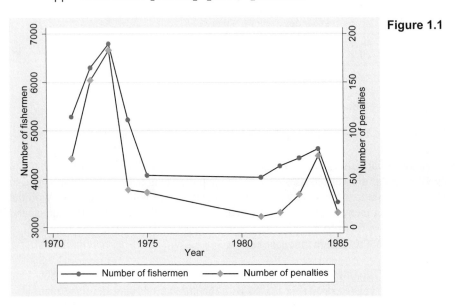

Figure 1.1

Because the years 1976 to 1980 are missing in these data, Figure 1.1 shows 1975 connected to 1981. For some purposes, we might hesitate to do this. Instead, we could either find the missing values or leave the gap unconnected — for example, by typing years 1976–1980 into the dataset, then including **cmissing(n)** options in the command after each **yaxis( )**.

To print this graph, go to the Graph window and click its print icon 🖨 or File > Print. To copy the graph directly into a word processor or other document, right-click on the graph, and select Copy Graph . Switch to your word processor, go to the desired insertion point, and issue an appropriate "paste" command such as Edit > Paste, Edit > Paste Special (Metafile) , or click a paste icon (different word processors will handle this differently).

To save the graph for future use, either right-click and Save Graph , click 💾 in the Graph window, or select File > Save As from the Graph window's top menu bar. The Save as type submenu offers several different file formats. On a Windows system, the choices include

    Stata graph (*.gph)   (A "live" graph, containing enough information for Stata to edit)
    As-is graph (*.gph)   (A more compact Stata graph format)
    Windows Metafile (*.wmf)
    Enhanced Metafile (*.emf)
    Portable Network Graphics (*.png)
    TIFF (*.tif)
    PostScript (*.ps)
    Encapsulated PostScript with or without TIFF preview (*.eps)

Other platforms such as Macintosh or Linux offer different choices for graph file formats. Regardless of which format we want, it often is worthwhile to save one copy of our graph in "live" .gph format. Such live .gph-format graphs can later be retrieved, combined, recolored, or reformatted using the **graph use** or **graph combine** commands, or edited in more detail using the Graph Editor (Chapter 3).

Through all of the preceding analyses, the log file *monday1.smcl* has been storing our results. An easy way to review this file to see what we have done is to open the file in its own Viewer window by selecting

    File > Log > View > OK

We could print this log file by clicking the 🖨 icon on the top bar of the log file's Viewer window. Log files close automatically at the end of a Stata session, or earlier if instructed by 📓 > Close log file, typing the command **log close** , or by choosing

    File > Log > Close

Once closed, the file *monday1.smcl* could be opened to view again through File > Log > View or 📓 during a subsequent Stata session. To create an output file that can be opened easily by your word processor, either translate the log file from .smcl (a Stata format) to .log (standard ASCII text format) by typing

. translate *monday1.smcl monday1.log*

or start out by creating the file in .log instead of .smcl format. You can also start and stop a log file temporarily, any number of times:

File > Log > Suspend

File > Log > Resume

The log icon 📓 on Stata's main icon menu bar can also perform all these tasks.

Stata's Documentation and Help Files

The complete Stata 10 Documentation Set includes over 7,000 pages in 15 volumes: a slim *Getting Started* manual (for example, *Getting Started with Stata for Windows*), the more extensive *User's Guide*, the encyclopedic three-volume *Base Reference Manual*, and separate reference manuals on data management, graphics, longitudinal and panel data, matrix programming (Mata), multivariate statistics, programming, survey data, survival analysis and epidemiological tables, and time series analysis. *Getting Started* helps you do just that, with the basics of installation, window management, data entry, printing, and so on. The *User's Guide* contains an extended discussion of general topics, including resources and troubleshooting. Of particular note for new users is the *User's Guide* section on "Commands everyone should know." The *Base Reference Manual* lists all Stata commands alphabetically. Entries for each command include the full command syntax, descriptions of all available options, examples, technical notes regarding formulas and rationale, and references for further reading. Data management, graphics, panel data, etc. are covered in the general references, but these complicated topics get more detailed treatment and examples in their own specialized manuals. A *Quick Reference and Index* volume rounds out the whole collection.

When we are in the midst of a Stata session, it is often simpler to ask for onscreen help instead of consulting the manuals. Selecting Help from the top menu bar invokes a drop-down menu of further choices, including specific commands, what's new, online updates, the *Stata Journal* and user-written programs, or connections to Stata's website (www.stata.com). Choosing Search allows keyword searching of Stata's documentation, of Net resources, or both. Alternatively, choosing Contents (or typing **help**) allows us to look up how to do things by category. The **help** command is particularly useful when used with a command name. Typing **help correlate**, for example, causes a description of that command to appear in a Viewer window. Like the reference manuals, this onscreen help provides command syntax diagrams and complete lists of options. It also includes some examples, although often less detailed and without the technical discussions found in the manuals. The onscreen help has several advantages over the manuals, however. The Viewer allows searching for keywords in the documentation or on Stata's website. Hypertext links take you directly to related entries. Onscreen help can also include material about recent updates, or the "unofficial" Stata programs that you have downloaded from Stata's website or from other users.

Searching for Information

Selecting Help > Search > Search documentation and FAQs provides a direct way to search for information in Stata's documentation or in the website's FAQs (frequently asked questions) and other pages. Alternatively, we can search net resources including the *Stata Journal*. Search results in the Viewer window contain clickable hyperlinks leading to further information or original citations.

The **search** command can do similar things. One specialized use for a quick **search** command is to provide more information on those occasions when our command does not succeed as planned, but instead results in one of Stata's cryptic numerical error messages. For example, **tabulate** is a Stata command, but suppose we spelled this incorrectly as "tabluate." Stata responds with the error message and "return code" r(199):

```
. tabluate penalty
unrecognized command:  tabluate
r(199);
```

We can learn more about what a return code such as r(199) refers to by clicking on the blue "r(199);" link in the error message. This brings up a more informative note telling us that "Stata failed to recognize the command, program, or ado-file name, probably because of a typographical or abbreviation error." We could also find this note by typing **search rc 199**. Type **help search** for more about this command.

StataCorp

The mailing or physical address is
> StataCorp
> 4905 Lakeway Drive
> College Station, TX 77845 USA

Telephone access includes an easy-to-remember 800 number.
> telephone: 1-800-782-8272 (or 1-800-STATAPC) U.S. and Canada
> 1-979-696-4600 International
> fax: 1-979-696-4601

For orders, licensing, and upgrade information, you can contact StataCorp by e-mail at
> service@stata.com

or visit their website at
> http://www.stata.com

Stata Press also has its own website, containing information about Stata publications including the datasets used for examples.
> http://www.stata-press.com

The refereed *Stata Journal* has become an important resource as well.
> http://www.stata-journal.com

Stata's main website, www.stata.com, provides extensive user resources, starting with pages describing Stata products in detail, how to order Stata, and many kinds of user support such as:

FAQs — Frequently asked questions and their answers. If you are puzzled by something and can't find the answer in the manuals, check here next — it might be a FAQ. Example questions range from basic questions such as "How can I convert other packages' files to Stata format data files?" to more technical queries like "How do I impose the restriction that rho is zero using the heckman command with full ml?"

Updates — Online updates within major versions are free to registered Stata users. These provide a fast, simple way to obtain the latest enhancements, bug fixes, etc. for your current version. Instead of going to the website you can ask within Stata whether updates exist for your version, and initiate the update process by typing the command

```
. update query
```

Technical support — Technical support can be obtained by sending e-mail messages to

> tech-support@stata.com

Responses tend to be prompt and helpful. Before writing for technical help, though, you should check whether your question is a FAQ.

Training — Enroll in web-based NetCourses on selected topics such as Introduction to Stata, Introduction to Stata Programming, or Advanced Stata Programming.

Stata News — The *Stata News* contains information about software features, current NetCourses, recent issues of the *Stata Journal*, and other topics.

Publications — Links to information about the *Stata Journal*, documentation and manuals, a bookstore selling books about Stata and other up-to-date statistical references, and Stata's author support program for people writing new books about Stata. The following sections have more to say about the *Stata Journal* and Stata books.

The *Stata Journal*

From 1991 through 2001, a bimonthly publication called the *Stata Technical Bulletin* (*STB*) served as a means of distributing new commands and Stata updates, both user-written and official. Accumulated *STB* articles were published in book form each year as *Stata Technical Bulletin Reprints*, which can be ordered directly from StataCorp. With the growth of the Internet, instant communication among users became possible. Program files could easily be downloaded from distant sources. A bimonthly printed journal and disk no longer provided the best avenues either for communicating among users, or for distributing updates and user-written programs. To adapt to a changing world, the *STB* had to evolve into something new.

The *Stata Journal* was launched to meet this challenge and the needs of Stata's broadening user base. Like the old *STB*, the *Stata Journal* contains articles describing new commands by users along with unofficial commands written by StataCorp employees. New commands are not its primary focus, however. The *Stata Journal* also contains refereed expository articles about statistics, book reviews, tips on using Stata, and a number of interesting columns, including "Speaking Stata" by Nicholas J. Cox, on effective use of the Stata programming language. The *Stata Journal* is intended for novice as well as experienced Stata users. For example, here are the contents from one recent issue.

"Robust standard errors for panel regressions with cross-sectional dependence"(D. Hoechle)
"Estimating parameters of dichotomous and ordinal item response models with gllamm" (X. Zheng and S. Rabe-Hesketh)
"Simulation-based sensitivity analysis for matching estimators" (T. Nannicini)
"Modeling of the cure fraction in survival studies" (P.C. Lambert)
"Profile likelihood for estimation and confidence intervals" (P. Royston)
"Fitting mixed logit models by using maximum simulated likelihood" (A.R. Hole)
"An exact and a Monte Carlo proposal to the Fisher Pitman permutation tests for paired replicates and for independent samples" (J. Kaiser)
"Speaking Stata: Turning over a new leaf" (N.J. Cox)
Stata tips on "Discrete uses for uniform()" (M.L. Buis), "Range frame plots" (S. Merryman), "Efficient use of summarize" (N.J. Cox), "Events in intervals" (N.J. Cox)
"Software Updates"

The *Stata Journal* is published quarterly. Subscriptions can be purchased by visiting www.stata-journal.com. The www.stata-journal.com archives also list contents of back issues, which you can order individually. Of historical interest, a special issue on the occasion of Stata's 20th anniversary (5(1), 2005) contains articles about the early development of Stata, and one about the first Stata book, "A short history of *Statistics with Stata.*"

Books Using Stata

In addition to Stata's own reference manuals, a growing library of books describe Stata, or use Stata to illustrate analytical techniques. These books include general introductions; disciplinary applications such as social science, biostatistics, or econometrics; and focused texts concerning survey analysis, experimental data, categorical dependent variables, and other subjects.

The Bookstore pages on Stata's website have up-to-date lists, with descriptions of content:
     http://www.stata.com/bookstore/
This online bookstore provides a central place to learn about and order Stata-relevant books from many different publishers. Examples below illustrate the wide range of choices.

A Gentle Introduction to Stata (A.C. Acock)
Using Stata for Principles of Econometrics (L.C. Adkins, R.C. Hill)
An Introduction to Modern Econometrics Using Stata (C.F. Baum)
Applied Microeconometrics Using Stata (A.C. Cameron, P.K. Trivedi)
Event History Analysis with Stata (H-P. Blossfeld, K. Golsch, G.Rohwer)
An Introduction to Survival Analysis Using Stata (M. Cleves, W. Gould, R. Gutierrez, Y. Marchenko)
Statistical Modeling for Biomedical Researchers (W.D. Dupont)
Maximum Likelihood Estimation with Stata (W. Gould, J. Pitblado, W. Sribney)
Statistics with Stata (L.C. Hamilton)
Generalized Linear Models and Extensions (J.W. Hardin, J.M. Hilbe)
Negative Binomial Regression (J.M. Hilbe)
A Short Introduction to Stata for Biostatistics (M. Hills, B.L. De Stavola)
Applied Survival Analysis: Regression Modeling of Time to Event Data (D.W. Hosmer, S. Lemeshow, S. May)
Applied Econometrics for Health Economists (A. Jones)
Applied Health Economics (A. Jones, N. Rice, T.B. d'Uva, S. Balia)
An Introduction to Stata for Health Researchers (S. Juul)
Data Analysis Using Stata (U. Kohler, F. Kreuter)
Sampling of Populations: Methods and Applications (P.S. Levy, S. Lemeshow)
Workflow in Data Analysis Using Stata (J.S. Long)
Regression Models for Categorical Dependent Variables Using Stata (J.S. Long, J. Freese)
A Visual Guide to Stata Graphics (M. Mitchell)
Thirty-three Stata Tips (H.J. Newton, N. J. Cox editors)
Analyzing Health Equity Using Household Survey Data (O. O'Donnell and others)
A Stata Companion to Political Analysis (P.H. Pollock III)
A Handbook of Statistical Analyses Using Stata (S. Rabe-Hesketh, B. Everitt)
Multilevel and Longitudinal Modeling Using Stata (S. Rabe-Hesketh, A. Skrondal)

Managing Your Patients? Data in the Neonatal and Pediatric ICU (J. Schulman)
Epidemiology: Study Design and Data Analysis (M. Woodward)

2

Data Management

The first steps in data analysis involve organizing the raw data into a format usable by Stata. We can bring new data into Stata in several ways: type the data from the keyboard; read a text or ASCII file containing the raw data; paste data from a spreadsheet into the Editor; or, using a third-party data transfer program, translate the dataset directly from a system file created by another spreadsheet, database, or statistical program. Once Stata has the data in memory, we can save the data in Stata format for easy retrieval and updating in the future.

Data management encompasses the initial tasks of creating a dataset, editing to correct errors, identifying the missing values, and adding internal documentation such as variable and value labels. It also encompasses many other jobs required by ongoing projects, such as adding new observations or variables; reorganizing, simplifying, or sampling from the data; separating, combining, or collapsing datasets; converting variable types; and creating new variables through algebraic or logical expressions. When data-management tasks become complex or repetitive, Stata users can write their own programs to automate the work. Although Stata is best known for its analytical capabilities, it possesses a broad range of data-management features as well. This chapter introduces some of the basics.

The *User's Guide* provides an overview of the different methods for inputting data, followed by eight rules for determining which input method to use. Input, editing, and many other operations discussed in this chapter can be accomplished through the Data menus. Data menu subheadings refer to the general category of task:
Describe data
Data Editor
Data Browser (read-only Editor)
Create or change variables
Sort
Combine datasets
Labels
Notes
Variable utilities
Matrices
Other utilities

Example Commands

. **append using** *olddata*
Reads previously-saved dataset *olddata.dta* and adds all its observations to the data currently in memory. Subsequently typing **save** *newdata*, **replace** will save the combined dataset as *newdata.dta*.

. **browse**
Opens the spreadsheet-like Data Browser for viewing the data. The Browser looks similar to the Data Editor, but it has no editing capability, so there is no risk of inadvertently changing your data. Alternatively, use the Data menu or click

. **browse** *boats men* **if** *year* > 1980
Opens the Data Browser showing only the variables *boats* and *men* for observations in which *year* is greater than 1980. This example illustrates the **if** qualifier, which can be used to focus the operation of many Stata commands.

. **compress**
Automatically converts all variables to their most efficient storage types to conserve memory and disk space. Subsequently typing the command **save** *filename*, **replace** will make these changes permanent.

. **drawnorm** *z1 z2 z3*, **n(5000)**
Creates an artificial dataset with 5,000 observations and three random variables, *z1*, *z2*, and *z3*, sampled from uncorrelated standard normal distributions. Options could specify other means, standard deviations, and correlation or covariance matrices.

. **dropmiss**
Automatically drops from the dataset in memory any *variables* that have missing values for every observation. This can be useful when working with a subset from a larger dataset, where some of the original variables are not applicable to any of the remaining observations. Typing **dropmiss, obs** will instead drop from memory any *observations* that have missing values for every variable.

. **edit**
Opens the spreadsheet-like Data Editor where data can be entered or edited. Alternatively, use the Data menu or click

. **edit** *boats year men*
Opens the Data Editor with only the variables *boats*, *year*, and *men* (in that order) visible and available for editing.

. **encode** *stringvar*, **gen(***numvar***)**
Creates a new variable named *numvar*, with labeled numeric values based on the string (non-numeric) variable *stringvar*.

. **format** *rainfall* **%8.2f**
Establishes a fixed (**f**) display format for numeric variable *rainfall*: 8 columns wide, with two digits always shown after the decimal. This affects only how values are displayed.

. **generate** *newvar* **= (x + y)/100**
Creates a new variable named *newvar*, equal to *x* plus *y* divided by 100.

. **generate** *newvar* **= uniform()**
Creates a new variable with values sampled from a uniform random distribution over the interval ranging from 0 to nearly 1, written [0,1).

. **infile x y z using** *data.raw*
Reads an ASCII file named *data.raw* containing data on three variables: *x*, *y*, and *z*. The values of these variables are separated by one or more white-space characters — blanks, tabs, and newlines (carriage return, linefeed, or both) — or by commas. With white-space delimiters, missing values for numerical variables are represented by periods, not blanks. With comma-delimited data, missing values are represented by a period or by two consecutive commas. Stata also provides for extended missing values, as discussed later in this chapter. Other commands are better suited for reading tab-delimited, comma-delimited, or fixed-column raw data; type **help infiling** for more infomation.

. **list**
Lists the data in default or "table" format. With large datasets, table format becomes hard to read, and **list, display** produces better results. See **help list** for other options. The Data Editor or Data Browser provide more useful views for many purposes.

. **list x y z in 5/20**
Lists the *x*, *y*, and *z* values of the 5th through 20th observations, as the data are presently sorted. The **in** qualifier works in similar fashion with most other Stata commands as well.

. **merge** *id* **using** *olddata*
Reads the previously-saved dataset *olddata.dta* and matches observations from *olddata* with observations in memory that have identical *id* values. Both *olddata* (the "using" data) and the data currently in memory (the "master" data) must already be sorted by *id*.

. **mvdecode** *var3-var62*, **mv(97=. \ 98=.a \ 99=.b)**
For variables *var3* through *var62*, recode the numerical values 97, 98, and 99 as missing. In this example we use three separate missing value codes, which Stata represents as a period, .a, and .b. These could represent different reasons the values are missing, such as responses of "Not applicable," "Don't know," and "Refused to answer" on a survey. If only one missing-value code is required, we can instead specify an option such as
mv(97 98 99=.)

. **replace** *oldvar* **= 100 * oldvar**
Replaces the values of *oldvar* with 100 times their previous values.

. **sample 10**

   Drops all the observations in memory except for a 10% random sample. Instead of selecting a certain percentage, we could select a certain number of cases. For example, **sample 55, count** would drop all but a random sample of size *n* = 55.

. **save** *newfile*

   Saves the data currently in memory, as a file named *newfile.dta*. If *newfile.dta* already exists, and you want to write over the previous version, type **save** *newfile*, **replace**. Alternatively, use the File menu. To save *newfile.dta* in the format of Stata version 9, type **saveold** *newfile* or select File > Save As > Save as type .

. **set memory 24m**

   (Windows or Unix systems only) Allocates 24 megabytes of memory for Stata data. The amount set could be greater or less than the current allocation. Virtual memory (disk space) is used if the request exceeds physical memory. Type **clear** to drop the current data from memory before using **set memory** .

. **sort** *x*

   Sorts the data from lowest to highest values of *x*. Observations with missing *x* values appear last after sorting because Stata views missing values as very high numbers. Type **help gsort** for a more general sorting command that can arrange values in either ascending or descending order and can optionally place the missing values first.

. **tabulate** *x* **if** *y* **> 65**

   Produces a frequency table for *x* using only those observations that have *y* values above 65. The **if** qualifier works similarly with most other Stata commands.

. **use** *oldfile*

   Retrieves previously-saved Stata-format dataset *oldfile.dta* from disk, and places it in memory. If other data are currently in memory and you want to discard those data without saving them, type **use** *oldfile*, **clear**. Alternatively, these tasks can be accomplished through File > Open or by clicking

Creating a New Dataset

Data that were previously saved in Stata format can be retrieved into memory either by typing a command of the form **use** *filename*, or by menu selections. This section describes basic methods for creating a Stata-format dataset in the first place, using as our example the 1995 data on Canadian provinces and territories listed in Table 2.1. (From the Federal, Provincial and Territorial Advisory Committee on Population Health, 1996. Canada's newest territory, Nunavut, is not listed here because it was part of the Northwest Territories until 1999.)

Table 2.1: Data on Canada and Its Provinces

Place	1995 Pop. (1000's)	Unemployment Rate (percent)	Male Life Expectancy	Female Life Expectancy
Canada	29606.1	10.6	75.1	81.1
Newfoundland	575.4	19.6	73.9	79.8
Prince Edward Island	136.1	19.1	74.8	81.3
Nova Scotia	937.8	13.9	74.2	80.4
New Brunswick	760.1	13.8	74.8	80.6
Quebec	7334.2	13.2	74.5	81.2
Ontario	11100.3	9.3	75.5	81.1
Manitoba	1137.5	8.5	75.0	80.8
Saskatchewan	1015.6	7.0	75.2	81.8
Alberta	2747.0	8.4	75.5	81.4
British Columbia	3766.0	9.8	75.8	81.4
Yukon	30.1	—	71.3	80.4
Northwest Territories	65.8	—	70.2	78.0

The simplest way to create a dataset from printed information like Table 2.1 is through the Data Editor, invoked by clicking ▣ , selecting Window > Data Editor from the menu bar, or by typing the command **edit**. Then begin typing values for each variable, in columns initially labeled *var1*, *var2*, etc. Thus, *var1* contains place names, *var2*, populations, and so forth.

We can assign more descriptive variable names by double-clicking on the column headings (such as *var1*) and then typing a new name in the resulting dialog box; eight characters or fewer works best, although names with up to 32 characters are allowed. We can also create variable labels that contain a brief description. For example, *var2* (population) might be renamed *pop*, and given the variable label "Population in 1000s, 1995".

Renaming and labeling variables can also be done outside of the Data Editor through the **rename** and **label variable** commands:

```
. rename var2 pop
. label variable pop "Population in 1000s, 1995"
```

Cells left empty, such as unemployment rates for the Yukon and Northwest Territories, will automatically be assigned Stata's default missing value code, a period. At any time, we can close the Data Editor and then save the dataset to disk. Clicking 🗋 or Data > Data Editor , or typing the command **edit**, brings the Editor back.

If the first value entered for a variable is a number, as with population, unemployment, and life expectancy, then Stata assumes that this column is a "numeric variable" and it will thereafter permit only numbers as values. Numeric values can also begin with a plus or minus sign, include decimal points, or be expressed in scientific notation. For example, we could represent Canada's population as 2.96061e+7, which means 2.96061×10^7 or about 29.6 million people. Numbers *should not include any commas*, such as 29,606,100 (or using commas as a decimal separator). If we did happen to put commas within the first value typed in a column, Stata would interpret this as a "string variable" (next paragraph) rather than as a number.

If the first value entered for a variable includes non-numeric characters, as did place names above (or "1,000" with the comma), then Stata thereafter considers this column to be a string or text variable. String variable values can be almost any combination of letters, numbers, symbols, or spaces up to 80 characters long in Small Stata, and up to 244 characters in Stata/IC, Stata/SE, or Stata/MP. They can store names, quotations, or other descriptive information. String variable values could be tabulated and counted, but not analyzed using means, correlations, or most other statistics. In the Data Editor or Data Browser, string variable values appear in red, distinguishing them from numeric (black) or labeled numeric (blue) variables.

After typing in the information from Table 2.1 in this fashion, we close the Data Editor and save our data, perhaps with the name *canada0.dta*:

```
. save canada0
```

Stata automatically adds the extension .dta to any dataset name, unless we tell it to do otherwise. If we already had saved and named an earlier version of this file, it is possible to write over that with the newest version by typing

```
. save, replace
```

At this point, our new dataset looks like this:

```
. describe
Contains data from C:\data\canada0.dta
  obs:            13
  vars:            5                            15 Mar 2008 10:47
  size:          585 (99.9% of memory free)

              storage  display     value
variable name   type   format      label    variable label

var1          str21    %21s
pop           float    %9.0g                 Population in 1000s, 1995
var3          float    %9.0g
var4          float    %9.0g
var5          float    %9.0g

Sorted by:
```

```
. list
```

		var1	pop	var3	var4	var5
1.	Canada		29606.1	10.6	75.1	81.1
2.	Newfoundland		575.4	19.6	73.9	79.8
3.	Prince Edward Island		136.1	19.1	74.8	81.3
4.	Nova Scotia		937.8	13.9	74.2	80.4
5.	New Brunswick		760.1	13.8	74.8	80.6
6.	Quebec		7334.2	13.2	74.5	81.2
7.	Ontario		11100.3	9.3	75.5	81.1
8.	Manitoba		1137.5	8.5	75	80.8
9.	Saskatchewan		1015.6	7	75.2	81.8
10.	Alberta		2747	8.4	75.5	81.4
11.	British Columbia		3766	9.8	75.8	81.4
12.	Yukon		30.1	.	71.3	80.4
13.	Northwest Territories		65.8	.	70.2	78

```
. summarize
```

Variable	Obs	Mean	Std. Dev.	Min	Max
var1	0				
pop	13	4554.769	8214.304	30.1	29606.1
var3	11	12.10909	4.250048	7	19.6
var4	13	74.29231	1.673052	70.2	75.8
var5	13	80.71539	.9754027	78	81.8

Examining such output tables gives us a chance to look for errors that should be corrected. The **summarize** table, for instance, provides several numbers useful in proofreading, including the count of nonmissing numerical observations (always 0 for string variables) and the minimum and maximum for each variable. Substantive interpretation of the summary statistics would be premature at this point, because our dataset contains one observation (Canada) that represents a combination of the other 12 provinces and territories.

The next step is to make our dataset more self-documenting. The variables could be given more descriptive names, such as the following:

```
. rename var1 place
. rename var3 unemp
. rename var4 mlife
. rename var5 flife
```

Stata also permits us to add several kinds of labels to the data. **label data** describes the dataset as a whole, whereas **label variable** describes an individual variable. For example,

```
. label data "Canadian dataset 0"
. label variable place "Place name"
. label variable unemp "% 15+ population unemployed, 1995"
. label variable mlife "Male life expectancy years"
. label variable flife "Female life expectancy years"
```

By labeling data and variables, we obtain a dataset that is more self-explanatory:

`. describe`

```
Contains data from C:\data\canada0.dta
  obs:             13                          Canadian dataset 0
 vars:              5                          15 Mar 2008 10:57
 size:            585 (99.9% of memory free)
```

variable name	storage type	display format	value label	variable label
place	str21	%21s		Place name
pop	float	%9.0g		Population in 1000s, 1995
unemp	float	%9.0g		% 15+ population unemployed, 1995
mlife	float	%9.0g		Male life expectancy years
flife	float	%9.0g		Female life expectancy years

```
Sorted by:
    Note:  dataset has changed since last saved
```

Once labeling is completed, we should save the data to disk by using File > Save or typing

`. save, replace`

We could later retrieve these data any time through [icon] , File > Open, or by typing

`. use canada0`
(Canadian dataset 0)

Now we can proceed with analysis. We might notice, for instance, that male and female life expectancies correlate positively with each other and also negatively with the unemployment rate. The life expectancy–unemployment rate correlation is stronger for males.

`. correlate unemp mlife flife`
(obs=11)

	unemp	mlife	flife
unemp	1.0000		
mlife	-0.7440	1.0000	
flife	-0.6173	0.7631	1.0000

The order of observations within a dataset can be changed by the **sort** command. For example, to rearrange observations from smallest to largest in population, type

`. sort pop`

String variables are sorted alphabetically instead of numerically. Typing **sort** *place* will rearrange observations putting Alberta first, British Columbia second, and so on.

The **order** command controls the order of variables within a dataset. For example, we could make unemployment the second variable, and population last:

`. order place unemp mlife flife pop`

The Data Editor also has buttons that perform these functions. The Sort button applies to the column currently highlighted by the cursor. The << and >> buttons move the current variable to the beginning or end of the variable list, respectively. As with any other editing, these changes only become permanent if we subsequently save our data.

The Data Editor's Hide button does not rearrange the data, but rather makes a column temporarily invisible on the spreadsheet. This feature is convenient if, for example, we need to type in more variables and want to keep the province names or some other case identification column in view, adjacent to the "active" column where we are entering data.

We can also restrict the Data Editor beforehand to work only with certain variables, in a specified order, or with a specified range of values. For example,

```
. edit place mlife flife
```

or

```
. edit place unemp if pop > 100
```

The last example employs an **if** qualifier, an important tool described in the next section.

Specifying Subsets of the Data: `in` and `if` Qualifiers

Many Stata commands can be restricted to a subset of the data by adding an **in** or **if** qualifier. (Qualifiers are also available for many menu selections: look for an if/in or by/if/in tab along the top of the menu.) **in** specifies the observation numbers to which the command applies. For example, **list in 5** tells Stata to list only the 5th observation. To list the 1st through 10th observations, type

```
. list in 1/10
```

The letter **l** denotes the last case, and **−4** , for example, the fourth-from-last. Thus, we could list the four most populous Canadian places (which will include Canada itself) as follows:

```
. sort pop
. list place pop in -4/l
```

Note the important, although typographically subtle, distinction between **1** (number one, or first observation) and **l** (letter "el," or last observation). The **in** qualifier works in a similar way with most other analytical or data-editing commands. It always refers to the data *as presently sorted*.

The **if** qualifier also has broad applications, but it selects observations based on specific variable values. As noted, the observations in *canada0.dta* include not only 12 Canadian provinces or territories, but also Canada as a whole. For many purposes, we might want to exclude Canada from analyses involving the 12 territories and provinces. One way to do so is to restrict the analysis to only those places with populations below 20 million (20,000 thousand); that is, every place except Canada:

```
. summarize if pop < 20000
```

Variable	Obs	Mean	Std. Dev.	Min	Max
place	0				
pop	12	2467.158	3435.521	30.1	11100.3
unemp	10	12.26	4.44877	7	19.6
mlife	12	74.225	1.728965	70.2	75.8
flife	12	80.68333	1.0116	78	81.8

Compare this with the earlier **summarize** output to see how much has changed. The previous mean of population, for example, was grossly misleading because it counted every person twice.

The " < " (is less than) sign is one of six *relational operators*:

==	is equal to
!=	is not equal to (~= also works)
>	is greater than
<	is less than
>=	is greater than or equal to
<=	is less than or equal to

A double equals sign, " == ", denotes the logical test, "*Is* the value on the left side the same as the value on the right?" To Stata, a single equals sign means something different: "*Make* the value on the left side be the same as the value on the right." The single equals sign is not a relational operator and cannot be used within **if** qualifiers. Single equals signs have other meanings. They are used with commands that generate new variables, or replace the values of old ones, according to algebraic expressions. Single equals signs also appear in certain specialized applications such as weighting and hypothesis tests.

Any of these relational operators can be used to select observations based on their values for numeric variables. Only two operators, == and !=, make sense with string variables. To apply an **if** qualifier to string values, enclose the target value in double quotes. For example, we could get a summary excluding Canada (leaving in the 12 provinces and territories):

```
. summarize if place != "Canada"
```

Two or more relational operators can be combined within a single **if** expression by the use of *logical operators*. Stata's logical operators are the following:

&	and
|	or (symbol is a vertical bar, not the number one or letter "el")
!	not (~ also works)

The Canadian territories (Yukon and Northwest) both have fewer than 100,000 people. To find the mean unemployment and life expectancies for the 10 Canadian provinces only, excluding both the smaller places (territories) and the largest (Canada), we could use this command:

```
. summarize unemp mlife flife if pop > 100 & pop < 20000
```

Variable	Obs	Mean	Std. Dev.	Min	Max
unemp	10	12.26	4.44877	7	19.6
mlife	10	74.92	.6051633	73.9	75.8
flife	10	80.98	.586515	79.8	81.8

Parentheses allow us to specify the precedence among multiple operators. For example, we might list all the places that either have unemployment below 9, *or* have life expectancies of at least 75.4 for men *and* 81.4 for women:

`. list if unemp < 9 | (mlife >= 75.4 & flife >= 81.4)`

	place	pop	unemp	mlife	flife
8.	Manitoba	1137.5	8.5	75	80.8
9.	Saskatchewan	1015.6	7	75.2	81.8
10.	Alberta	2747	8.4	75.5	81.4
11.	British Columbia	3766	9.8	75.8	81.4

A note of caution regarding missing values: Stata ordinarily shows missing values as a period, but in some operations (notably **sort** and **if**, although not in statistical calculations such as means or correlations), these same missing values are treated as if they were large positive numbers. Watch what happens if we sort places from lowest to highest unemployment rate, and then ask to see places with unemployment rates above 15%:

`. sort unemp`
`. list if unemp > 15`

	place	pop	unemp	mlife	flife
2.	Newfoundland	575.4	19.6	73.9	79.8
3.	Prince Edward Island	136.1	19.1	74.8	81.3
12.	Yukon	30.1	.	71.3	80.4
13.	Northwest Territories	65.8	.	70.2	78

The two places with missing unemployment rates were included among those "greater than 15." In this instance the result is obvious, but with a larger dataset we might not notice. Suppose that we were analyzing a political opinion poll. A command such as the following would tabulate the variable *vote* not only for people age 65 and older, as intended, but also for any people whose *age* values were missing:

`. tabulate vote if age >= 65`

Where missing values exist, we might have to deal with them explicitly as part of the **if** expression.

`. tabulate vote if age >= 65 & age < .`

A less-than inequality such as *age* < . is a general way to select observations with nonmissing values. As shown later in this chapter, Stata permits up to 27 different missing values codes, although so far we have used only the default " . " The other 26 codes are represented internally as numbers even larger than " . ", so an if **age** < . qualifier sets them all aside. Type **help missing** for more details.

An alternative way to screen out missing values uses the **missing( )** function, which evaluates to 1 if a value is missing, and 0 if it is not. For example, to tabulate *vote* only for those observations that have nonmissing values of *age*, *income*, and *education,* type

```
. tabulate vote if missing(age, income, education)==0
```

The **in** and **if** qualifiers set observations aside temporarily so that a particular command does not apply to them. These qualifiers have no effect on the data in memory, and the next command will apply to all observations unless it too has an **in** or **if** qualifier. To drop variables from the data in memory, use the **drop** command (or use the Data Editor). Returning to our Canadian data (*canada0.dta*), we could drop *mlife* and *flife* from memory by typing

```
. drop mlife flife
```

Either **in** or **if** qualifiers can be used to select which observations to drop. Following our earlier **sort unemp** command, the two territories occupy the 12th and 13th positions in the data. Canada itself is 6th. **drop in 12/13** means "drop the 12th through the 13th observations."

```
. list
```

	place	unemp	pop
1.	Saskatchewan	7	1015.6
2.	Alberta	8.4	2747
3.	Manitoba	8.5	1137.5
4.	Ontario	9.3	11100.3
5.	British Columbia	9.8	3766
6.	Canada	10.6	29606.1
7.	Quebec	13.2	7334.2
8.	New Brunswick	13.8	760.1
9.	Nova Scotia	13.9	937.8
10.	Prince Edward Island	19.1	136.1
11.	Newfoundland	19.6	575.4
12.	Yukon	.	30.1
13.	Northwest Territories	.	65.8

```
. drop in 12/13
(2 observations deleted)
. drop in 6
(1 observation deleted)
```

The same change could have been accomplished through an **if** qualifier, with a command that says "drop if *place* equals Canada or population is less than 100." (We first retrieve the saved dataset, because so we can show how to drop these three observations a different way.)

```
. use canada0, clear
(Canadian dataset 0)
. drop if place == "Canada" | pop < 100
(3 observations deleted)
```

After dropping Canada, the territories, and the variables *mlife* and *flife*, we have the following reduced dataset:

```
. list
```

	place	pop	unemp	mlife	flife
1.	Newfoundland	575.4	19.6	73.9	79.8
2.	Prince Edward Island	136.1	19.1	74.8	81.3
3.	Nova Scotia	937.8	13.9	74.2	80.4
4.	New Brunswick	760.1	13.8	74.8	80.6
5.	Quebec	7334.2	13.2	74.5	81.2
6.	Ontario	11100.3	9.3	75.5	81.1
7.	Manitoba	1137.5	8.5	75	80.8
8.	Saskatchewan	1015.6	7	75.2	81.8
9.	Alberta	2747	8.4	75.5	81.4
10.	British Columbia	3766	9.8	75.8	81.4

We can also drop selected variables or observations with the Delete button in the Data Editor.

Instead of telling Stata which variables or observations to drop, it sometimes is simpler to specify which to keep. The same reduced dataset could have been obtained as follows:

```
. keep place pop unemp
. keep if place != "Canada" & pop >= 100
(3 observations deleted)
```

Like any other changes to the data in memory, none of these reductions affect disk files until we save the data. At that point, we will have the option of writing over the old dataset (**save, replace**) and thus destroying it, or just saving the newly modified dataset with a new name (by choosing File > Save As , or by typing a command with the form **save *newname***) so that both versions exist on disk.

Generating and Replacing Variables

The **generate** and **replace** commands allow us to create new variables or change the values of existing variables. For example, in Canada, as in most industrial societies, women tend to live longer than men. To analyze regional variations in this gender gap, we might retrieve dataset *canada1.dta* and generate a new variable equal to female life expectancy (*flife*) minus male life expectancy (*mlife*). In the main part of a **generate** or **replace** statement (unlike **if** qualifiers) we use a single equals sign.

```
. use canada1, clear
(Canadian dataset 1)
. generate gap = flife - mlife
. label variable gap "Female-male life expectancy gap"
. describe gap
```

variable name	storage type	display format	value label	variable label
gap	float	%9.0g		Female-male life expectancy gap

. list *place flife mlife gap*

	place	flife	mlife	gap
1.	Canada	81.1	75.1	6
2.	Newfoundland	79.8	73.9	5.900002
3.	Prince Edward Island	81.3	74.8	6.5
4.	Nova Scotia	80.4	74.2	6.200005
5.	New Brunswick	80.6	74.8	5.799995
6.	Quebec	81.2	74.5	6.699997
7.	Ontario	81.1	75.5	5.599998
8.	Manitoba	80.8	75	5.800003
9.	Saskatchewan	81.8	75.2	6.600006
10.	Alberta	81.4	75.5	5.900002
11.	British Columbia	81.4	75.8	5.599998
12.	Yukon	80.4	71.3	9.099998
13.	Northwest Territories	78	70.2	7.800003

For the province of Newfoundland, the true value of *gap* should be 79.8 – 73.9 = 5.9 years, but the output shows this value as 5.900002 instead. Like all computer programs, Stata stores numbers in binary form, and 5.9 has no exact binary representation. The small inaccuracies that arise from approximating decimal fractions in binary are unlikely to affect statistical calculations much, but they appear disconcerting in data lists. We can change the display format so that Stata shows only a rounded-off version. The following command specifies a fixed display format four numerals wide, with one digit to the right of the decimal:

. format *gap* %4.1f

Even when the display shows 5.9, however, a command such as the following will return no observations:

. list if *gap* == 5.9

This occurs because Stata believes the value does not exactly equal 5.9. (More technically, Stata stores *gap* values in single precision but does all calculations in double precision, and the single- and double-precision approximations of 5.9 are not identical.)

Display formats, as well as variables names and labels, can also be changed by double-clicking on a column in the Data Editor. Fixed numeric formats such as **%4.1f** are one of the three most common numeric display format types. These are

%w.dg   General numeric format, where *w* specifies the total width or number of columns displayed and *d* the minimum number of digits that must follow the decimal point. Exponential notation (such as 1.00e+07, meaning 1.00×10^7 or 10 million) and shifts in the decimal-point position will be used automatically as needed, to display values in an optimal (but varying) fashion.

%*w*.*d*f   Fixed numeric format, where *w* specifies the total width or number of columns displayed and *d* the fixed number of digits that must follow the decimal point.

%*w*.*d*e   Exponential numeric format, where *w* specifes the total width or number of columns displayed and *d* the fixed number of digits that must follow the decimal point.

For example, as we saw in Table 2.1, the 1995 population of Canada was approximately 29,606,100 people, and the Yukon Territory population was 30,100. The table below shows how those two numbers appear under several different display formats.

format	Canada	Yukon
%9.0g	2.96e+07	30100
%9.1f	29606100.0	30100.0
%12.5e	2.96061e+07	3.01000e+04

Although the displayed values look different, their internal values are identical. Calculations remain unaffected by display formats. Other numeric display formatting options include the use of commas, left- and right-justification, or leading zeroes. There also exist special formats for dates, time series variables, and string variables. Type **help format** for more information.

replace can make the same sorts of calculations as **generate**, but it changes values of an existing variable instead of creating a new variable. For example, suppose that we had questionnaire data that included household income in dollars. We decide it would be more convenient to work with income in thousands of dollars. To convert income in dollars to income in thousands of dollars, we would just divide all values by 1,000:

```
. replace income = income/1000
```

replace can make such wholesale changes, or it can be used with **in** or **if** qualifiers to selectively edit the data. Suppose our survey variables include *age* and year born (*born*). A command such as the following would correct one or more typos where a subject's age had been incorrectly typed as 299 instead of 29:

```
. replace age = 29 if age == 299
```

Alternatively, the following command could correct an error in the value of *age* for observation number 1453:

```
. replace age = 29 in 1453
```

For a more complicated example,

```
. replace age = 2008-born if age >= .  |  age < 2008-born
```

This replaces values of variable *age* with 2008 minus the year of birth if *age* is missing or if the reported age is less than 2008 minus the year of birth.

generate and **replace** provide tools to create categorical variables as well. We noted earlier that our Canadian dataset includes several types of observations: 2 territories, 10 provinces, and one country combining them all. Although **in** and **if** qualifiers allow us to separate these, and **drop** can eliminate observations from the data, it might be most convenient to have a categorical variable that indicates the observation's "type." The following example shows one way to create such a variable. We start by generating *type* as a constant, equal to 1 for each observation. Next, we replace this with the value 2 for the Yukon and Northwest Territories, and with 3 for Canada. The final steps involve labeling new variable *type* and defining labels for values 1, 2, and 3.

```
. use canada2, clear
(Canadian dataset 2)
. generate type = 1
. replace type = 2 if place == "Yukon" | place == "Northwest
      Territories"
(2 real changes made)
. replace type = 3 if place == "Canada"
(1 real change made)
. label variable type "Province, territory or nation"
. label values type typelbl
. label define typelbl 1 "Province" 2 "Territory" 3 "Nation"
. list place flife mlife gap type
```

	place	flife	mlife	gap	type
1.	Canada	81.1	75.1	6.0	Nation
2.	Newfoundland	79.8	73.9	5.9	Province
3.	Prince Edward Island	81.3	74.8	6.5	Province
4.	Nova Scotia	80.4	74.2	6.2	Province
5.	New Brunswick	80.6	74.8	5.8	Province
6.	Quebec	81.2	74.5	6.7	Province
7.	Ontario	81.1	75.5	5.6	Province
8.	Manitoba	80.8	75	5.8	Province
9.	Saskatchewan	81.8	75.2	6.6	Province
10.	Alberta	81.4	75.5	5.9	Province
11.	British Columbia	81.4	75.8	5.6	Province
12.	Yukon	80.4	71.3	9.1	Territory
13.	Northwest Territories	78	70.2	7.8	Territory

As illustrated, labeling the values of a categorical variable requires two commands. The **label define** command specifies what labels go with what numbers. The **label values** command specifies to which variable these labels apply. One set of labels (created through one **label define** command) can apply to any number of variables (that is, be referenced in any number of **label values** commands). Value labels can have up to 32,000 characters, but work best for most purposes if they are not too long.

generate can create new variables, and **replace** can produce new values, using any mixture of old variables, constants, random values, and expressions. For numeric variables, the following *arithmetic operators* apply:

+   add
−   subtract
*   multiply
/   divide
^   raise to power

Parentheses will control the order of calculation. Without them, the ordinary rules of precedence apply. Of the arithmetic operators, only addition, "+", works with string variables, where it connects two string values into one.

Although their purposes differ, **generate** and **replace** have similar syntax. Either can use any mathematically or logically feasible combination of Stata operators and **in** or **if** qualifiers. These commands can also employ Stata's broad array of special functions, introduced later.

Missing Value Codes

Examples seen so far involve only a single missing-value code, Stata's default: a large number which Stata displays as a period. In some datasets, however, values might be missing for several different reasons. We could denote different kinds of missing values by using extended missing-value codes. These are even larger numbers, which Stata displays as ".a" through ".z". Unlike the default missing-value code ".", the extended missing-value codes can be labeled.

Different kinds of missing values often arise with surveys, where the question "In what year were you married?" might have no answer because the respondent has never been married, can't recall, or thinks it's none of your business. Dataset *Granite_06_10s.dta*, contains data from a political opinion survey, New Hampshire's Granite State Poll. This poll, conducted in October 2006, helps to illustrate Stata's extended missing-value codes.

```
Contains data from C:\data\Granite_06_10s.dta
  obs:           515                          Granite State Poll 10/2006 subset
 vars:            11                          15 Mar 2008 13:16
 size:        11,845 (99.9% of memory free)
```

variable name	storage type	display format	value label	variable label
respnum_	int	%8.0g		respondent ID number
censuswt	float	%9.0g		survey weight
sex	byte	%8.0g	sex	Sex of respondent
age	byte	%8.0g	age	Age of respondent
edlevel	byte	%8.0g	d3	Highest level of education
hincome	byte	%8.0g	d13	Household income
marstat	byte	%8.0g	d1	Marital status
partyaf	byte	%8.0g	d4	Political party affiliation
iraq2	byte	%8.0g	iraq2	Support/oppose US war in Iraq
favbush	byte	%8.0g	nhfav1	Favorability rating Pres. Bush
novint	byte	%20.0g	novint	Interest in Nov 2006 election

```
Sorted by:
```

One variable in this poll, *novint*, asked respondents about their level of interest in the upcoming November 2006 elections.

```
. tab novint
```

Interest in Nov 2006 election	Freq.	Percent	Cum.
Extremely interested	102	19.81	19.81
Very interested	174	33.79	53.59
Somewhat interested	171	33.20	86.80
Not very interested	60	11.65	98.45
Don't know	5	0.97	99.42
No answer	3	0.58	100.00
Total	515	100.00	

Like many other surveys, the Granite State Poll employs particular numbers to represent different kinds of non-answers. In this case, the number 98 means the answer was "Don't know," and 99 means no answer was given. We can see these numerical values if we ask for the same table without value labels.

```
. tab novint, nolabel
```

Interest in Nov 2006 election	Freq.	Percent	Cum.
1	102	19.81	19.81
2	174	33.79	53.59
3	171	33.20	86.80
4	60	11.65	98.45
98	5	0.97	99.42
99	3	0.58	100.00
Total	515	100.00	

The problem here is that Stata views 98 and 99 as legitimate numbers, and includes them when it calculates a mean or absurdly high standard deviation. The 98s and 99s are less visible, but similarly distort our results, when we calculate correlations or other statistics. Interest in the elections shows a very weak negative correlation (–0.0699) with *age*.

```
. summarize novint
```

Variable	Obs	Mean	Std. Dev.	Min	Max
novint	515	3.864078	11.91984	1	99

```
. correlate novint age
(obs=497)
```

	novint	age
novint	1.0000	
age	-0.0699	1.0000

If we replace the 98 and 99 values with ".", that would solve our problem with means and correlations, but at the cost of throwing away information — namely, the distinction between "Don't know" and "No answer" responses, which might be important for some polling calculations. A better solution is to define a new variable *novint2*, then recode its 98 and 99 as separate extended missing values. The **mvdecode** command recodes numerical values into missing, as specified in its **mv( )** option. In this example, we recode 98 as ".a" and 99 as ".b".

```
. generate novint2 = novint
. mvdecode novint2, mv(98=.a \ 99=.b)
    novint2: 8 missing values generated
. tabulate novint2, miss
```

novint2	Freq.	Percent	Cum.
1	102	19.81	19.81
2	174	33.79	53.59
3	171	33.20	86.80
4	60	11.65	98.45
.a	5	0.97	99.42
.b	3	0.58	100.00
Total	515	100.00	

Extended missing values such as .a and .b can be labeled, and these labels will show up if we ask for a tabulation that includes missing values. The missing values do not enter into calculations such as means or correlations, however, so we now see more reasonable summary statistics. With this correction, the correlation between *novint* and *age* (–0.2641) is stronger than it previously appeared. Younger respondents expressed less interest in this election.

```
. label variable novint2 "Interest in Nov election, v.2"
. label values novint2 novint2
. label define novint2 1 "Extremely interested" 2 "Very interested"
        3 "Somewhat interested" 4 "Not very interested"
        .a "Don't know" .b "No answer"
. tabulate novint2, miss
```

Interest in Nov election, v.2	Freq.	Percent	Cum.
Extremely interested	102	19.81	19.81
Very interested	174	33.79	53.59
Somewhat interested	171	33.20	86.80
Not very interested	60	11.65	98.45
Don't know	5	0.97	99.42
No answer	3	0.58	100.00
Total	515	100.00	

```
. summ novint2
```

Variable	Obs	Mean	Std. Dev.	Min	Max
novint2	507	2.372781	.9351971	1	4

```
. correlate novint2 age
(obs=489)
```

	novint2	age
novint2	1.0000	
age	-0.2641	1.0000

A glance at the summary statistics for these data reveals that three other variables also have maximum values of 98 or 99, with similar meanings of "Don't know" or "No answer." Further investigation would uncover a third code number, 97, which the pollsters describe as "refused."

```
. summarize
```

Variable	Obs	Mean	Std. Dev.	Min	Max
respnum_	515	301.6951	173.5092	1	599
censuswt	515	1.000057	.4479156	.2206591	2.511936
sex	515	1.576699	.4945626	1	2
age	497	3.547284	1.397823	1	6
edlevel	513	5.403509	4.362812	1	98
hincome	489	31.33129	41.8643	1	98
marstat	511	1.870841	1.423525	1	6
partyaf	504	3.89881	2.061596	1	8
iraq2	510	3.32549	1.695926	1	5
favbush	512	2.621094	6.052687	1	98
novint	515	3.864078	11.91984	1	99
novint2	507	2.372781	.9351971	1	4

We could designate values of 97, 98, and 99 as missing for more than one variable with a single **mvdecode** command. Note that household income, always a sensitive survey question, has 140 more missing values after this treatment, whereas education level has only one more.

```
. mvdecode novint edlevel hincome favbush , mv(97=. \ 98=.a \ 99=.b)
    novint: 8 missing values generated
   edlevel: 1 missing value generated
   hincome: 140 missing values generated
   favbush: 2 missing values generated
```

Further commands using **label define** would be needed to supply appropriate labels for .a and .b along with the nonmissing values for each variable. With or without value labels, however, the **mvdecode** command results in correct statistics and counts. Note that the mean household income has dropped from 31.33 (on a 1–7 scale!) to a more reasonable 4.86.

```
. summarize
```

Variable	Obs	Mean	Std. Dev.	Min	Max
respnum_	515	301.6951	173.5092	1	599
censuswt	515	1.000057	.4479156	.2206591	2.511936
sex	515	1.576699	.4945626	1	2
age	497	3.547284	1.397823	1	6
edlevel	512	5.222656	1.503172	1	7
hincome	349	4.862464	1.867291	1	7
marstat	511	1.870841	1.423525	1	6
partyaf	504	3.89881	2.061596	1	8
iraq2	510	3.32549	1.695926	1	5
favbush	510	2.247059	.9453335	1	3
novint	507	2.372781	.9351971	1	4
novint2	507	2.372781	.9351971	1	4

As usual, the changes we have made do not become permanent until our dataset is saved. After so much recoding, it makes sense to save these data with a new name — in case, for some future reason, we want to take another look at the original "raw" data.

```
. save Granite_06_10s2, replace
file C:\data\Granite_06_10s2.dta saved
```

Using Functions

This section lists many of the functions available for use with **generate** or **replace**. For example, we could create a new variable named *loginc*, equal to the natural logarithm of *income*, by using the natural log function **ln** in a **generate** command:

```
. generate loginc = ln(income)
```

ln is one of Stata's *mathematical functions*. See **help math functions** for a complete list with details. Some example math functions are listed below.

abs(*x*)	Absolute value of *x*.
acos(*x*)	Arc-cosine returning radians. Because 360 degrees = 2π radians, **acos**(*x*)***180/_pi** would give the arc-cosine returning degrees (_pi denotes the mathematical constant π).
comb(*n*,*k*)	Combinatorial function (number of possible combinations of *n* things taken *k* at a time).
cos(*x*)	Cosine of radians. To find the cosine of *y* degrees, type **generate y = cos(*y* *_pi/180)**
exp(*x*)	Exponential (*e* to power).
int(*x*)	Integer obtained by truncating *x* towards zero.
ln(*x*)	Natural (base *e*) logarithm. For any other base number B, to find the base B logarithm of *x*, type **generate y = ln(*x*)/ln(B)**
lnfactorial(*x*)	Natural log of factorial. To find *x* factorial, type **generate y = round(exp(lnfact(*x*),1)**
log10(*x*)	Base 10 logarithm.
logit(*x*)	Log of odds ratio of *x*: ln(*x* /(1–*x*))
max(*x1*,*x2*,..,*xn*)	Maximum of *x1*, *x2*, ..., *xn*.
min(*x1*,*x2*,..,*xn*)	Minimum of *x1*, *x2*, ..., *xn*
round(*x*)	Round *x* to nearest whole number.
round(*x*,*y*)	Round *x* in units of *y*.
sign(*x*)	–1 if *x*<0, 0 if *x*=0, +1 if *x*>0
sin(*x*)	Sine of radians.
sqrt(*x*)	Square root.
sum(x)	Running sum of *x* (also see **help egen**)
tan(*x*)	Tangent of radians.

Many *probability functions* exist as well. Consult **help density functions** and the reference manuals for a full list and details such as definitions, constraints on parameters, and the treatment of missing values. A few examples are listed on the following page.

betaden(*a,b,x*)   Probability density of the beta distribution.

binomialtail(*n,k,p*) Probability of *k* or more successes in *n* trials when the probability of a success on a single trial is *p*.

chi2(*n,x*)   Cumulative chi-squared distribution with *n* degrees of freedom.

chi2tail(*n,x*)   Reverse cumulative (upper-tail, survival) chi-squared distribution with *n* degrees of freedom. chi2tail(*n,x*) = 1 – chi2(*n,x*)

F(*n1,n2,f*)   Cumulative *F* distribution with *n1* numerator and *n2* denominator degrees of freedom.

gammaden(*a,b,g,x*)   Probability density function for the gamma family, where gammaden(*a*,1,0,*x*) = the probability density function for the cumulative gamma distribution gammap(*a,x*).

invbinomial(*n,k,P*)   Inverse binomial. For $P \leq 0.5$, probability *p* such that the probability of observing *k* or more successes in *n* trials is *P*; for $P > 0.5$, probability *p* such that the probability of observing *k* or fewer successes in *n* trials is $1 - P$.

invchi2(*n,p*)   Inverse of **chi2( )**. If chi2(*n,x*) = *p*, then invchi2(*n,p*) = *x*

invF(*n1,n2,p*)   Inverse cumulative *F* distribution. If   F(*n1,n2,f*) = *p*, then invF(*n1,n2,p*) = *f*

invnormal(*p*)   Inverse cumulative standard normal distribution. If normal(*z*) = *p*, then invnormal(*p*) = *z*

normal(*z*)   Cumulative standard normal distribution.

normalden(*x,m,s*) Normal density, mean *m* and standard deviation *s*.

tden(*n,t*)   Probability density function of Student's *t* distribution with *n* degrees of freedom.

ttail(*n,t*)   Reverse cumulative (upper-tail) Student's *t* distribution with *n* degrees of freedom. This function returns probability $T > t$.

uniform()   Pseudo-random number generator, returning values from a uniform distribution theoretically ranging from 0 to nearly 1, written [0,1).

Nothing goes inside the parentheses with **uniform()**. Optionally, we can control the pseudo-random generator's starting seed, and hence the stream of "random" numbers, by first issuing a **set seed** # command, where # could be any integer from 0 to $2^{31} - 1$ inclusive. Omitting the **set seed** command corresponds to **set seed 123456789**. If we want to generate identical streams of random numbers in different Stata sessions, or different parts of one session, specifying the same **set seed** value can do this.

Stata provides many *date functions*, date-related *time series functions* and special formats for displaying time or date variables. Lists and details can be found in the *User's Guide*, or by typing **help dates and times**. Below are some examples of date functions. "Elapsed date" in these functions refers to the number of days since January 1, 1960.

date(*s_1,s_2*[,y])   Elapsed date corresponding to s_1. s_1 is a string variable indicating the date in virtually any format. Months can be spelled out, abbreviated to three characters, or given as numbers; years can include or exclude the century; blanks and punctuation are allowed. s_2 is any permutation of M, D, and [##]Y, such as "MDY" (in quotes) — in the order that month, day, and year occur in s_1. ## gives the century if years in s_1 have only two digits.

mdy(*M,D,Y*)   Elapsed date corresponding to *M*, *D*, and *Y*.

day(*e*)   Numeric day of the month corresponding to *e*, the elapsed date.
month(*e*)   Numeric month corresponding to *e*, the elapsed date.
year(*e*)   Numeric year corresponding to *e*, the elapsed date.
dow(*e*)   Numeric day of the week corresponding to *e*, the elapsed date.
doy(*e*)   Numeric day of the year corresponding to *e*, the elapsed date.
week(*e*)   Numeric week of the year corresponding to *e*, the elapsed date.

Some useful special functions include the following:

autocode(*x,n,xmin,xmax*) Forms categories from *x* by partitioning the interval from *xmin* to *xmax* into *n* equal-length intervals and returning the upper bound of the interval that contains *x*.

cond(*x,a,b*)   Returns *a* if *x* evaluates to "true" and *b* if *x* evaluates to "false."
                          generate *y* = cond(*inc1* > *inc2*, *inc1*, *inc2*)
                          creates the variable *y* as the maximum of *inc1* and *inc2* (assuming neither is missing).

sum(*x*)   Returns the running sum of *x*, treating missing values as zero.

String functions, not described here, help to manipulate and evaluate string variables. Type **help string functions** for a complete list of string functions. The reference manuals and *User's Guide* give examples and details of these and other functions.

Multiple functions, operators, and qualifiers can be combined in one command as needed. The functions and algebraic operators just described can also be used in another way that does not create or change any dataset variables. The **display** command performs a single calculation and shows the results onscreen. For example:

```
. display 2+3
5
```

```
. display log10(10^83)
83
```

```
. display invttail(120,.025) * 34.1/sqrt(975)
2.1622305
```

Thus, **display** can serve as an onscreen statistical calculator.

Unlike a calculator, **display**, **generate**, and **replace** have direct access to Stata's statistical results. For example, suppose that we summarized the unemployment rates from dataset *canada1.dta*:

```
. summarize unemp
```

Variable	Obs	Mean	Std. Dev.	Min	Max
unemp	11	12.10909	4.250048	7	19.6

After **summarize**, Stata temporarily stores the mean as a scalar named r(mean) .

```
. display r(mean)
12.109091
```

We could use this result to create variable *unempDEV*, defined as deviations from the mean:

```
. gen unempDEV = unemp - r(mean)
(2 missing values generated)
. summ unemp unempDEV
```

Variable	Obs	Mean	Std. Dev.	Min	Max
unemp	11	12.10909	4.250048	7	19.6
unempDEV	11	4.33e-08	4.250048	-5.109091	7.49091

Stata temporarily saves results after many analyses, such as r(mean) after **summarize**. These can be valuable for subsequent calculations or programming. To see a complete list of the names and values currently saved, type **return list**. In this example, saved values named r(N), f(sum_w), r(mean), and so forth describe the most recent **summarize** results for *unempDEV*.

```
. return list
```

scalars:
```
             r(N) =  11
         r(sum_w) =  11
          r(mean) =  4.33488325639e-08
           r(Var) =  18.0629105746097
            r(sd) =  4.250048302620771
           r(min) =  -5.109090805053711
           r(max) =  7.490909576416016
           r(sum) =  4.76837158203e-07
```

Stata also provides another variable-creation command, **egen** ("extensions to **generate**"), which has its own set of functions to accomplish tasks not easily done by **generate**. These include such things as creating new variables from the sums, maxima, minima, medians, interquartile ranges, standardized values, or moving averages of existing variables or expressions. For example, the following command creates a new variable named *zscore*, equal to the standardized (mean 0, variance 1) values of *x*:

```
. egen zscore = std(x)
```

Or, the following command creates new variable *avg*, equal to the row mean of each observation's values on *x*, *y*, *z*, and *w*, ignoring any missing values.

```
. egen avg = rowmean(x,y,z,w)
```

To create a new variable named *total*, equal to the row sum of each observation's values on *x*, *y*, *z*, and *w*, treating missing values as zeroes, type

```
. egen total = rowtotal(x,y,z,w)
```

The following command creates new variable *xrank*, holding ranks corresponding to values of *x*: *xrank* = 1 for the observation with highest *x*. *xrank* = 2 for the second highest, and so forth.

```
. egen xrank = rank(x)
```

Consult **help egen** for a complete list of **egen** functions, or the reference manuals for further examples.

Converting Between Numeric and String Formats

Dataset *canada3.dta* contains one string variable, *place*. It also has a labeled categorical variable, *type*. Both seem to have nonnumeric values.

```
. use canada2, clear
(Canadian dataset 3)

. list place type
```

		place	type
1.	Canada		Nation
2.	Newfoundland		Province
3.	Prince Edward Island		Province
4.	Nova Scotia		Province
5.	New Brunswick		Province
6.	Quebec		Province
7.	Ontario		Province
8.	Manitoba		Province
9.	Saskatchewan		Province
10.	Alberta		Province
11.	British Columbia		Province
12.	Yukon		Territory
13.	Northwest Territories		Territory

Beneath the labels, however, *type* remains a numeric variable, indicated by a blue font in the Data Editor or Browser. Clicking on that cell will show the underlying numbers, or we can **list** these asking for the **nolabel** option:

```
. list place type, nolabel
```

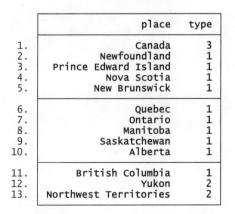

		place	type
1.	Canada		3
2.	Newfoundland		1
3.	Prince Edward Island		1
4.	Nova Scotia		1
5.	New Brunswick		1
6.	Quebec		1
7.	Ontario		1
8.	Manitoba		1
9.	Saskatchewan		1
10.	Alberta		1
11.	British Columbia		1
12.	Yukon		2
13.	Northwest Territories		2

String and labeled numeric variables behave differently when analyzed. Most statistical operations and algebraic relations are not defined for string variables, so we might want to have both string and labeled-numeric versions of the same information in our data. The **encode** command generates a labeled-numeric variable from a string variable. The number 1 is given to the alphabetically first value of the string variable, 2 to the second, and so on. The following example creates a labeled numeric variable named *placenum* from the string variable *place*:

```
. encode place, gen(placenum)
```

An opposite conversion is possible, too: The **decode** command generates a string variable using the values of a labeled numeric variable. Here we create string variable *typestr* from numeric variable *type*:

```
. decode type, gen(typestr)
```

When listed, the new numeric variable *placenum*, and the new string variable *typestr*, look similar to the originals:

```
. list place placenum type typestr
```

	place	placenum	type	typestr
1.	Canada	Canada	Nation	Nation
2.	Newfoundland	Newfoundland	Province	Province
3.	Prince Edward Island	Prince Edward Island	Province	Province
4.	Nova Scotia	Nova Scotia	Province	Province
5.	New Brunswick	New Brunswick	Province	Province
6.	Quebec	Quebec	Province	Province
7.	Ontario	Ontario	Province	Province
8.	Manitoba	Manitoba	Province	Province
9.	Saskatchewan	Saskatchewan	Province	Province
10.	Alberta	Alberta	Province	Province
11.	British Columbia	British Columbia	Province	Province
12.	Yukon	Yukon	Territory	Territory
13.	Northwest Territories	Northwest Territories	Territory	Territory

But with the **nolabel** option, the differences become visible. Stata views *placenum* and *type* basically as numbers.

```
. list place placenum type typestr, nolabel
```

		place	placenum	type	typestr
1.	Canada		3	3	Nation
2.	Newfoundland		6	1	Province
3.	Prince Edward Island		10	1	Province
4.	Nova Scotia		8	1	Province
5.	New Brunswick		5	1	Province
6.	Quebec		11	1	Province
7.	Ontario		9	1	Province
8.	Manitoba		4	1	Province
9.	Saskatchewan		12	1	Province
10.	Alberta		1	1	Province
11.	British Columbia		2	1	Province
12.	Yukon		13	2	Territory
13.	Northwest Territories		7	2	Territory

Most statistical analyses, such as finding means and standard deviations, work only with numeric variables. For calculation purposes, their labels do not matter.

```
. summarize place placenum type typestr
```

Variable	Obs	Mean	Std. Dev.	Min	Max
place	0				
placenum	13	7	3.89444	1	13
type	13	1.307692	.6304252	1	3
typestr	0				

Occasionally we encounter a string variable where the values are all or mostly numbers. To convert these string values into their numeric counterparts, use the **real** function. For example, in the artificial dataset below, the variable *siblings* is a string variable, although it only has one value, "4 or more," that could not be represented just as easily by a number.

```
. describe siblings
```

variable name	storage type	display format	value label	variable label
siblings	str9	%9s		Number of siblings (string)

```
. list
```

	siblings
1.	1
2.	3
3.	0
4.	2
5.	4 or more

```
. generate sibnum = real(siblings)
(1 missing value generated)
```

The new variable *sibnum* is numeric, with a missing value where *siblings* had "4 or more."

```
. list
```

	siblings	sibnum
1.	1	1
2.	3	3
3.	0	0
4.	2	2
5.	4 or more	.

The **destring** command provides a more flexible method for converting string variables to numeric. In the example above, we could have accomplished the same thing by typing

```
. destring siblings, generate(sibnum) force
```

See **help destring** for information about syntax and options.

Creating New Categorical and Ordinal Variables

A previous section illustrated how to construct a categorical variable called *type* to distinguish among territories, provinces, and nation in our Canadian dataset. You can create categorical or ordinal variables in many other ways. This section gives a few examples.

type has three categories:

```
. tabulate type
```

Province, territory or nation	Freq.	Percent	Cum.
Province	10	76.92	76.92
Territory	2	15.38	92.31
Nation	1	7.69	100.00
Total	13	100.00	

Suppose we want to re-express *type* as a set of dichotomies or "dummy variables," each coded 0 or 1. **tabulate** will create dummy variables automatically if we add the **generate** option. In the following example, this results in a set of variables called *type1*, *type2*, and *type3*, each representing one of the three categories of *type*:

```
. tabulate type, generate(type)
```

Province, territory or nation	Freq.	Percent	Cum.
Province	10	76.92	76.92
Territory	2	15.38	92.31
Nation	1	7.69	100.00
Total	13	100.00	

```
. describe
```

Contains data from **c:\data\canada3.dta**
  obs: 13   Canadian dataset 3
  vars: 10   15 Apr 2008 18:23
  size: 689 (99.9% of memory free)

variable name	storage type	display format	value label	variable label
place	str21	%21s		Place name
pop	float	%9.0g		Population in 1000s, 1995
unemp	float	%9.0g		% 15+ population unemployed, 1995
mlife	float	%9.0g		Male life expectancy years
flife	float	%9.0g		Female life expectancy years
gap	float	%4.1f		Female-male life expectancy gap
type	byte	%9.0g	typelbl	Province, territory or nation
type1	byte	%8.0g		type==Province
type2	byte	%8.0g		type==Territory
type3	byte	%8.0g		type==Nation

Sorted by:
    Note: dataset has changed since last saved

```
. list place type type1-type3
```

	place	type	type1	type2	type3
1.	Canada	Nation	0	0	1
2.	Newfoundland	Province	1	0	0
3.	Prince Edward Island	Province	1	0	0
4.	Nova Scotia	Province	1	0	0
5.	New Brunswick	Province	1	0	0
6.	Quebec	Province	1	0	0
7.	Ontario	Province	1	0	0
8.	Manitoba	Province	1	0	0
9.	Saskatchewan	Province	1	0	0
10.	Alberta	Province	1	0	0
11.	British Columbia	Province	1	0	0
12.	Yukon	Territory	0	1	0
13.	Northwest Territories	Territory	0	1	0

Re-expressing categorical information as a set of dummy variables involves no loss of information; in this example, *type1* through *type3* together tell us exactly as much as *type* itself does. Occasionally, however, analysts choose to re-express a measurement variable in categorical or ordinal form, even though this *does* result in a substantial loss of information. For example, *unemp* in *canada2.dta* gives a measure of the unemployment rate. Excluding Canada itself from the data, we see that *unemp* ranges from 7% to 19.6%, with a mean of 12.26:

```
. summarize unemp if type != 3
```

Variable	Obs	Mean	Std. Dev.	Min	Max
unemp	10	12.26	4.44877	7	19.6

Having Canada in the data becomes a nuisance at this point, so we drop it:

```
. drop if type == 3
(1 observation deleted)
```

Two commands create a dummy variable named *unemp2* with values of 0 when unemployment is below average (12.26), 1 when unemployment is equal to or above average, and missing when *unemp* is missing. In reading the second command, recall that Stata's sorting and relational operators treat missing values as very large numbers.

```
. generate unemp2 = 0 if unemp < 12.26
(7 missing values generated)
. replace unemp2 = 1 if unemp >= 12.26 & unemp < .
(5 real changes made)
```

We might want to group the values of a measurement variable, thereby creating an ordered-category or ordinal variable. The **autocode** function (see "Using Functions") provides automatic grouping of measurement variables. To create new ordinal variable *unemp3*, which groups values of *unemp* into three equal-width groups over the interval from 5 to 20, type

```
. generate unemp3 = autocode(unemp,3,5,20)
(2 missing values generated)
```

A list of the data shows how the new dummy (*unemp2*) and ordinal (*unemp3*) variables correspond to values of the original measurement variable *unemp*.

```
. list place unemp unemp2 unemp3
```

	place	unemp	unemp2	unemp3
1.	Newfoundland	19.6	1	20
2.	Prince Edward Island	19.1	1	20
3.	Nova Scotia	13.9	1	15
4.	New Brunswick	13.8	1	15
5.	Quebec	13.2	1	15
6.	Ontario	9.3	0	10
7.	Manitoba	8.5	0	10
8.	Saskatchewan	7	0	10
9.	Alberta	8.4	0	10
10.	British Columbia	9.8	0	10
11.	Yukon	.	.	.
12.	Northwest Territories	.	.	.

Using Explicit Subscripts with Variables

When Stata has data in memory, it also defines certain system variables that describe those data. For example, _N represents the total number of observations. _n represents the observation number: _n = 1 for the first observation, _n = 2 for the second, and so on to the last observation (_n = _N). If we issue a command such as the following, it creates a new variable, *caseID*, equal to the number of each observation as presently sorted:

```
. generate caseID = _n
```

Sorting the data another way will change each observation's value of _n , but its *caseID* value will remain unchanged. Thus, if we do sort the data another way, we can later return to the earlier order by typing

```
. sort caseID
```

Creating and saving unique case identification numbers that store the order of observations at an early stage of dataset development can greatly facilitate later data management.

We can use explicit subscripts with variable names, to specify particular observation numbers. For example, the 6th observation in dataset *canada1.dta* (if we have not dropped or re-sorted anything) is Quebec. Consequently, *pop[6]* refers to Quebec's population, 7334.2 thousand.

```
. display pop[6]
7334.2002
```

Similarly, *pop[12]* is the Yukon's population:

```
. display pop[12]
30.1
```

Explicit subscripting and the _n system variable have additional relevance when our data form a series. If we had the daily stock market price of a particular stock as a variable named *price*, for instance, then either *price* or, equivalently, *price*[_n] denotes the value of the _nth observation or day. *price*[_n–1] denotes the previous day's price, and *price*[_n+1] denotes the next. Thus, we might define a new variable *difprice*, which is equal to the change in *price* since the previous day:

```
. generate difprice = price - price[_n-1]
```

Chapter 13, on time series analysis, returns to this topic.

Importing Data from Other Programs

Previous sections illustrated how to enter and edit data using the Data Editor. If our original data reside in an appropriately formatted spreadsheet, a simpler way is to copy and paste blocks of data from the spreadsheet directly into the empty Data Editor. Pasting directly from a spreadsheet requires some care and perhaps experimentation because Stata will decide whether each data column represents a numeric or string variable based on the first value in that column. If the first (top) value in a column is a number, Stata assumes that column is a numeric variable; any later non-numeric values are coded as "missing." Conversely if the first value in the column appears non-numeric (such as words, or even numbers containing commas such as "1,500"), Stata views that as a string variable, so any numbers appearing later in that column are treated just as text, rather than actual numbers we could average or correlate. Black (numeric) or red (string) fonts in the Data Editor or Data Browser show at once which is which.

Spreadsheets created for other purposes might contain labels, explanations, row separators, or other features that we don't want to paste into a Stata file. Before selecting the block of data to copy, careful editing of the spreadsheet might be needed. One nice trick is to insert a row of variable names (single words) above the top column of data in our spreadsheet. When these names are copied into an empty Data Editor along with the rest of the data, Stata should automatically recognize that first row as the variable names.

These Data Editor methods are quick and easy, but for larger projects it is important to have tools that work directly with computer files created by other programs. Such files fall into two general categories: raw-data ASCII (text) files, which can be read into Stata with the appropriate Stata commands; and system files, which must be translated to Stata format by a special third-party program before Stata can read them.

To illustrate ASCII file methods, we return to the Canadian data of Table 2.1. Suppose that, instead of typing these data into Stata's Data Editor, we typed them into our word processor, with at least one space between each value. String values must be in double quotes if they contain internal spaces, as does "Prince Edward Island". For other string values, quotes are optional. Word processors allow the option of saving documents as ASCII (text) files, a simpler and more universal type than the word processor's usual saved-file format. We can thus create an ASCII file named *canada.raw* that looks something like this:

```
"Canada"  29606.1  10.6  75.1  81.1
"Newfoundland"  575.4  19.6  73.9  79.8
"Prince Edward Island"  136.1  19.1  74.8  81.3
"Nova Scotia"  937.8  13.9  74.2  80.4
"New Brunswick"  760.1  13.8  74.8  80.6
"Quebec"  7334.2  13.2  74.5  81.2
"Ontario"  11100.3  9.3  75.5  81.1
"Manitoba"  1137.5  8.5  75  80.8
"Saskatchewan"  1015.6  7  75.2  81.8
"Alberta"  2747  8.4  75.5  81.4
"British Columbia"  3766  9.8  75.8  81.4
"Yukon"  30.1  .  71.3  80.4
"Northwest Territories"  65.8  .  70.2  78
```

Note the use of periods, not blanks, to indicate missing values for the Yukon and Northwest Territories. If the dataset should have five variables, then for every observation, exactly five values (including periods for missing values) must exist.

infile reads into memory an ASCII file, such as *canada.raw*, in which the values are separated by one or more whitespace characters — blanks, tabs, and newlines (carriage return, line feed, or both) — or by commas. Its basic form is

```
. infile variable-list using filename.raw
```

With purely numeric data, the variable list could be omitted, in which case Stata assigns the names *var1*, *var2*, *var3*, and so forth. On the other hand, we might want to give each variable a distinctive name. We also need to identify string variables individually. For *canada.raw*, the **infile** command might be

```
. infile str30 place pop unemp mlife flife using canada.raw, clear
(13 observations read)
```

The **infile** variable list specifies variables in the order that they appear in the data file. The **clear** option drops any current data from memory before reading in the new file.

If any string variables exist, their names must each be preceded by a **str#** statement. **str30**, for example, informs Stata that the next-named variable (*place*) is a string variable with as many as 30 characters. Actually, none of the Canadian place names require more than 21 characters, but we do not need to know that in advance. It is often easier to overestimate string variable lengths. Then, once data are in memory, use **compress** to ensure that no variable takes up more space than it needs. The **compress** command automatically changes all variables to their most memory-efficient storage type.

```
. compress
place was str30 now str21
```

```
. describe
```

```
Contains data
  obs:           13
  vars:           5
  size:         585 (99.9% of memory free)
```

variable name	storage type	display format	value label	variable label
place	str21	%21s		
pop	float	%9.0g		
unemp	float	%9.0g		
mlife	float	%9.0g		
flife	float	%9.0g		

```
Sorted by:
    Note:  dataset has changed since last saved
```

We can now proceed to label variables and data as described earlier. At any point, the commands **save** *canada0* (or **save** *canada0*, **replace**) would save the new dataset in Stata format, as file *canada0.dta*. The original raw-data file, *canada.raw*, remains unchanged on disk.

If our variables have non-numeric values (for example, "male" and "female") that we want to store as labeled numeric variables, then adding the option **automatic** will accomplish this. For example, we might read in raw survey data through this **infile** command:

```
. infile gender age income vote using survey.raw, automatic
```

Spreadsheet and database programs commonly write ASCII files that have only one observation per line, with values separated by tabs or commas. To read these files into Stata, use **insheet**. Its general syntax resembles that of **infile**, with options telling Stata whether the data are delimited by tabs, commas, or other characters. For example, assuming tab-delimited data,

```
. insheet variable-list using filename.raw, tab
```

Or, assuming comma-delimited data with the first row of the file containing variable names (also comma-delimited),

```
. insheet variable-list using filename.raw, comma names
```

With **insheet** we do not need to separately identify string variables. If we include no variable list, and do not have variable names in the file's first row, Stata automatically assigns the variable names *var1*, *var2*, *var3*, ... Errors will occur if some values in our ASCII file are not separated by tabs, commas, or some other delimiter as specified in the **insheet** command.

Raw data files created by other statistical packages can be in "fixed-column" format, where the values are not necessarily delimited at all, but do occupy predefined column positions. Both **infile** and the more specialized command **infix** permit Stata to read such files. In the command syntax itself, or in a "data dictionary" existing in a separate file or as the first part of the data file, we have to specify exactly how the columns should be read.

Here is a simple example. Data exist in an ASCII file named *nfresour.raw*:

```
198624087641691000
198725247430001044
198825138637481086
198925358964371140
1990    8615731195
1991    7930001262
```

These data concern natural resource production in Newfoundland. The four variables occupy fixed column positions: columns 1–4 are the years (1986...1991); columns 5–8 measure forestry production in thousands of cubic meters (2408...missing); columns 9–14 measure mine production in thousands of dollars (764,169...793,000); and columns 15–18 are the consumer price index relative to 1986 (1000...1262). Notice that in fixed-column format, unlike space or tab-delimited files, blanks indicate missing values, and the raw data contain no decimal points. To read *nfresour.raw* into Stata, we specify each variable's column position:

```
. infix year 1-4 wood 5-8 mines 9-14 CPI 15-18
      using nfresour.raw, clear
(6 observations read)
```

```
. list
```

	year	wood	mines	CPI
1.	1986	2408	764169	1000
2.	1987	2524	743000	1044
3.	1988	2513	863748	1086
4.	1989	2535	896437	1140
5.	1990	.	861573	1195
6.	1991	.	793000	1262

More complicated fixed-column formats might require a data "dictionary." Data dictionaries can be straightforward, but they offer many possible choices. Type **help infiling** to see an outline of these commands. For more examples and explanation, consult the *User's Guide* and reference manuals. Stata also can load, write, or view data from ODBC (Open Database Connectivity) sources; see **help odbc**.

What if we need to export data from Stata to some other, non-ODBC program? The **outfile** command writes ASCII files to disk. A command such as the following will create a space-delimited ASCII file named *canada6.raw*, containing whatever data were in memory:

```
. outfile using canada6
```

The **infile**, **insheet**, **infix**, and **outfile** commands just described all manipulate raw data in ASCII files. A second, very quick, possibility is to copy your data from Stata's Browser and paste this directly into a spreadsheet such as Excel. Often the best option, however, is to transfer data directly between the specialized system files saved by various spreadsheet, database, or statistical programs. Several third-party programs perform such translations. Stat/Transfer, for example, will transfer data across many different formats including dBASE, Excel, FoxPro, Gauss, JMP, MATLAB, Minitab, OSIRIS, Paradox, S-Plus, SAS, SPSS, SYSTAT, and Stata. Even large datasets hundreds of megabytes in size can be translated or excerpted quickly with this program. It is available through StataCorp (www.stata.com) or from its maker, Circle Systems (www.stattransfer.com). Transfer programs prove indispensable for analysts working in multi-program environments or exchanging data with colleagues.

One distinguishing feature of Stata is worth mentioning here. Stata datasets saved on one Stata platform (whether Windows, Macintosh, or Unix) can be read without translation by Stata on any of the other platforms.

Combining Two or More Stata Files

We can combine Stata datasets in two general ways: **append** a second dataset that contains additional observations, or **merge** with other datasets that contain new variables or values. In keeping with this chapter's Canadian theme, we will illustrate these procedures using data on Newfoundland. File *newf1.dta* records the province's population for 1985 to 1989.

```
. use newf1, clear
(Newfoundland 1985-89)

. describe

Contains data from c:\data\newf1.dta
  obs:             5                        Newfoundland 1985-89
  vars:            2                        19 Jan 2008 19:02
  size:           70  (99.9% of memory free)
```

variable name	storage type	display format	value label	variable label
year	int	%9.0g		Year
pop	float	%9.0g		Population

```
Sorted by:

. list
```

	year	pop
1.	1985	580700
2.	1986	580200
3.	1987	568200
4.	1988	568000
5.	1989	570000

File *newf2.dta* has population and unemployment counts for some later years:

```
. use newf2
(Newfoundland 1990-95)

. describe

Contains data from c:\data\newf2.dta
  obs:             6                        Newfoundland 1990-95
  vars:            3                        19 Jan 2008 19:03
  size:          108  (99.9% of memory free)
```

variable name	storage type	display format	value label	variable label
year	int	%9.0g		Year
pop	float	%9.0g		Population
jobless	float	%9.0g		Number of people unemployed

```
Sorted by:

. list
```

	year	pop	jobless
1.	1990	573400	42000
2.	1991	573500	45000
3.	1992	575600	49000
4.	1993	584400	49000
5.	1994	582400	50000
6.	1995	575449	.

To combine these datasets, with *newf2.dta* already in memory, we use the **append** command:

. **append using** *newf1*

. **list**

	year	pop	jobless
1.	1990	573400	42000
2.	1991	573500	45000
3.	1992	575600	49000
4.	1993	584400	49000
5.	1994	582400	50000
6.	1995	575449	.
7.	1985	580700	.
8.	1986	580200	.
9.	1987	568200	.
10.	1988	568000	.
11.	1989	570000	.

Because variable *jobless* occurs in *newf2* (1990 to 1995) but not in *newf1*, its 1985 to 1989 values are missing in the combined dataset. We can now put the observations in order from earliest to latest and save these combined data as a new file, *newf3.dta*:

. **sort** *year*
. **list**

	year	pop	jobless
1.	1985	580700	.
2.	1986	580200	.
3.	1987	568200	.
4.	1988	568000	.
5.	1989	570000	.
6.	1990	573400	42000
7.	1991	573500	45000
8.	1992	575600	49000
9.	1993	584400	49000
10.	1994	582400	50000
11.	1995	575449	.

. **save** *newf3*

append might be compared to lengthening a sheet of paper (that is, the dataset in memory) by taping a second sheet with new observations (rows) to its bottom. **merge**, in its simplest form, corresponds to "widening" our sheet of paper by taping a second sheet to its right side, thereby adding new variables (columns). For example, dataset *newf4.dta* contains further Newfoundland time series: the numbers of births and divorces over the years 1980 to 1994. Thus it has some observations in common with our earlier dataset *newf3.dta*, as well as one variable (*year*) in common, but it also has two new variables not present in *newf3.dta*.

```
. use newf4
(Newfoundland 1980-94)

. describe

Contains data from c:\data\newf4.dta
  obs:            15                          Newfoundland 1980-94
  vars:            3                          19 Jan 2008 19:03
  size:          210 (99.9% of memory free)
```

variable name	storage type	display format	value label	variable label
year	int	%9.0g		Year
births	int	%9.0g		Number of births
divorces	int	%9.0g		Number of divorces

```
Sorted by:

. list
```

	year	births	divorces
1.	1980	10332	555
2.	1981	11310	569
3.	1982	9173	625
4.	1983	9630	711
5.	1984	8560	590
6.	1985	8080	561
7.	1986	8320	610
8.	1987	7656	1002
9.	1988	7396	884
10.	1989	7996	981
11.	1990	7354	973
12.	1991	6929	912
13.	1992	6689	867
14.	1993	6360	930
15.	1994	6295	933

We want to merge *newf3* with *newf4*, matching observations according to *year* wherever possible. To accomplish this, both datasets must be sorted by an index or key variable (which in this example is *year*). We earlier issued a **sort *year*** command before saving *newf3.dta*, so we now do the same with *newf4.dta*. Then we merge the two, specifying *year* as the key variable to match.

```
. sort year
. merge year using newf3
. describe
```

```
Contains data from c:\data\newf4.dta
  obs:             16                       Newfoundland 1980-94
  vars:             6                       19 Jan 2008 19:03
  size:           368 (99.9% of memory free)
```

variable name	storage type	display format	value label	variable label
year	int	%9.0g		Year
births	int	%9.0g		Number of births
divorces	int	%9.0g		Number of divorces
pop	float	%9.0g		Population
jobless	float	%9.0g		Number of people unemployed
_merge	byte	%8.0g		

Sorted by:

. **list**

	year	births	divorces	pop	jobless	_merge
1.	1980	10332	555	.	.	1
2.	1981	11310	569	.	.	1
3.	1982	9173	625	.	.	1
4.	1983	9630	711	.	.	1
5.	1984	8560	590	.	.	1
6.	1985	8080	561	580700	.	3
7.	1986	8320	610	580200	.	3
8.	1987	7656	1002	568200	.	3
9.	1988	7396	884	568000	.	3
10.	1989	7996	981	570000	.	3
11.	1990	7354	973	573400	42000	3
12.	1991	6929	912	573500	45000	3
13.	1992	6689	867	575600	49000	3
14.	1993	6360	930	584400	49000	3
15.	1994	6295	933	582400	50000	3
16.	1995	.	.	575449	.	2

In this example, we simply used **merge** to add new variables to our data, matching observations. By default, whenever the same variables are found in both datasets, those of the "master" data (the file already in memory) are retained and those of the "using" data are ignored. The **merge** command has several options, however, that override this default. A command of the following form would allow any *missing values* in the master data to be replaced by corresponding nonmissing values found in the using data (here, *newf5.dta*):

. **merge** *year* **using** *newf5*, **update**

Or, a command such as the following causes *any values* from the master data to be replaced by nonmissing values from the using data, if the latter are different:

. **merge** *year* **using** *newf5*, **update replace**

Values of the key variable can occur more than once in either dataset. For example, suppose the year 1990 occurs twice in the master data. Then values from the using data with *year* = 1990 are matched with each occurrence of *year* = 1990 in the master data. You can use this capability for many purposes, such as combining background data on individual patients with data on any number of separate doctor visits they made. Although **merge** makes this and many other data-management tasks straightforward, analysts should look closely at the results to be certain that the command is accomplishing what they intend.

As a diagnostic aid, **merge** automatically creates a new variable called *_merge*. Unless **update** was specified, *_merge* codes have the following meanings:
1   Observation from the master dataset only.
2   Observation from the using dataset only.
3   Observation from both master and using data (using values ignored if different).

If the **update** option was specified, *_merge* codes convey what happened:
1   Observation from the master dataset only.
2   Observation from the using dataset only.
3   Observation from both, master data agrees with using.
4   Missing in master updated.
5   Master disagrees with using (replaced if **replace** option specified).

It is an important step to review our *_merge* values carefully after each **merge** operation, making sure things turned out as planned. Before performing another **merge** operation, we can **drop _merge** to make way for the new version.

It is possible to merge multiple datasets with a single **merge** command. For example, if *newf5.dta* through *newf8.dta* are four datasets, each sorted by the variable *year*, then merging all four with the master dataset could be accomplished as follows.

```
. merge year using newf5 newf6 newf7 newf8, update replace
```

Other **merge** options include checks on whether the merging-variable values are unique, and the ability to specify which variables to keep. Type **help merge** for details. Merging and appending data can be accomplished through Data > Combine datasets menus, as well.

Transposing, Reshaping, or Collapsing Data

Long after a dataset has been created, we might discover that for some analytical purposes it has the wrong organization. Fortunately, several commands facilitate drastic restructuring of datasets. We will illustrate these using data (*growth1.dta*) on recent population growth in five eastern provinces of Canada. In these data, unlike our previous examples, province names are represented by a numeric variable (*provinc2*) with eight-character labels.

```
. use growth1, clear
(Eastern Canada growth)
```

```
. describe
```

```
Contains data from c:\data\growth1.dta
  obs:             5                          Eastern Canada growth
  vars:            5                          19 Jan 2008 19:09
  size:          125 (99.9% of memory free)
```

variable name	storage type	display format	value label	variable label
provinc2	byte	%8.0g	provinc2	Eastern Canadian province
grow92	float	%9.0g		Pop. gain in 1000s, 1991-92
grow93	float	%9.0g		Pop. gain in 1000s, 1992-93
grow94	float	%9.0g		Pop. gain in 1000s, 1993-94
grow95	float	%9.0g		Pop. gain in 1000s, 1994-95

Sorted by:

```
. list
```

	provinc2	grow92	grow93	grow94	grow95
1.	New Brun	10	2.5	2.2	2.4
2.	Newfound	4.5	.8	-3	-5.8
3.	Nova Sco	12.1	5.8	3.5	3.9
4.	Ontario	174.9	169.1	120.9	163.9
5.	Quebec	80.6	77.4	48.5	47.1

In this organization, population growth for each year is stored as a separate variable. We could analyze changes in the mean population growth from year to year. But given this organization, Stata could not readily draw a time plot of population growth against year, nor find the correlation between population growth in New Brunswick and Newfoundland. All the necessary information is here, but such analyses require different organizations of the data.

One simple reorganization involves transposing variables and observations. In effect, the dataset rows become its columns, and vice versa. This is accomplished by the **xpose** command. The option **clear** is required with this command, because it always clears the present data from memory. Including the **varname** option creates an additional variable (named *_varname*) in the transposed dataset, containing original variable names as strings.

```
. xpose, clear varname
. describe
```

```
Contains data
  obs:             5
  vars:            6
  size:          180 (99.9% of memory free)
```

variable name	storage type	display format	value label	variable label
v1	float	%9.0g		
v2	float	%9.0g		
v3	float	%9.0g		
v4	float	%9.0g		
v5	float	%9.0g		
_varname	str8	%9s		

Sorted by:

```
. list
```

	v1	v2	v3	v4	v5	_varname
1.	1	2	3	4	5	provinc2
2.	10	4.5	12.1	174.9	80.6	grow92
3.	2.5	.8	5.8	169.1	77.4	grow93
4.	2.2	-3	3.5	120.9	48.5	grow94
5.	2.4	-5.8	3.9	163.9	47.1	grow95

Value labels are lost along the way, so provinces in the transposed dataset are indicated only by their numbers (1 = New Brunswick, 2 = Newfoundland, and so on). The second through last values in each column are the population gains for that province, in thousands. Thus, variable *v1* has a province identification number (1, meaning New Brunswick) in its first row, and New Brunswick's population growth values for 1992 to 1995 in its second through fifth rows. We can now find correlations between population growth in different provinces, for instance, by typing a **correlate** command with **in 2/5** (second through fifth observations only) qualifier:

```
. correlate v1-v5 in 2/5
(obs=4)
```

	v1	v2	v3	v4	v5
v1	1.0000				
v2	0.8058	1.0000			
v3	0.9742	0.8978	1.0000		
v4	0.5070	0.4803	0.6204	1.0000	
v5	0.6526	0.9362	0.8049	0.6765	1.0000

The strongest correlation appears between the growth of neighboring maritime provinces New Brunswick (*v1*) and Nova Scotia (*v3*): $r = .9742$. Newfoundland's (*v2*) growth has a much weaker correlation with that of Ontario (*v4*): $r = .4803$.

More sophisticated restructuring is possible through the **reshape** command. This command switches datasets between two basic configurations termed "wide" and "long." Dataset *growth1.dta* is initially in wide format.

```
. use growth1, clear
(Eastern Canada growth)
```

```
. list
```

	provinc2	grow92	grow93	grow94	grow95
1.	New Brun	10	2.5	2.2	2.4
2.	Newfound	4.5	.8	-3	-5.8
3.	Nova Sco	12.1	5.8	3.5	3.9
4.	Ontario	174.9	169.1	120.9	163.9
5.	Quebec	80.6	77.4	48.5	47.1

A **reshape** command switches this to long format.

```
. reshape long grow, i(provinc2) j(year)
(note: j = 92 93 94 95)
```

Data	wide	->	long
Number of obs.	5	->	20
Number of variables	5	->	3
j variable (4 values)		->	year
xij variables:			
grow92 grow93 ... grow95		->	grow

Listing the data shows their new shape. A **sepby( )** option with the **list** command gives us an output table with horizontal lines visually separating the provinces, instead of every five observations (the default).

```
. list, sepby(provinc2)
```

	provinc2	year	grow
1.	New Brun	92	10
2.	New Brun	93	2.5
3.	New Brun	94	2.2
4.	New Brun	95	2.4
5.	Newfound	92	4.5
6.	Newfound	93	.8
7.	Newfound	94	-3
8.	Newfound	95	-5.8
9.	Nova Sco	92	12.1
10.	Nova Sco	93	5.8
11.	Nova Sco	94	3.5
12.	Nova Sco	95	3.9
13.	Ontario	92	174.9
14.	Ontario	93	169.1
15.	Ontario	94	120.9
16.	Ontario	95	163.9
17.	Quebec	92	80.6
18.	Quebec	93	77.4
19.	Quebec	94	48.5
20.	Quebec	95	47.1

```
. label data "Eastern Canadian growth--long"
. label variable grow "Population growth in 1000s"
. save growth2
file c:\data\growth2.dta saved
```

The **reshape** command above began by stating that we want to put the dataset in **long** form. Next, it named the new variable to be created, *grow*. The **i(*provinc2*)** option specified the observation identifier, or the variable whose unique values denote logical observations. In this example, each province forms a logical observation. The **j(*year*)** option specifies the sub-observation identifier, or the variable whose unique values (within each logical observation) denote sub-observations. Here, the sub-observations are years within each province.

Figure 2.1 visualizes the structure of this long-format dataset. With one **graph** command, we can now produce time plots comparing the population changes in New Brunswick,

Newfoundland, and Nova Scotia (observations for which *provinc2* < 4). The **graph** command below calls for connected-line plots of *grow* (as *y*-axis variable) against *year* (*x* axis) if *province2* < 4, with horizontal lines at *y* = 0 (zero population growth), and separate plots for each value of *provinc2*.

```
. graph twoway connected grow year if provinc2 < 4, yline(0)
      by(provinc2)
```

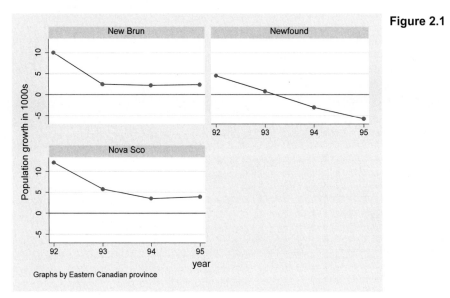

Figure 2.1

Graphs by Eastern Canadian province

Declines in their fisheries during the early 1990s contributed to economic hardships in these three provinces. Growth slowed dramatically in New Brunswick and Nova Scotia, while Newfoundland (the most fisheries-dependent province) actually lost population.

reshape works equally well in reverse, to switch data from "long" to "wide" format. Dataset *growth3.dta* serves as an example of long format.

```
. use growth3, clear
(Eastern Canadian growth--long)


. list, sepby(provinc2)
```

	provinc2	grow	year
1.	New Brun	10	92
2.	New Brun	2.5	93
3.	New Brun	2.2	94
4.	New Brun	2.4	95
5.	Newfound	4.5	92
6.	Newfound	.8	93
7.	Newfound	-3	94
8.	Newfound	-5.8	95

```
  9.  │  Nova Sco    12.1      92
 10.  │  Nova Sco     5.8      93
 11.  │  Nova Sco     3.5      94
 12.  │  Nova Sco     3.9      95
      │
 13.  │  Ontario    174.9      92
 14.  │  Ontario    169.1      93
 15.  │  Ontario    120.9      94
 16.  │  Ontario    163.9      95
      │
 17.  │  Quebec      80.6      92
 18.  │  Quebec      77.4      93
 19.  │  Quebec      48.5      94
 20.  │  Quebec      47.1      95
```

To convert this to wide format, we use **reshape wide**:

```
. reshape wide grow, i(provinc2) j(year)
(note: j = 92 93 94 95)

Data                                  long    ->    wide
─────────────────────────────────────────────────────────────────────
Number of obs.                          20    ->       5
Number of variables                      3    ->       5
j variable (4 values)                 year    ->    (dropped)
xij variables:

                                      grow    ->    grow92 grow93 ... grow95
─────────────────────────────────────────────────────────────────────
```

```
. list

      │  provinc2   grow92   grow93   grow94   grow95
      ├────────────────────────────────────────────────
  1.  │  New Brun       10      2.5      2.2      2.4
  2.  │  Newfound      4.5       .8       -3     -5.8
  3.  │  Nova Sco     12.1      5.8      3.5      3.9
  4.  │  Ontario     174.9    169.1    120.9    163.9
  5.  │  Quebec       80.6     77.4     48.5     47.1
```

With this step, we have recreated the organization of dataset *growth1.dta*.

Another important tool for restructuring datasets is the **collapse** command, which creates an aggregated dataset of statistics (for example, means, medians, or sums). The long *growth3* dataset has four observations for each province:

```
. use growth3, clear
(Eastern Canadian growth--long)
```

```
. list, sepby(provinc2)

      │  provinc2    grow    year
      ├───────────────────────────
  1.  │  New Brun      10      92
  2.  │  New Brun     2.5      93
  3.  │  New Brun     2.2      94
  4.  │  New Brun     2.4      95
      ├───────────────────────────
```

```
 5. │ Newfound    4.5    92
 6. │ Newfound     .8    93
 7. │ Newfound     -3    94
 8. │ Newfound   -5.8    95

 9. │ Nova Sco   12.1    92
10. │ Nova Sco    5.8    93
11. │ Nova Sco    3.5    94
12. │ Nova Sco    3.9    95

13. │ Ontario   174.9    92
14. │ Ontario   169.1    93
15. │ Ontario   120.9    94
16. │ Ontario   163.9    95

17. │  Quebec    80.6    92
18. │  Quebec    77.4    93
19. │  Quebec    48.5    94
20. │  Quebec    47.1    95
```

We might want to aggregate the different years into a mean growth rate for each province. In the collapsed dataset, each observation will correspond to one value of the **by( )** variable.

```
. collapse (mean) grow, by(provinc2)
. list
```

```
    │ provinc2       grow
 1. │ New Brun      4.275
 2. │ Newfound  -.8750001
 3. │ Nova Sco      6.325
 4. │  Ontario      157.2
 5. │   Quebec       63.4
```

For a slightly more complicated example, suppose we had a dataset similar to *growth3.dta* but also containing the variables *births*, *deaths*, and *income*. We want an aggregate dataset with each province's total numbers of births and deaths over these years, the mean income (to be named *meaninc*), and the median income (to be named *medinc*). If we do not specify a new variable name, as with *grow* in the previous example, or *births* and *deaths*, the collapsed variable takes on the same name as the old variable.

```
. collapse (sum) births deaths (mean) meaninc = income
      (median) medinc = income, by(provinc2)
```

collapse can create variables based on any of the following summary statistics:

mean	Means (default, if statistic is not specified)
median	Medians
p1	1st percentiles
p2	2nd percentiles (and so forth to **p99**)
sd	Standard deviations
sum	Sums
rawsum	Sums, ignoring optionally specified weight
count	Number of nonmissing observations
max	Maximums

min	Minimums
iqr	Interquartile range
first	First values
last	Last values
firstnm	First nonmissing values
lastnm	Last nonmissing values

A wider range of statistics can be collected using the flexible **statsby** command, which works as a prefix for other analyses. For example, to *create a new dataset* containing all **summarize** statistics describing growth in each Canadian province,

```
. use growth3, clear
(Eastern Canadian growth--long)


. statsby, by(provinc2): summarize grow
(running summarize on estimation sample)

       command:  summarize grow
             N:  r(N)
         sum_w:  r(sum_w)
          mean:  r(mean)
           Var:  r(Var)
            sd:  r(sd)
           min:  r(min)
           max:  r(max)
           sum:  r(sum)
            by:  provinc2

Statsby groups
———+——— 1 ———+——— 2 ———+——— 3 ———+——— 4 ———+——— 5
.....


. describe

Contains data
  obs:           5                          statsby: summarize
  vars:          9
  size:        205 (99.9% of memory free)

              storage  display   value
variable name   type   format    label    variable label

provinc2       byte    %8.0g     provinc2 Eastern Canadian province
N              float   %9.0g              r(N)
sum_w          float   %9.0g              r(sum_w)
mean           float   %9.0g              r(mean)
Var            float   %9.0g              r(Var)
sd             float   %9.0g              r(sd)
min            float   %9.0g              r(min)
max            float   %9.0g              r(max)
sum            float   %9.0g              r(sum)

Sorted by:
    Note:  dataset has changed since last saved
```

If we wanted a dataset containing only the mean growth for each province, similar to that created using a **collapse** command earlier, we could have specified a new variable name such as *meangrow*, equal to the r(mean) result:

```
. statsby meangrow = r(mean), by(provinc2): summarize grow
. describe
```

```
Contains data
  obs:            5                          statsby: summarize
  vars:           2
  size:          65  (99.9% of memory free)
```

variable name	storage type	display format	value label	variable label
provinc2	byte	%8.0g	provinc2	Eastern Canadian province
meangrow	float	%9.0g		r(mean)

```
Sorted by:
    Note:  dataset has changed since last saved
```

statsby can also make a dataset of results from regression models or other analyses. For illustration, the following commands estimate time trends by regressing *grow* on *year* for each province, and build a new dataset containing the slope (which becomes a variable named *_b_year*), y-intercept (*_b_cons*), and their standard errors for each province.

```
. use growth3, clear
(Eastern Canadian growth--long)
```

```
. statsby _b _se, by(provinc2):  regress grow year
(running regress on estimation sample)
```

```
        command:  regress grow year
             by:  provinc2
```

```
Statsby groups
———+—— 1 ——+—— 2 ——+—— 3 ——+—— 4 ——+—— 5
.....
```

```
. describe
```

```
Contains data
  obs:            5                          statsby: regress
  vars:           5
  size:         125  (99.9% of memory free)
```

variable name	storage type	display format	value label	variable label
provinc2	byte	%8.0g	provinc2	Eastern Canadian province
_b_year	float	%9.0g		_b[year]
_b_cons	float	%9.0g		_b[_cons]
_se_year	float	%9.0g		_se[year]
_se_cons	float	%9.0g		_se[_cons]

```
Sorted by:
    Note:  dataset has changed since last saved
```

Type **help statsby** or consult the *Data Management Reference Manual* for more information and examples. Selecting

> Statistics > Other > Collect statistics for a command across a by list

from the menus brings up the dialog box for this command.

Using Weights

Stata understands four types of weighting:

aweight   Analytical weights, used in weighted least squares (WLS) regression and similar procedures.

fweight   Frequency weights, counting the number of duplicated observations. Frequency weights must be integers.

iweight   Importance weights, however you define "importance."

pweight   Probability or sampling weights, proportional to the inverse of the probability that an observation is included due to sampling strategy.

Not all types of weighting have been defined for all types of analyses. We cannot, for example, use **pweight** with the **tabulate** command. Using weights effectively requires a clear understanding of what we want them to accomplish in a particular analysis.

When researchers speak of "weighted data," they often refer to methods of compensating for originally disproportionate or complex sampling designs — a common feature of surveys. **pweight** provides one way to adjust for disproportionate sampling, using probability weights proportional to 1/(probability of selection). Analysis of survey data using probability weights is a particular strength of Stata. Chapter 14 provides examples and more discussion.

In some instances, "weighted data" can mean something simpler — an aggregate dataset, perhaps constructed from a frequency table or cross-tabulation, or from means based on many individual observations. In that case, we use **fweight**. The small dataset *nfschool.dta*, containing results from a survey of 1,381 rural Newfoundland high school students, illustrates this idea.

```
. describe
```

```
Contains data from c:\data\nfschool.dta
  obs:             6                          Newf.school/univer.(Seyfrit 93)
  vars:            3                          19 Jan 2008 19:53
  size:           72 (99.9% of memory free)
```

variable name	storage type	display format	value label	variable label
univers	byte	%8.0g	yes	Expect to attend university?
year	byte	%8.0g		What year of school now?
count	int	%8.0g		observed frequency

```
Sorted by:
```

```
. list, sep(3)
```

	univers	year	count
1.	no	10	210
2.	no	11	260
3.	no	12	274
4.	yes	10	224
5.	yes	11	235
6.	yes	12	178

At first glance, the dataset seems to contain only 6 observations. When we cross-tabulate whether students expect to attend a university (*univers*) by their current year in high school (*year*), we get a table with one observation per cell.

. **tabulate** *univers year*

Expect to attend university ?	What year of school now? 10	11	12	Total
no	1	1	1	3
yes	1	1	1	3
Total	2	2	2	6

To understand these data, we need to apply frequency weights. The variable *count* gives frequencies: 210 of these students are tenth graders who said they did not expect to attend a university, 260 are eleventh graders who said no, and so on. Specifying **[fweight = *count*]** obtains a cross-tabulation showing responses of all 1,381 students.

. **tabulate** *univers year* **[fweight = *count*]**

Expect to attend university ?	What year of school now? 10	11	12	Total
no	210	260	274	744
yes	224	235	178	637
Total	434	495	452	1,381

Carrying the analysis further, we might add options asking for a table with column percentages (**col**), no cell frequencies (**nof**), and a χ^2 test of independence (**chi2**). This reveals a statistically significant relationship (P = .001). The percentage of students expecting to go to college declines with each year of high school, from about 52% among 10th graders to just 39% among 12th graders.

. **tabulate** *univers year* **[fw = *count*], col nof chi2**

Expect to attend university ?	What year of school now? 10	11	12	Total
no	48.39	52.53	60.62	53.87
yes	51.61	47.47	39.38	46.13
Total	100.00	100.00	100.00	100.00

Pearson chi2(2) = 13.8967   Pr = 0.001

Probability weights (**pweight**) will get more attention in Chapter 14. Analytical weights (**aweight**) are useful in graphing (Chapters 3, 7) and for weighted least squares (Chapters 6, 9), among other things. Importance weights (**iweight**) have no fixed definition, but could be applied in programs written for special purposes.

Creating Random Data and Random Samples

The pseudo-random number function **uniform()** lies at the heart of Stata's ability to generate random data or to sample randomly from the data at hand. The *Base Reference Manual* (Functions) provides a technical description of this 32-bit pseudo-random generator. If we presently have data in memory, then a command such as the following creates a new variable named *randnum*, having apparently random 16-digit values over the interval [0,1) for each case in the data.

```
. generate randnum = uniform()
```

We could also create a random dataset from scratch. Suppose we want to start a new dataset containing 10 random values. We first clear any other data from memory (if they were valuable, **save** them first). Next, set the number of observations desired for the new dataset. Explicitly setting the seed number makes it possible to later reproduce the same "random" results. Finally, we generate our random variable.

```
. clear
. set obs 10
obs was 0, now 10
. set seed 12345
. generate randnum = uniform()
. list
```

	randnum
1.	.309106
2.	.6852276
3.	.1277815
4.	.5617244
5.	.3134516
6.	.5047374
7.	.7232868
8.	.4176817
9.	.6768828
10.	.3657581

In combination with Stata's algebraic, statistical, and special functions, **uniform()** can simulate values sampled from a variety of theoretical distributions. If we want *newvar* sampled from a uniform distribution over [0,428) instead of the usual [0,1), we type

```
. generate newvar = 428 * uniform()
```

These will still be 16-digit values. Perhaps we want only integers from 1 to 428 (inclusive):

```
. generate newvar = 1 + int(428 * uniform())
```

To simulate 1,000 throws of a six-sided die, type

```
. clear
. set obs 1000
obs was 0, now 1000
. generate roll = 1 + int(6 * uniform())
. tabulate roll
```

roll	Freq.	Percent	Cum.
1	170	17.00	17.00
2	167	16.70	33.70
3	149	14.90	48.60
4	171	17.10	65.70
5	166	16.60	82.30
6	177	17.70	100.00
Total	1,000	100.00	

We theoretically expect 16.67% ones, 16.67% twos, and so on, but in any one sample like these 1,000 "throws," the observed percentages will vary randomly around their expected values.

To simulate 1,000 throws of a pair of six-sided dice, type

```
. generate dice = 2 + int(6 * uniform()) + int(6 * uniform())
. tabulate dice
```

dice	Freq.	Percent	Cum.
2	27	2.70	2.70
3	62	6.20	8.90
4	78	7.80	16.70
5	120	12.00	28.70
6	154	15.40	44.10
7	147	14.70	58.80
8	145	14.50	73.30
9	97	9.70	83.00
10	89	8.90	91.90
11	52	5.20	97.10
12	29	2.90	100.00
Total	1,000	100.00	

We can use _n to begin an artificial dataset as well. The following commands create a new 5,000-observation dataset with one variable named *index*, containing values from 1 to 5,000.

```
. set obs 5000
obs was 1000, now 5000
. generate index = _n
. summarize
```

Variable	Obs	Mean	Std. Dev.	Min	Max
roll	1000	3.527	1.732129	1	6
dice	1000	6.948	2.428414	2	12
index	5000	2500.5	1443.52	1	5000

It is possible to generate variables from a normal (Gaussian) distribution using **uniform()**. The following example creates a dataset with 2,000 observations and 2 variables: z from an N(0,1) population, and x from N(500,75).

```
. clear
. set obs 2000
obs was 0, now 2000
. generate z = invnormal(uniform())
. generate x = 500 + 75*invnormal(uniform())
```

The actual sample means and standard deviations differ slightly from their theoretical values:

```
. summarize
```

Variable	Obs	Mean	Std. Dev.	Min	Max
z	2000	.0194897	1.000984	-3.280097	3.840595
x	2000	500.3424	74.45384	258.1613	717.5144

If z follows a normal distribution, $v = e^z$ follows a lognormal distribution. To form a lognormal variable v based upon a standard normal z,

```
. generate v = exp(invnormal(uniform()))
```

To form a lognormal variable w based on an N(100,15) distribution,

```
. generate w = exp(100 + 15*invnormal(uniform()))
```

Taking logarithms, of course, normalizes a lognormal variable.

To simulate y values drawn randomly from an exponential distribution with mean and standard deviation $\mu = \sigma = 3$,

```
. generate y = -3 * ln(uniform())
```

For other means and standard deviations, substitute other values for 3.

$X1$ follows a χ^2 distribution with one degree of freedom, which is the same as a squared standard normal:

```
. generate X1 = (invnormal(uniform()))^2
```

By similar logic, $X2$ follows a χ^2 with two degrees of freedom:

```
. generate X2 = (invnormal(uniform()))^2 + (invnormal(uniform()))^2
```

Other statistical distributions, including t and F, can be simulated along the same lines. In addition, programs have been written for Stata to generate random samples following distributions such as binomial, Poisson, gamma, and inverse Gaussian.

Although **invnormal(uniform())** can be adjusted to yield normal variates with particular correlations, a much easier way to do this is through the **drawnorm** command. To generate 5,000 observations from N(0,1), type

```
. clear
. drawnorm z, n(5000)
(obs 5000)
. summ
```

Variable	Obs	Mean	Std. Dev.	Min	Max
z	5000	.0232413	1.008863	-3.496124	3.907439

Below, we will create three further variables. Variable *x1* is from an N(0,1) population, variable *x2* is from N(100,15), and *x3* is from N(500,75). Furthermore, we define these variables to have the following population correlations:

	x1	**x2**	**x3**
x1	1.0	0.4	-0.8
x2	0.4	1.0	0.0
x3	-0.8	0.0	1.0

The procedure for creating such data requires first defining the correlation matrix *C*, and then using *C* in the **drawnorm** command:

```
. mat C = (1, .4, -.8 \ .4, 1, 0 \ -.8, 0, 1)
. drawnorm x1 x2 x3, means(0,100,500) sds(1,15,75) corr(C)
. summarize x1-x3
```

Variable	Obs	Mean	Std. Dev.	Min	Max
x1	5000	-.0040733	1.011686	-3.933968	3.169514
x2	5000	100.1839	14.92045	41.14071	151.774
x3	5000	500.1485	75.64395	227.187	772.7881

```
. correlate x1-x3
(obs=5000)
```

	x1	x2	x3
x1	1.0000		
x2	0.4070	1.0000	
x3	-0.8034	-0.0115	1.0000

Compare the sample variables' correlations and means with the theoretical values given earlier. Random data generated in this fashion can be viewed as samples drawn from theoretical populations. We should not expect the samples to have exactly the theoretical population parameters (in this example, an *x3* mean of 500, *x1–x2* correlation of 0.4, *x1–x3* correlation of –.8, and so forth). Artificial uncorrelated or correlated datasets also can be created via menus and dialog boxes, under

Statistics > Other > Draw a sample from a normal distribution

or

Statistics > Other > Create a dataset with specified correlation structure

The command **sample** makes unobtrusive use of **uniform**'s random generator to obtain random samples of the data in memory. For example, to discard all but a 10% random sample of the original data, type

```
. sample 10
```

When we add an **in** or **if** qualifier, **sample** applies only to those observations meeting our criteria. For example,

```
. sample 10 if age < 26
```

would leave us with a 10% sample of those observations with *age* less than 26, plus 100% of the original observations with $age \geq 26$.

We could also select random samples of a particular size. To discard all but 90 randomly-selected observations from the dataset in memory, type

```
. sample 90, count
```

The sections in Chapter 16 on bootstrapping and Monte Carlo simulations provide further examples of random sampling and random variable generation.

Writing Programs for Data Management

Data management on larger projects often involves repetitive or error-prone tasks that are best handled by writing specialized Stata programs. Advanced programming can become very technical, but we can also begin by writing simple programs that consist of nothing more than a sequence of Stata commands, typed and saved as an ASCII file. ASCII files can be created using your favorite word processor or text editor, which should offer "ASCII text file" among its options under File > Save As. An even easier way to create such text files is through Stata's Do-file Editor, which is brought up by clicking Window > Do-file Editor or the icon 📓. Alternatively, bring up the Do-file Editor by typing the command **doedit**, or **doedit** *filename* if *filename* exists. Commands in the Review window can be highlighted and sent directly to the Do-File Editor (right-click to get this menu choice). Commands can also be copied and pasted into the Do-File Editor from other sources such as log files or the Results window.

Using the Do-file Editor any of these ways, we might create a file named *canada.do* (which contains the commands to read in a raw data file named *canada.raw*), then label the dataset and its variables, compress it, and save it in Stata format. The commands in this file are identical to those seen earlier when we went through the example step by step.

```
infile str30 place pop unemp mlife flife using canada.raw
label data "Canadian dataset 1"
label variable pop "Population in 1000s, 1995"
label variable unemp "% 15+ population unemployed, 1995"
label variable mlife "Male life expectancy years"
label variable flife "Female life expectancy years"
compress
save canada1, replace
```

Once this file has been written and saved, we can run it by selecting File > Do and opening *canada.do* from the menus; or just by typing a command such as

. do *canada*

Such batch-mode programs, termed "do-files," are usually saved with a .do extension. More elaborate programs (defined either by do-files or "automatic" ado-files) can be stored in memory, and can call other programs in turn, creating new Stata commands and opening a world of possibilities for the adventurous. The Do-file Editor has several other features that you might find useful. Chapter 3 describes a simple way to use do-files in building graphs. For more information, see the *Getting Started* manual on Using the Do-file Editor.

Stata ordinarily interprets the end of a command line as the end of that command. This is reasonable onscreen, where the line can be arbitrarily long, but does not work as well when we are typing commands in a text file. One way to avoid line-length problems is through the **#delimit ;** command, which can set a semicolon as the end-of-command delimiter. In the following example, we make a semicolon the delimiter; then type two long commands that do not end until a semicolon appears; and then finally reset the delimiter to its usual value, a carriage return (cr):

```
#delimit ;
infile str30 place pop unemp mlife flife births deaths
    marriage medinc mededuc using newcan.raw;
order place pop births deaths marriage medinc mededuc
    unemp mlife flife;
#delimit cr
```

Stata normally pauses each time the Results window becomes full of information, and waits to proceed until we press the space bar or any other key (or click 🔘). Instead of pausing, we can ask Stata to continue scrolling until the output is complete. Typed in the Command window or as part of a program, the command

. set more off

calls for continuous scrolling. This is convenient if our program produces much screen output that we don't want to see, or if it is writing to a log file that we will examine later. Typing

. set more on

returns to the usual mode of waiting for keyboard input before scrolling.

Managing Memory

When we select File > Open or type a **use** command to open a dataset, Stata reads the disk file and loads it into memory. Loading into memory permits rapid analysis, but it is only possible if the dataset can fit within the amount of memory currently allocated to Stata. If we try to open a dataset that is too large, we get an error message saying "no room to add more observations" or "no room to add more variables," and advising what to do next. For example, PUMS_1.dta contains Public Use Microsample (PUMS) data on 2.8 million people — roughly 1% of the U.S. population from the 2000 Census. With its default settings Stata won't open such a large file, but reports an error message containing helpful links to the relevant commands.

```
. use PUMS_1.dta
(1% PUMS sample, 2000)
no room to add more observations
    An attempt was made to increase the number of observations beyond what is
    currently possible.  You have the following alternatives:

    1.  Store your variables more efficiently; see help compress.  (Think of
        Stata's data area as the area of a rectangle; Stata can trade off width
        and length.)

    2.  Drop some variables or observations; see help drop.

    3.  Increase the amount of memory allocated to the data area using the set
        memory command; see help memory.
r(901);
```

How much memory do we need for *PUMS_1.dta*? We can **describe** the file even though we can't yet open it, by typing

```
. describe using PUMS_1.dta, short

Contains data                         1% PUMS sample, 2000
  obs:      2,818,644                  20 Jan 2008 12:55
  vars:           107
  size:   518,630,496
Sorted by:
```

The dataset contains 107 variables, 2,818,644 observations, and is more than 518 megabytes in size. It will require even more than 518 megabytes to analyze. Small Stata allocates a fixed amount of memory to data, and this limit cannot be changed. Stata/IC, Stata/SE, and Stata/MP versions are flexible, however. With those versions we can increase Stata's memory allocation, within limits of our computer's physical memory, by issuing a **set memory** command. To allocate 700 megabytes to data, type

```
. set memory 700m

Current memory allocation
```

settable	current value	description	memory usage (1M = 1024k)
set maxvar	5000	max. variables allowed	1.909M
set memory	700M	max. data space	700.000M
set matsize	400	max. RHS vars in models	1.254M
			703.163M

If there are data already in memory, first type the command **clear** to remove them. To reset the memory allocation "permanently," so it will be the same next time we start up, type

```
. set memory 700m, permanently
```

Most users have no need to set their default memory allocation this high, however.

In the example above, *PUMS_1.dta* is a 500-megabyte dataset that would not fit into the original memory allocation. Asking for a 700-megabyte allocation has now given us more than enough room for the 2.8 million observations and 107 variables in these data. Stata reports that 29.3% of the allocated memory remains free.

```
. use PUMS_1.dta
. describe, short
Contains data from C:\data\PUMS_1.dta
  obs:      2,818,644                     1% PUMS sample, 2000
  vars:           107                     20 Jan 2008 12:55
  size:   529,905,072 (29.3% of memory free)
Sorted by:
```

Stata's analytical methods require memory space and also temporarily create new variables as calculations proceed, so the remaining 29.3% of memory could become critical. Stata is among the faster statistical programs; calculating means and standard deviations for all 107 variables in this 2.8 million observation dataset requires less than a minute on an ordinary computer. For more complicated calculations or graphics, however, desktop analysis of very large datasets can feel slow. One work-around strategy is to draw a random sample from the large dataset. For example, typing

```
. sample 1
```

takes an approximately 1% random sample from the data in memory, dropping *PUMS_1.dta* down to a more manageable 28,000 observations. We could save this smaller dataset as *PUMS_01.dta*, then use it for most of our exploratory work. Once we have figured out exactly which analyses we want, and how to recode or create any new variables that are needed, these successful commands can be copied from the Results window and edited into a do-file that will implement all commands, and record their results in a log file. Test the do-file with the sample dataset, then apply it to the original complete dataset and step out for coffee.

It is possible to **set memory** to values higher than the computer's available physical memory. In that case, Stata uses "virtual memory," which is really disk storage. Although virtual memory allows bypassing hardware limitations, it can be terribly slow. If you regularly work with datasets that push the limits of your computer, you might soon conclude that it is time to buy more memory.

Type **help limits** to see a list of limitations in Stata, not only on dataset size but also other dimensions including matrix size, command lengths, lengths of names, and numbers of variables in commands. Some of these limitations can be adjusted by the user.

3

Graphs

Graphs appear in every chapter of this book — one indication of their value and integration with other analyses in Stata. Analytical graphics have always been one of Stata's strengths, and reason enough for many users to choose Stata over other packages. Attractive and publishable basic graphs are easy to draw, using commands or choices from the menus under Graphics . Users who imagine more elaborate or creative graphs will find their efforts supported by an impressive array of tools and options, described in the 600-page *Graphics Reference Manual*, and illustrated by many examples in *A Visual Guide to Stata Graphics* (Mitchell 2008).

In the short space of this chapter, we cover the spectrum from basic to creative graphing taking an example- rather than syntax-oriented approach (see the *Graphics Reference Manual* or **help graph** for more details about syntax). We begin with the seven basic types of graphs:

histogram	histograms
graph twoway	two-variable scatterplots, line plots, and many others
graph matrix	scatterplot matrices
graph box	box plots
graph pie	pie charts
graph bar	bar charts
graph dot	dot plots

For each of these basic types, there exist many options. That is especially true for the versatile **twoway** type.

More specialized graphs such as symmetry plots, quantile plots, and quantile–normal plots allow detailed inpection of variable distributions. A few examples appear in this chapter. Type **help graph other** for details.

Finally, the chapter concludes with techniques particularly useful in building data-rich, self-contained graphics for publication. Stata provides tools to add text to graphs, overlay multiple twoway plots, retrieve and reformat saved graphs, and combine multiple graphs into one. As our graphing commands grow more complicated, simple batch programs (do-files) can help to write, modify, and re-use them. The Graph Editor, a new feature with Stata 10, can add arrows or text and even make design changes to previously-saved graphs.

The full range of graphical choices goes far beyond what this book can cover, but these examples point out a few of the possibilities. Later chapters supply further illustrations. The Graphics menu provides point-and-click access to most graphing procedures. Experimenting with dialog boxes and using the Submit button is a very good way to learn what is available, and particularly the many choices for **twoway** graphs.

Example Commands

. `histogram y, frequency`
Draws histogram of variable y, showing frequencies on the vertical axis.

. `histogram y, start(0) width(10) norm fraction`
Draws histogram of y with bins 10 units wide, starting at 0. Adds a normal curve based on the sample mean and standard deviation, and shows fraction of the data on the vertical axis.

. `histogram y, by(x, total) percent`
In one figure, draws separate histograms of y for each value of x, and also a "total" histogram for the sample as a whole. Shows percentages on the vertical axis.

. `kdensity x, generate(xpoints xdensity) width(20) biweight`
Produces and graphs kernel density estimate of the distribution of x. Two new variables are created: *xpoints* containing the x values at which the density is estimated, and *xdensity* with the density estimates themselves. **width(20)** specifies the halfwidth of the kernel, in units of the variable x. If **width( )** is not specified, the default follows a simple formula for "optimal." The **biweight** option in this example calls for a biweight kernel, instead of the default **epanechnikov**.

. `graph twoway scatter y x`
Displays a basic two-variable scatterplot of y against x. The **graph** part is optional for all **twoway** family commands; for example, we could type **twoway scatter** y x.

. `graph twoway lfit y x || scatter y x`
Visualizes the linear regression of y on x by overlaying two **twoway** graphs: the regression (linear fit or **lfit**) line, and the y vs. x scatterplot To include a 95% confidence band for the regression line, replace **lfit** with **lfitci**.

. `graph twoway scatter y x, xlabel(0(10)100) ylabel(-3(1)6, horizontal)`
Constructs scatterplot of y vs. x, with x axis labeled at 0, 10, ..., 100. y axis is labeled at -3, -2, ..., 6, with labels written horizontally instead of vertically (the default).

. `graph twoway scatter y x, mlabel(country)`
Constructs scatterplot of y vs. x, with data points (markers) labeled by the values of variable *country*.

. `graph twoway scatter y x1, by(x2)`
In one figure, draws separate y vs. *x1* scatterplots for each value of *x2*.

. `graph twoway scatter y x1 [fweight = population], msymbol(Oh)`
Draws a scatterplot of y vs. *x1*. Marker symbols are hollow circles **(Oh)**, with their size (area) proportional to frequency-weight variable *population*.

. **graph twoway connected** *y time*
A basic time plot of *y* against *time*. Data points are shown connected by line segments. To show line segments but no data-point markers, use **line** instead of **connected**.
. **graph twoway line** *y time*

. **graph twoway line** *y1 y2 time*
Draws a time plot (in this example, a line plot) with two *y* variables that both have the same scale, and are graphed against an *x* variable named *time*.

. **graph twoway line** *y1 time*, **yaxis(1)** || **line** *y2 time*, **yaxis(2)**
Draws a time plot with two *y* variables that have different scales, by overlaying two individual line plots. The left-hand *y* axis, **yaxis(1)**, gives the scale for *y1*, while the right-hand *y* axis, **yaxis(2)**, gives the scale for *y2*.

. **graph matrix** *x1 x2 x3 x4 y*
Constructs a scatterplot matrix, showing all possible scatterplot pairs among the variables listed.

. **graph box** *y1 y2 y3*
Constructs box plots of variables *y1*, *y2*, and *y3*.

. **graph box** *y*, **over(x) yline(23)**
Constructs box plots of *y* for each value of *x*, and draws a horizontal line at *y* = 23.

. **graph pie** *a b c*
Draws one pie chart with slices indicating the relative amounts of variables *a*, *b*, and *c*. The variables must have similar units.

. **graph bar (sum)** *a b c*
Shows the sums of variables *a*, *b*, and *c* as side-by-side bars in a bar chart. To obtain means instead of sums, type **graph bar (mean)** *a b c*. Further options include bars representing medians, percentiles, counts, or other statistics for each variable (same options as **collapse**).

. **graph bar (mean)** *a*, **over(x)**
Draws a bar chart showing the mean of variable *a* at each value of variable *x*.

. **graph bar (asis)** *a b c*, **over(x) stack**
Draws a bar chart in which the values ("as is") of variables *a*, *b*, and *c* are stacked on top of one another, at each value of variable *x*.

. **graph dot (median)** *y*, **over(x)**
Draws a dot plot, in which dots along a horizontal scale mark the median value of *y* at each level of *x*. **graph dot** supports the same statistical options as **graph bar** or **collapse**.

. **qnorm** *y*
Draws a quantile–normal plot (normal probability plot) showing quantiles of *y* versus corresponding quantiles of a normal distribution.

```
. rchart x1 x2 x3 x4 x5, connect(1)
```
Constructs a quality-control R chart graphing the range of values represented by variables *x1 – x5*. Type **help qc** to see the full range of quality-control graphs Stata offers. These can also be accessed through menus: Graphics > Quality control .

Graph options, such as those controlling titles, labels, and tick marks on the axes are common across graph types wherever this makes sense. Moreover, the underlying logic of Stata's graph commands is consistent from one type to the next. These common elements are the key to gaining graph-building fluency, as the basics fall into place.

Histograms

Histograms, displaying the distribution of measurement variables, are most easily produced with their own command **histogram**. For examples, we turn to *states.dta*, which contains selected environment and education measures on the 50 U.S. states plus the District of Columbia (data from the League of Conservation Voters 1991; National Center for Education Statistics 1992, 1993; World Resources Institute 1993).

```
. use states
(U.S. states data 1990-91)

. describe
```
```
Contains data from c:\data\states.dta
  obs:            51                          U.S. states data 1990-91
  vars:           21                          14 Sep 2003 18:34
  size:       4,590 (99.9% of memory free)
```

variable name	storage type	display format	value label	variable label
state	str20	%20s		State
region	byte	%9.0g	region	Geographical region
pop	float	%9.0g		1990 population
area	float	%9.0g		Land area, square miles
density	float	%7.2f		People per square mile
metro	float	%5.1f		Metropolitan area population, %
waste	float	%5.2f		Per capita solid waste, tons
energy	int	%8.0g		Per capita energy consumed, Btu
miles	float	%8.0g		Per capita miles/year, 1,000
toxic	float	%5.2f		Per capita toxics released, lbs
green	float	%5.2f		Per capita greenhouse gas, tons
house	byte	%8.0g		House '91 environ. voting, %
senate	byte	%8.0g		Senate '91 environ. voting, %
csat	int	%9.0g		Mean composite SAT score
vsat	int	%8.0g		Mean verbal SAT score
msat	int	%8.0g		Mean math SAT score
percent	byte	%9.0g		% HS graduates taking SAT
expense	int	%9.0g		Per pupil expenditures prim&sec
income	double	%10.0g		Median household income, $1,000
high	float	%9.0g		% adults HS diploma
college	float	%9.0g		% adults college degree

```
Sorted by:  state
```

Figure 3.1 shows a simple histogram of *college*, the percentage of a state's over-25 population with a bachelor's degree or higher. It was produced by the following command:

```
.  histogram college, frequency title("Figure 3.1")
```

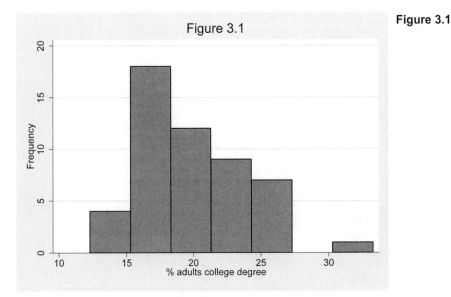

Figure 3.1

Figure 3.1

Under the Prefs > Graph Preferences menus, we have the choice of several pre-designed "schemes" for the default colors and shading of our graphs. Custom schemes can be defined as well. The examples in this book employ the s2mono (monochrome) scheme, which among other things calls for shaded margins around each graph. The s1mono scheme does not have such margins. Experimenting with the different monochrome and color schemes helps to determine which works best for a particular purpose. A graph drawn and saved under one scheme can subsequently be retrieved and re-saved under a different one, as described later.

Options can be listed in any order following the comma in a graph command. Figure 3.1 illustrates two options: frequency (instead of density, the default) is shown on the vertical axis; and the title "Figure 3.1" appears over the graph. Once a graph is onscreen, menu choices provide the easiest way to print it, save it to disk, or cut and paste it into another program such as a word processor.

Figure 3.1 reveals the positive skew of this distribution, with a mode above 15 and an outlier around 35. It is hard to describe the graph more specifically because the bars do not line up with *x*-axis tick marks. Figure 3.2 contains a version with several improvements (based on some quick experiments to find the right values):

xlabel(12(2)34)	The *x* axis is labeled from 12 to 34, in increments of 2.
ylabel(0(2)12)	The *y* axis is labeled from 0 to 12, in increments of 2.
ytick(1(2)13)	Tick marks are drawn on the *y* axis from 1 to 13, in increments of 2.
start(12)	The histogram's first bar (bin) starts at 12.
width(2)	The width of each bar (bin) is 2.

```
. histogram college, frequency title("Figure 3.2") xlabel(12(2)34)
    ylabel(0(2)12) ytick(1(2)13) start(12) width(2)
```

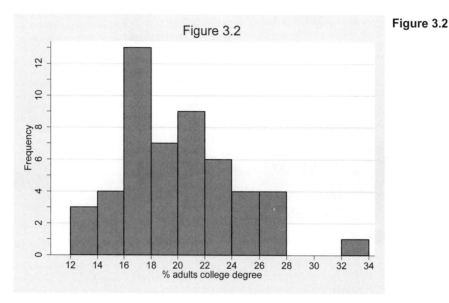

Figure 3.2

Figure 3.2 helps us to describe the distribution more specifically. For example, we now see that in 13 states, between 16 and 18 percent of adults have college degrees.

Other useful **histogram** options include the following.

bin(#)   Draw a histogram with # bins (bars). We can specify either **bin(#)** or, as in Figure 3.2, **start(#)** and **width(#)** — but not both.

percent   Show percentages on the vertical axis. **ylabel** and **ytick** then refer to percentage values. Another possibility, **frequency**, is illustrated in Figure 3.2. We could also ask for **fraction** of the data. The default histogram shows **density**, meaning that bars are scaled so that the sum of their *areas* equals 1.

gap(#)   Leave a gap between bars. # is relative, $0 \leq \# < 100$; experiment to find a suitable value.

addlabels Label the heights of histogram bars. A separate option, **addlabopts**, controls how the labels look.

discrete   Specify discrete data, requiring one bar for each value of *x*.

norm   Overlay a normal curve on the histogram, based on sample mean and standard deviation.

kdensity   Overlay a kernal-density estimate on the histogram. The option **kdenopts** controls density computation; see **help kdensity** for details.

There also exists a separate command, **twoway histogram**, that draws histograms allowing other options common to the **twoway** family of graphs. This command is not shown here, but you can learn about it by typing **help twoway histogram**.

With histograms or most other graphs, we can override the defaults and specify our own titles for the horizontal and vertical axes. The option **ytitle** controls *y*-axis titles, and **xtitle** controls *x*-axis titles. Figure 3.3 illustrates such titles, together with some other histogram options. Note the incremental buildup from basic (Figure 3.1) to more elaborate (Figure 3.3) graphs. These are the usual steps for graph construction in Stata: we start simply, then experimentally add options to earlier commands retrieved from the Review window, as we work toward an image that most clearly represents our findings. Figure 3.3 is perhaps over-elaborate, but drawn here to show off multiple options.

```
. histogram college, frequency title("Figure 3.3") ylabel(0(2)12)
     ytick(1(2)13) xlabel(12(2)34) start(12) width(2) addlabel
     norm gap(15)
```

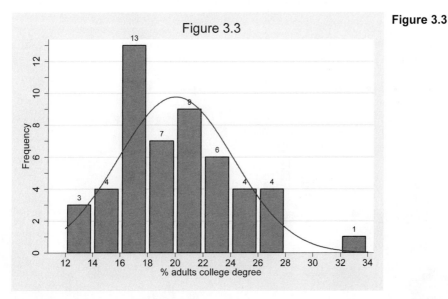

Figure 3.3

Suppose we want to see how the distribution of *college* varies by *region*. The **by** option obtains a separate histogram for each value of *region*. Other options work as they do for single histograms. Figure 3.4 shows an example in which we ask for percentages on the vertical axis, and the data grouped into 8 bins.

```
. histogram college, by(region) percent bin(8)
```

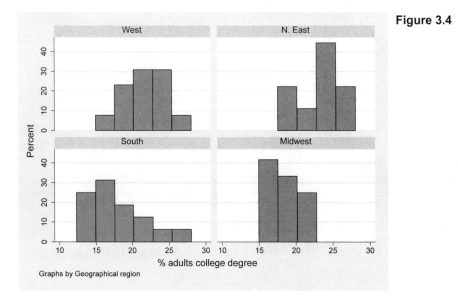

Figure 3.4

Figure 3.5, below, contains a similar set of four regional histograms, but also includes a fifth "total" histogram showing the distribution for all regions combined.

```
. histogram college, percent bin(8) by(region, total)
```

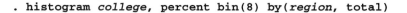
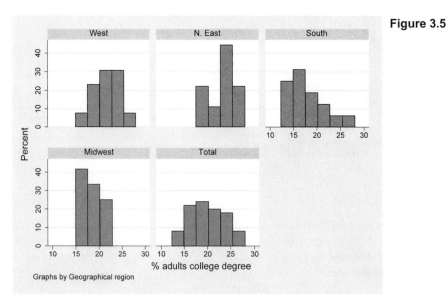

Figure 3.5

Axis labeling, tick marks, titles, and the **by(***varname***)** or **by(***varname***, total)** options work in a similar fashion with other Stata graphing commands.

Scatterplots

Basic scatterplots are obtained through commands of the general form

```
. graph twoway scatter y x
```

where *y* is the vertical or *y*-axis variable, and *x* is the horizontal or *x*-axis one. (The initial **graph** part of this command is optional.) For example, again using the *states.dta* dataset, we could plot *waste* (per capita solid waste) against *metro* (percent population in metropolitan areas), with the result shown in Figure 3.6. Each point in Figure 3.6 represents one of the 50 U.S. states (or Washington DC).

```
. graph twoway scatter waste metro
```

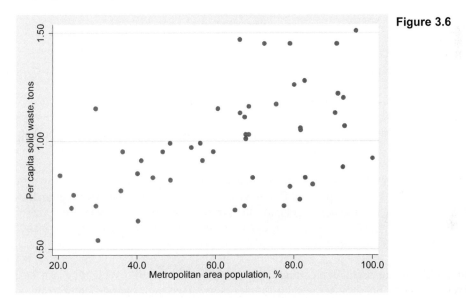

Figure 3.6

As with histograms, we can use **xlabel**, **xtick**, **xtitle**, etc. to control axis labels, tick marks, or titles. Scatterplots also allow control of the shape, color, size, and other attributes of markers. Figure 3.6 employs the default markers, which are solid circles. The same effect would result if we included the option **msymbol(circle)**, or wrote this option in abbreviated form as **msymbol(O)**. **msymbol(diamond)** or **msymbol(D)** would produce a graph with diamond markers, and so forth. The following table lists possible shapes:

msymbol( )	Abbreviation	Description
circle	**O**	circle, solid
diamond	**D**	diamond, solid
triangle	**T**	triangle, solid
square	**S**	square, solid
plus	**+**	plus sign
x	**X**	letter x

smcircle	**o**	small circle, solid
smdiamond	**d**	small diamond, solid
smsquare	**s**	small square, solid
smtriangle	**t**	small triangle, solid
smplus	**smplus**	small plus sign
smx	**x**	small letter x
circle_hollow	**Oh**	circle, hollow
diamond_hollow	**Dh**	diamond, hollow
triangle_hollow	**Th**	triangle, hollow
square_hollow	**Sh**	square, hollow
smcircle_hollow	**oh**	small circle, hollow
smdiamond_hollow	**dh**	small diamond, hollow
smtriangle_hollow	**th**	small triangle, hollow
smsquare_hollow	**sh**	small square, hollow
point	**p**	very small dot
none	**i**	invisible

The **mcolor** option controls marker colors. For example, the command

```
. graph twoway scatter waste metro, msymbol(S) mcolor(purple)
```

would produce a scatterplot in which the symbols are large purple squares. Type **help colorstyle** for a list of available colors.

One interesting possibility with scatterplots is to make symbol size (area) proportional to a third variable, thereby giving the data points different visual weight. For example, we might redraw the scatterplot of *waste* against *metro*, but make the symbol size reflect each state's population (*pop*). This can be done as shown in Figure 3.7, using the **fweight[ ]** or frequency weight feature. Hollow circles, **msymbol(Oh)**, provide a suitable shape.

Frequency weights are useful with some other graph types as well. Weighting can be a deceptively complex topic because "weights" come in several types and have different meanings in different contexts. For an overview of weighting in Stata, type **help weight**.

```
. graph twoway scatter waste metro [fweight = pop], msymbol(Oh)
```

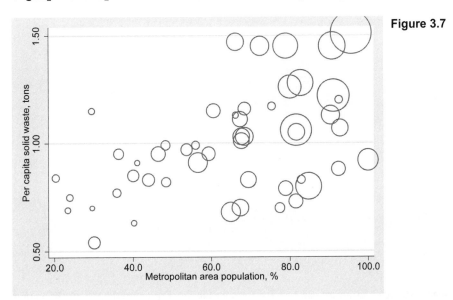

Figure 3.7

Density-distribution sunflower plots provide an alternative to scatterplots with high-density data. They resemble scatterplots in which some of the individual data points are replaced with sunflower-like symbols to indicate more than one observation at that location. Figure 3.8 shows a sunflower-plot version of Figure 3.6, in which some of the flower symbols (those with four "petals") represent up to four individual data points, or states. Sunflower plots are most helpful with datasets where many observations plot at similar coordinates. A table printed after the **sunflower** command provides a key regarding how many observations each flower represents. The number of petals and the darkness of the flower convey data density.

```
. sunflower waste metro
```

```
Bin width          =   11.3714
Bin height         =   .295096
Bin aspect ratio   = .0224739
Max obs in a bin   =         4
Light              =         3
Dark               =        13
X-center           =     67.55
Y-center           =       .96
Petal weight       =         1
```

flower type	petal weight	No. of petals	No. of flowers	estimated obs.	actual obs.
none				23	23
light	1	3	5	15	15
light	1	4	3	12	12
				50	50

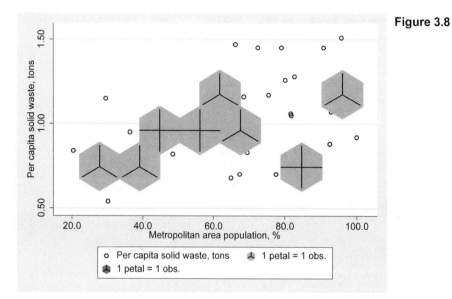

Figure 3.8

Markers in an ordinary scatterplot can be identified by labels. For example, we might want to name the states in a scatterplot such as Figure 3.6. Fifty state names, however, would turn the graph into a visual jumble. Concentrating on one region such as the West seems more promising. An **if** qualifier accomplishes this, producing the results seen in Figure 3.9.

```
. graph twoway scatter waste metro if region==1, mlabel(state)
```

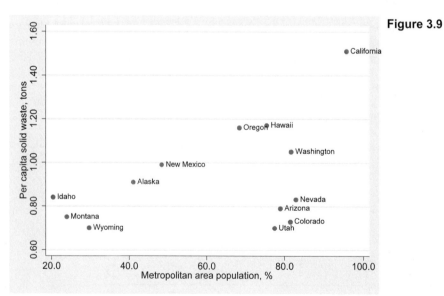

Figure 3.9

Figure 3.10 shows separate *waste – metro* scatterplots for each region. The relationship between these two variables appears noticeably steeper in the South and Midwest than it does in the West and Northeast, an impression we will later confirm. The **ylabel** and **xlabel** options in this example give the *y*- and *x*-axis labels three-digit (maximum) fixed display formats with no decimals, making them easier to read in the small subplots.

```
. graph twoway scatter waste metro, by(region)
     ylabel(, format(%3.0f)) xlabel(, format(%3.0f))
```

Figure 3.10

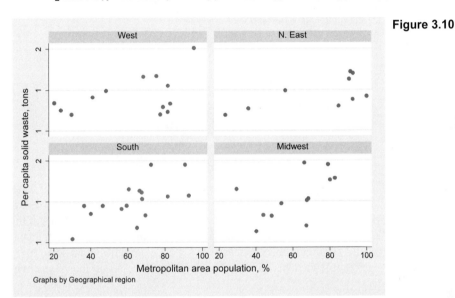

The two-line **graph** command shown with Figure 3.10 should actually be typed as if it were one continuous line (without hitting "Enter") in the Command window. In this book, however, long commands appear broken up into two or more lines for readability. Multiline commands are possible within do-files or programs through use of the **#delimit** command, or by typing /// to indicate continuation at the end of a physical, but not logical, line.

Scatterplot matrices, produced by **graph matrix**, prove useful in multivariate analysis. They provide a compact display of the relationships between a number of variable pairs, allowing the analyst to scan for signs of nonlinearity, outliers, or clustering that might affect statistical modeling. Figure 3.11 shows a scatterplot matrix involving three variables from *states.dta*.

```
. graph matrix miles metro income waste, half msymbol(oh)
```

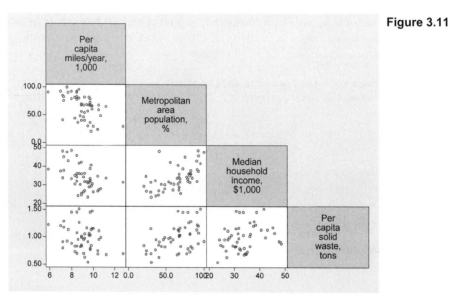

Figure 3.11

The **half** option specified that Figure 3.11 should include only the lower triangular part of the matrix. The upper triangular part is symmetrical and, for many purposes, redundant. **msymbol(oh)** called for small hollow circles as markers, just as we might with a scatterplot. Control of the axes is more complicated, because there are as many axes as variables; type **help graph matrix** for details.

When the variables of interest include one dependent or "effect" variable, and several independent or "cause" variables, it helps to list the dependent variable last in the **graph matrix** command's variable list. That results in a neat row of dependent-versus-independent variable (*y* vs. *x*) graphs across the bottom row of the matrix.

Line Plots

Mechanically, line plots are scatterplots in which the points are connected by line segments. Like scatterplots, the various types of line plots belong to Stata's versatile **graph twoway** family. The scatterplot options that control axis labeling and markers work much the same with line plots, too. Further options control characteristics of the lines themselves.

Line plots tend to have different uses than scatterplots. For example, as time plots they depict changes in a variable over time. Dataset *cod.dta* contains time-series data reflecting the unhappy story of Newfoundland's Northern Cod fishery. This fishery, which had been among the world's richest, collapsed in 1992 due to overfishing and environmental change.

```
Contains data from c:\data\cod.dta
  obs:            38                      Newfoundland's Northern Cod
                                             fishery, 1960-1997
  vars:             5                     4 Jul 2003 15:02
  size:           836 (99.9% of memory free)
```

variable name	storage type	display format	value label	variable label
year	int	%8.0g		Year
cod	float	%8.0g		Total landings, 1000t
canada	int	%8.0g		Canadian landings, 1000t
TAC	int	%8.0g		Total Allowable Catch, 1000t
biomass	float	%9.0g		Estimated biomass, 1000t

Sorted by: **year**

A simple time plot showing Canadian and total landings can be constructed by drawing line plots of both variables against *year*. Figure 3.12 does this, showing the "killer spike" of international overfishing in the late 1960s, followed by a decade of Canadian fishing pressure in the 1980s, leading up to the 1992 collapse of the Northern Cod.

```
. graph twoway line cod canada year
```

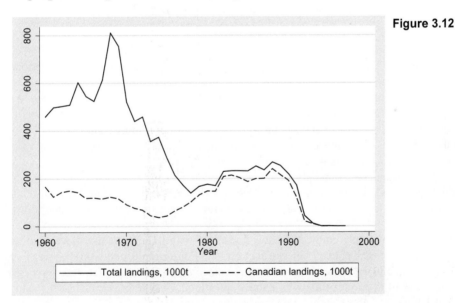

Figure 3.12

In Figure 3.12 above, Stata automatically chose a solid line for the first-named *y* variable, *cod*, and a dashed line for the second, *canada*. (Different colors would have been chosen instead, under one of Stata's color schemes such as s2color.) A legend at the bottom explains the meaning of the lines. We could improve this graph by rearranging the legend, and suppressing the redundant *y*-axis title, as illustrated in Figure 3.13.

```
. graph twoway line cod canada year, legend(label (1 "all nations")
        label(2 "Canada") position(2) ring(0) rows(2)) ytitle("")
```

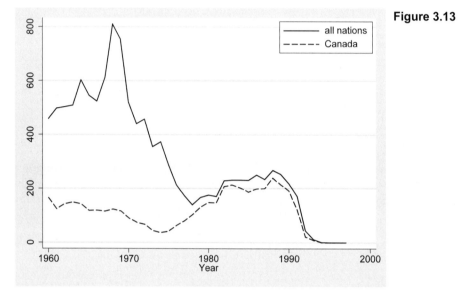

Figure 3.13

The **legend** option for Figure 3.13 breaks down as follows. Note that all of these suboptions occur *within the parentheses* following **legend**.

label(1 "all nations")	label first-named *y* variable "all nations"
label(2 "Canada")	label second-named *y* variable "Canada"
position(2)	place the legend at 2 o'clock position (upper right)
ring(0)	place the legend within the plot space
rows(2)	organize the legend to have two rows

By shortening the legend labels and placing them within the plot space, we leave more room to show the data and create a more attractive, readable figure. **legend** works similarly for other graph styles that have legends. Type **help legend option** to see a list of the many suboptions.

Figures 3.12 and 3.13 simply connected each data point with line segments. Several other connecting styles are possible, using the **connect** option. For example, **connect(stairstep)** or equivalently, **connect(J)** will cause points to be connected in stairstep (flat, then vertical) fashion. Figure 3.14 illustrates this with a stairstep time plot of the government-set fishing quota called Total Allowable Catch (*TAC*) in *cod.dta*.

. **graph twoway line** *TAC year,* **connect(stairstep)**

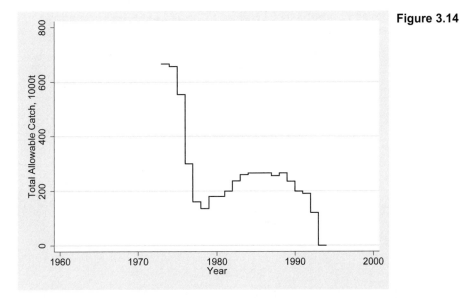

Figure 3.14

Other **connect** choices are listed below. The default, straight line segments, corresponds to **connect(direct)** or **connect(l)**. For more details, see **help connectstyle**.

connect( )	Abbreviation	Description
none	i	do not connect
direct	l (letter "el")	connect with straight lines
ascending	L	direct, but only if $x[i+1] > x[i]$
stairstep	J	flat, then vertical
stepstair		vertical, then flat

Figure 3.15 repeats this stairstep plot of *TAC*, but with some enhancements of axis labels and titles. The option **xtitle("")** requests no *x*-axis title (because "year" is obvious). We added tick marks at two-year intervals to the *x* axis, labeled the *y* axis at intervals of 100, and printed *y*-axis labels horizontally instead of vertically (the default).

```
. graph twoway line TAC year, connect(stairstep)  xtitle("")
    xtick(1960(2)2000)  ytitle("Thousands of tons")
    ylabel(0(100)800, angle(horizontal)) xtitle("")
    clpattern(dash)
```

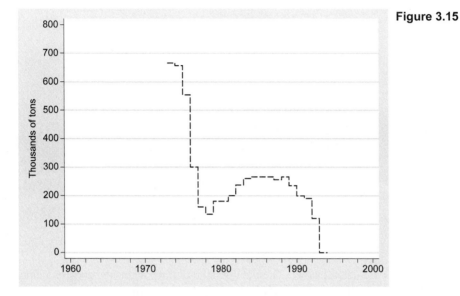

Figure 3.15

Instead of letting Stata determine the line patterns (solid, dashed, etc.) in Figure 3.15, we used the **clpattern(dash)** option to call for a dashed line. Possible line pattern choices are listed in the table below (also see **help linepatternstyle**).

clpattern( )	Description
solid	solid line
dash	dashed line
dot	dotted line
dash_dot	dash then dot
shortdash	short dash
shortdash_dot	short dash followed by dot
longdash	long dash
longdash_dot	long dash followed by dot
blank	invisible line
formula	for example, **clpattern(-.)** or **clpattern(—..)**

Before we move on to other examples and types, Figure 3.16 unites the three variables discussed in this section to create a single graphic showing the tragedy of the Northern Cod. Note how the **connect( )**, **clpattern( )**, and **legend( )** options work in this three-variable context.

```
. graph twoway line cod canada TAC year, connect(line line
    stairstep) clpattern(solid longdash dash) xtitle("")
    xtick(1960(2)2000) ytitle("Thousands of tons")
    ylabel(0(100)800, angle(horizontal))
    xtitle("") legend(label (1 "all nations") label(2 "Canada")
    label(3 "TAC") position(2) ring(0) rows(3))
```

Figure 3.16

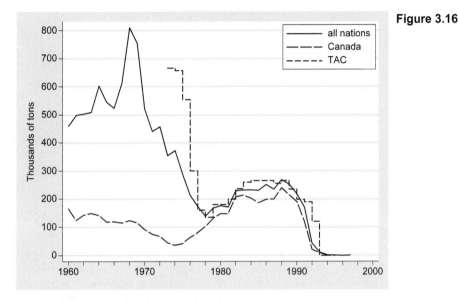

Connected-Line Plots

In the line plot examples of the previous section, data points are invisible and we see only the connecting lines. The **graph twoway connected** command creates connected-line plots in which the data points are marked by scatterplot symbols. The marker-symbol options described earlier for **graph twoway scatter**, and also the line-connecting options described for **graph twoway line**, both apply to **graph twoway connected** as well. Figure 3.17 on the next page shows a default example, a connected-line time plot of the cod biomass variable (*bio*) from *cod.dta*.

```
. graph twoway connected bio year
```

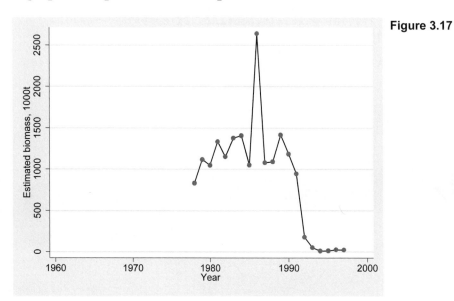

Figure 3.17

The dataset contains only biomass values for 1978 through 1997, resulting in much empty space in Figure 3.17. **if** qualifiers allow us to restrict the range of years. Figure 3.18, on the following page, does this. It also dresses up the image through control of marker symbols, line patterns, axes, and legends. With cod landings and biomass both in the same image, we see that the biomass began its crash in the late 1980s, several years before a crisis was officially recognized. Figure 3.17 also dramatizes one improbably high biomass estimate in 1986, a statistical outlier that contributed to official overconfidence about the state of the resource (Haedrich and Hamilton 2000).

```
. graph twoway connected bio cod year if year > 1977 & year < 1999,
    msymbol(T Oh) clpattern(dash solid) xlabel(1978(2)1996)
    xtick(1979(2)1997) ytitle("Thousands of tons") xtitle("")
    ylabel(0(500)2500, angle(horizontal))
    legend(label(1 "Estimated biomass") label(2 "Total landings")
    position(2) rows(2) ring(0))
```

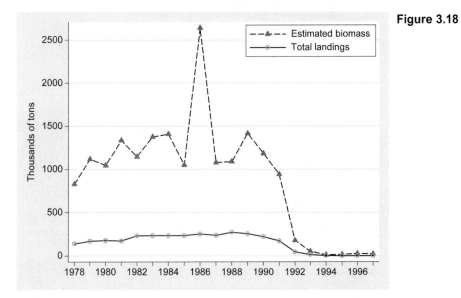

Figure 3.18

Other Twoway Plot Types

In addition to basic line plots and scatterplots, the **graph twoway** command can draw a wide variety of other types. The following table lists the possibilities.

graph twoway	Description
scatter	scatterplot
line	line plot
connected	connected-line plot
scatteri	scatter with immediate arguments (data given *in* the command line)
area	line plot with shading
bar	twoway bar plot (different from **graph bar**)
spike	twoway spike plot
dropline	dropline plot (spikes dropped vertically or horizontally to given value)
dot	twoway dot plot (different from **graph dot**)
rarea	range plot, shading the area between high and low values
rbar	range plot with bars between high and low values
rspike	range plot with spikes between high and low values
rcap	range plot with capped spikes
rcapsym	range plot with spikes capped with symbols

rscatter	range plot with scatterplot marker symbols
rline	range plot with lines
rconnected	range plot with lines and markers
pcspike	paired-coordinate plot with spikes
pccapsym	paired-coordinate plot with spikes capped with symbols
pcarrow	paired-coordinate plot with arrows
pcbarrow	paired-coordinate plot with arrows having two heads
pcscatter	paired-coordinate plot with markers
pci	**pcspike** with immediate arguments
pcarrowi	**pcarrow** with immediate arguments
tsline	time-series plot
tsrline	time-series range plot
mband	straight line segments connect the (x, y) cross-medians within bands
mspline	cubic spline curve connects the (x, y) cross-medians within bands
lowess	LOWESS (locally weighted scatterplot smoothing) curve
lfit	linear regression line
qfit	quadratic regression curve
fpfit	fractional polynomial plot
lfitci	linear regression line with confidence band
qfitci	quadratic regression curve with confidence band
fpfitci	fractional polynomial plot with confidence band
function	line plot of function
histogram	histogram plot
kdensity	kernel density plot
lpoly	local polynomial smooth plot
lpolyci	local polynomial smooth plot with confidence intervals

The usual options to control line patterns, marker symbols, and so forth work where appropriate with all **twoway** commands. For more information about a particular command, type **help twoway mband**, **help twoway function**, etc. (using any of the names above). Note that **graph twoway bar** is a different command from **graph bar**. Similarly, **graph twoway dot** differs from **graph dot**. The **twoway** versions provide various methods for plotting a measurement y variable against a measurement x variable, analogous to a scatterplot or a line plot. The non-twoway versions, on the other hand, provide ways to plot summary statistics (such as means or medians) of one or more measurement y variables against categories of one or more x variables. The **twoway** versions thus are comparatively specialized, although (as with all **twoway** plots) they can be overlaid with other **twoway** plots for more complex graphical effects.

Many of these plot types are most useful in composite figures, constructed by overlaying two or more simple plots as described later in this chapter. Others produce nice stand-alone graphs. For example, Figure 3.19 shows an area plot of the Newfoundland cod landings.

```
. graph twoway area cod canada year, ytitle("")
```

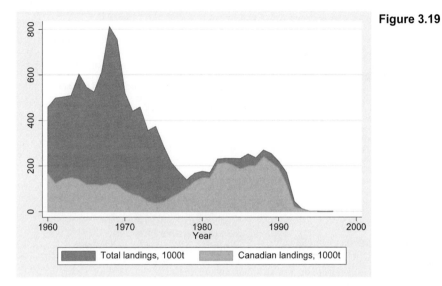

Figure 3.19

The shading in area graphs and other types with shaded regions can be controlled through the option **bcolor**. Type **help colorstyle** for a list of the available colors, which include gray scales. The darkest gray, gs0, is actually black. The lightest gray, gs16, is white. Other values are in between. For example, Figure 3.20 shows a light-gray version of this graph.

```
. graph twoway area cod canada year, ytitle("") bcolor(gs12 gs14)
```

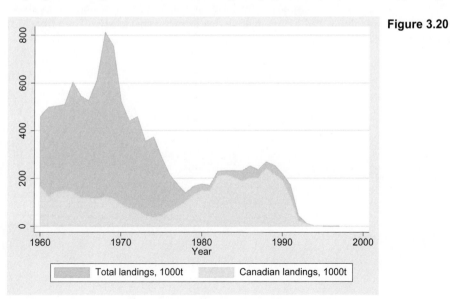

Figure 3.20

Unusually cold atmosphere/ocean conditions played a secondary role in Newfoundland's fisheries disaster, which involved not only the Northern Cod but also other species and populations. For example, key fish species in the neighboring Gulf of St. Lawrence declined during this period as well (Hamilton, Haedrich, and Duncan 2004). Dataset *gulf.dta* describes environment and Northern Gulf cod catches (raw data from DFO 2003).

```
Contains data from c:\data\gulf.dta
  obs:            56                      Gulf of St. Lawrence environment
                                            and cod fishery
  vars:            7                      2 Feb 2008 14:22
  size:         1,568 (99.9% of memory free)

              storage  display   value
variable name   type   format    label    variable label

winter          int    %8.0g              Winter
minarea         float  %9.0g              Minimum ice area, 1000 km^2
maxarea         float  %9.0g              Maximum ice area, 1000 km^2
mindays         byte   %8.0g              Minimum ice days
maxdays         byte   %8.0g              Maximum ice days
cil             float  %9.0g              Cold Intermediate Layer
                                            temperature minimum, C
cod             float  %9.0g              N. Gulf cod catch, 1000 tons

Sorted by:  winter
```

The maximum annual ice cover averaged 173,017 km^2 during these years.

```
. summarize maxarea
```

Variable	Obs	Mean	Std. Dev.	Min	Max
maxarea	38	173.0172	37.18623	47.8901	220.1905

Figure 3.21 uses this mean (173 thousand) as the base for a spike plot, in which spikes above and below the line show above and below-average ice cover, respectively. The **yline(173)** option draws a horizontal line at 173.

```
. graph twoway spike maxarea winter if winter > 1963, base(173)
     yline(173) ylabel(40(20)220, angle(horizontal))
     xlabel(1965(5)2000)
```

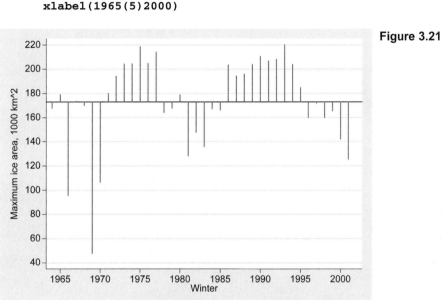

Figure 3.21

The **base( )** format of Figure 3.21 emphasizes the succession of unusually harsh winters (above-average maximum ice cover) during the late 1980s and early 1990s, around the time of Newfoundland's fisheries crisis. We also see an earlier spell of mild winters in the early 1980s, and hints of a recent warming trend.

A different view of the same data, in Figure 3.22, employs lowess regression to smooth the time series. The bandwidth option, **bwidth(.4)**, specifies a curve based on smoothed data points that are calculated from weighted regressions within a moving band containing 40% of the sample. Lower bandwidths such as **bwidth(.2)**, or 20% of the data, would give us a more jagged, less smoothed curve that more closely resembles the raw data. Higher bandwidths such as **bwidth(.8)**, the default, will smooth more radically. Regardless of the bandwidth chosen, smoothed points towards either extreme of the x values must be calculated from increasingly narrow bands, and therefore will show less smoothing. Chapter 8 contains more about lowess smoothing.

```
. graph twoway lowess maxarea winter if winter > 1963, bwidth(.4)
      yline(173) ylabel(40(20)220, angle(horizontal))
      xlabel(1965(5)2000)
```

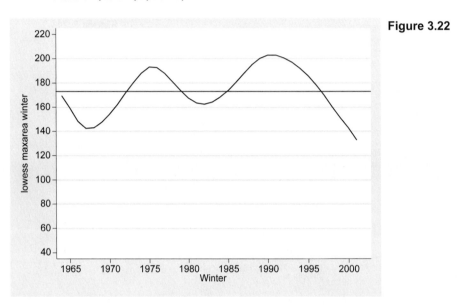

Figure 3.22

Range plots connect high and low *y* values at each level of *x*, using bars, spikes, or shaded areas. Daily stock market prices are often graphed in this way. Figure 3.23 shows a capped-spike range plot using the minimum and maximum ice cover variables from *gulf.dta*.

```
. graph twoway rcap minarea maxarea winter if winter > 1963,
      ylabel(0(20)220, angle(horizontal))
      ytitle("Ice area, 1000 km^2") xlabel(1965(5)2000)
```

Figure 3.23

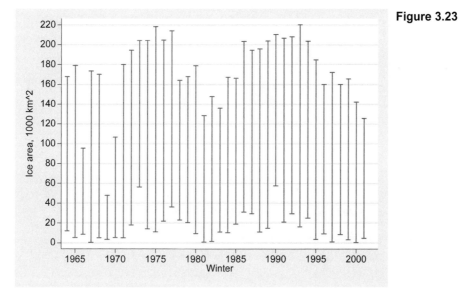

These examples by no means exhaust the possibilities for twoway graphs. Other applications appear throughout the book. Later in this chapter, we will see examples involving overlays of several twoway graphs, forming a single image.

Box Plots

Box plots convey information about center, spread, symmetry, and outliers at a glance. To obtain a single box plot, type a command of the form

```
. graph box y
```

If several different variables have similar scales, we can visually compare their distributions through commands of the form

```
. graph box w x y z
```

One of the most common applications for box plots involves comparing the distribution of one variable over categories of a second. Figure 3.24 compares the distribution of *college* (percent college graduates) across states of four U.S. regions, from dataset *states.dta*.

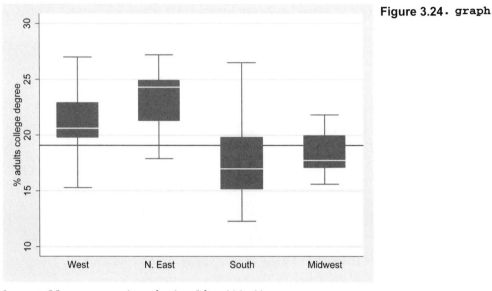

Figure 3.24. `graph`

`box` *`college`*`, over(`*`region`*`) yline(19.1)`

The median proportion of adults with college degrees tends to be highest in the Northeast, and lowest in the South. On the other hand, southern states are more varied. Regional medians, the lines within boxes in Figure 3.24, can be compared visually to the 50-state median indicated by the **yline(19.1)** option. (This 50-state median was obtained via a **summarize, detail** command, as described in Chapter 4.) The **if region < .** qualifier above restrict our analysis to observations that have nonmissing values of *region*; that is, to every place except Washington DC.

The box in a box plot extends from approximate first to third quartiles, a distance called the interquartile range (IQR). It therefore contains roughly the middle 50% of the data. Outliers, defined as observations more than 1.5(IQR) beyond the first or third quartile, are plotted individually in a box plot. No outliers appear among the four distributions in Figure 3.24. Stata's box plots define quartiles in the same manner as **summarize, detail**. This is not the same approximation used to calculate "fourths" for letter-value displays, **lv** (Chapter 4). See Frigge, Hoaglin, and Iglewicz (1989) and Hamilton (1992b) for more about quartile approximations and their role in identifying outliers.

Numerous options control the appearance, shading and details of boxes in a box plot; type **help graph box** for a list. Figure 3.25 demonstrates some of these options, and also the horizontal arrangement of **graph hbox**, using per capita energy consumption from *states.dta*. The option **over(region, sort(1))** calls for boxes sorted in ascending order according to their medians on the first-named (and in this case, the only) *y* variable. **intensity(30)** controls the intensity of shading in the boxes, setting this somewhat lower (less dark) than the default seen in Figure 3.24. Counterintuitively, the vertical line marking the overall median (320) in Figure 3.25 requires a **yline** option, rather than **xline**.

```
. graph hbox energy, over(region, sort(1)) yline(320)
      intensity(30) marker(1, mlabel(state) mlabpos(12))
```

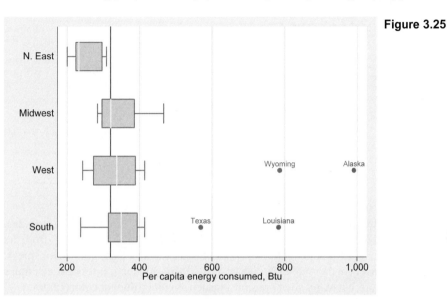

Figure 3.25

The energy box plots in Figure 3.25 make clear not only the differences among medians, but also the presence outliers — four very high-consumption states in the West and South. The option **marker(1, mlabel(*state*) mlabpos(12))** requested that marker symbols for the first (here, the only) *y* variable be labeled with values of *state*, and labels shown in the 12 o'clock position relative to the marker. The outliers are four oil-producing states: Wyoming, Alaska, Texas, and Louisiana. Box plots excel at drawing attention to outliers, which easily get overlooked (and can cause trouble) in other steps of statistical analysis.

Pie Charts

Pie charts are popular tools for "presentation graphics," although they tend to have less value for analytical work. Stata's basic pie chart command has the form

```
. graph pie w x y z
```

where the variables *w*, *x*, *y*, and *z* all measure quantities of something in similar units (for example, all are in dollars, hours, or people).

Dataset *AKethnic.dta*, on the ethnic composition of Alaska's population, provides an illustration. Alaska's indigenous Native population divides into three broad cultural/linguistic groups: Aleut, Indian (including Athabaska, Tlingit, and Haida), and Eskimo (Yupik and Inupiat). The variables *aleut*, *indian*, *eskimo*, and *nonnativ* are population counts for each group, taken from the 1990 U.S. Census. This dataset contains only three observations, representing three types or sizes of communities: cities of 10,000 people or more; towns of 1,000 to 10,000; and villages with fewer than 1,000 people.

```
Contains data from c:\data\AKethnic.dta
  obs:             3                         Alaska ethnicity 1990
  vars:            7                         6 Sep 2003 12:47
  size:           75 (99.9% of memory free)
```

variable name	storage type	display format	value label	variable label
comtype	byte	%8.0g	popcat	Community type (size)
pop	float	%9.0g		Population
n	int	%8.0g		number of communities
aleut	int	%8.0g		Aleut
indian	int	%8.0g		Indian
eskimo	int	%8.0g		Eskimo
nonnativ	float	%9.0g		Non-Native

```
Sorted by:
```

The majority of the state's population is non-Native, as clearly seen in a pie chart (Figure 3.26). The option **pie(3, explode)** causes the third-named variable, *eskimo*, to be "exploded" from the pie for emphasis. The fourth-named variable, *nonnativ*, is shaded a light gray color, **pie(4, color(gs13))**, for contrast with the smaller Native groups. (In this monochrome book, examples use only gray-scale colors, but many other possiblities such as **color(blue)** or **color(cranberry)** exist. Type **help colorstyle** for the list.) **plabel(3 percent, gap(20))** causes a percentage label to be printed by the *eskimo* (variable 3) slice, with a gap of 20 relative radial units from the center. We see that about 8% of the population is Eskimo (Inupiat or Yupik). The **legend** option specifies a four-row box at the 11 o'clock position within the plot space.

```
. graph pie aleut indian eskimo nonnativ, pie(3, explode)
       pie(4, color(gs13)) plabel(3 percent, gap(20))
       legend(position(11) rows(4) ring(0))
```

Figure 3.26

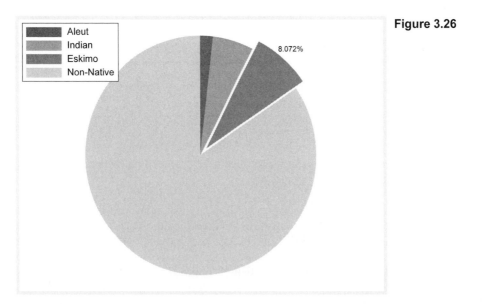

Non-Natives are the dominant group in Figure 3.26, but if we draw separate pies for each type of community by adding a **by(*comtype*)** option, new details emerge (Figure 3.27). The option **angle0( )** specifies the angle of the first slice of pie. Setting this first-slice angle at 0 (horizontal) orients the pies in Figure 3.27 in such a way that the labels are more readable. The figure shows that whereas Natives are only a small fraction of the population in Alaska cities, they constitute the majority among those living in villages. In particular, Eskimos make up a large fraction of villagers — 35% across all villages, and more than 90% in some. This gives Alaska villages a different character from Alaska cities.

```
. graph pie aleut indian eskimo nonnativ, pie(3, explode)
      pie(4, color(gs13)) plabel(3 percent, gap(8))
      legend(rows(1)) by(comtype) angle0(0)
```

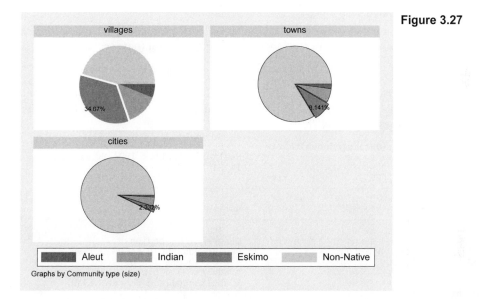

Figure 3.27

Bar Charts

Although they contain less information than box plots, bar charts provide simple and versatile displays for comparing sets of summary statistics such as means, medians, sums, or counts. To obtain vertical bars showing the mean of *y* across categories of *x*, for example, type

```
. graph bar (mean) y, over(x)
```

For horizontal bars showing the sum of *y* across categories of *x1* within categories of *x2*, type

```
. graph hbar (sum) y, over(x1) over(x2)
```

The bar chart could display any of the following statistics:
  mean   Means (default, if statistic is not specified)
  median Medians

p1	1st percentiles
p2	2nd percentiles (and so forth to **p99**)
sd	Standard deviations
sum	Sums
rawsum	Sums, ignoring optionally specified weight
count	Number of nonmissing observations
max	Maximums
min	Minimums
iqr	Interquartile range
first	First values
last	Last values
firstnm	First nonmissing values
lastnm	Last nonmissing values

This list of available summary statistics is the same as that for the **collapse** command (see Chapter 2), and also for certain other commands including **graph dot** (next section) and **table** (Chapter 4).

Dataset *statehealth.dta* contains further data on the U.S. states, combining socioeconomic measures from the 1990 Census with several health-risk indicators from the Centers for Disease Control (2003), averaged over 1994–98.

```
Contains data from c:\data\statehealth.dta
  obs:            51                          Health indicators 1994-98 (CDC)
  vars:           12                          14 Sep 2003 18:46
  size:        3,519 (99.9% of memory free)
```

variable name	storage type	display format	value label	variable label
state	str20	%20s		US State
region	byte	%9.0g	region	Geographical region
income	long	%10.0g		Median household income, 1990
income2	float	%11.0g	income2	Median income low or high
high	float	%9.0g		% adults HS diploma, 1990
college	float	%9.0g		% adults college degree, 1990
overweight	float	%9.0g		% overweight
inactive	float	%9.0g		% inactive in leisure time
smokeM	float	%9.0g		% male adults smoking
smokeF	float	%9.0g		% female adults smoking
smokeT	float	%9.0g		% adults smoking
motor	float	%9.0g		Age-adjusted motor-vehicle related deaths/100,000

```
Sorted by:  state
```

Figure 3.28 graphs the median percent of population inactive in leisure time (*inactive*) across four geographical regions (*region*). We see a pronounced regional difference: inactivity rates are highest in the South (36%), and lowest in the West (21%). Note that the vertical axis has automatically been labeled "p 50 of inactive," meaning the 50th percentile or median. The **blabel(bar)** option labels the bar heights (20.9, etc.). **bar(1, bcolor(gs10))** specifies that bars for the first-named *y* variable (*inactive*; there is only one) should be filled with a medium-light gray color.

```
. graph bar (median) inactive, over(region) blabel(bar)
      bar(1, bcolor(gs10))
```

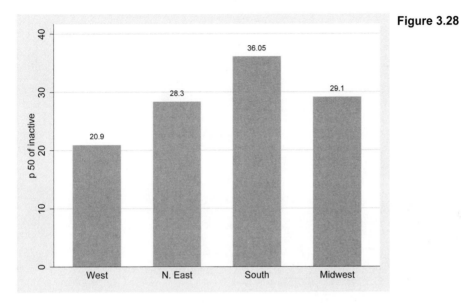

Figure 3.28

Figure 3.29 adds a second variable, *overweight*, and colors the bars a darker gray. Bar labels are **size(medium)** in Figure 3.29, larger than the defaults, **size(small)**, used in Figure 3.28. Other possibilities for **size( )** suboptions include labels that are **tiny** , **medsmall**, **medlarge**, or **large**. See **help textsizestyle** for a complete list. Figure 3.29 shows that regional differences in the prevalence of overweight individuals are less pronounced than differences in inactivity, although both variables' medians are highest in the South and Midwest.

```
. graph bar (median) inactive overweight, over(region)
     blabel(bar, size(medium))
     bar(1, bcolor(gs10)) bar(2, bcolor(gs7))
```

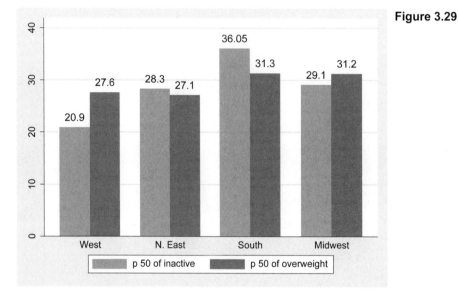

Figure 3.29

The risk indicators in *statehealth.dta* include motor-vehicle related fatalities per 100,000 population (*motor*). On the next page, Figure 3.30 breaks these down regionally, and then into subgroups of low- and high-income states (states having median household incomes below or above the national median), revealing a striking correlation with wealth. Within each region, the low-income states exhibit higher mean fatality rates. Across both income categories, fatality rates are higher in the South, and lower in the Northeast. The order of the two **over** options in the command controls their order in organizing the chart. For this example we chose a horizontal bar chart or **hbar** . In such horizontal charts, **ytitle**, **yline**, etc. refer counter-intuitively to the horizontal axis. **yline(17.2)** marks the overall mean with a vertical line.

```
. graph hbar (mean) motor, over(income2) over(region) yline(17.2)
     ytitle("Mean motor-vehicle related fatalities/100,000")
```

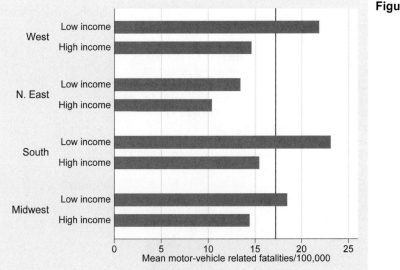

Figure 3.30

Bars also can be stacked, as shown in Figure 3.31. This plot, based on the Alaska ethnicity data (*AKethnic.dta*), employs all the defaults to display ethnic composition by type of community (village, town, or city).

```
. graph bar (sum) nonnativ aleut indian eskimo, over(comtyp) stack
```

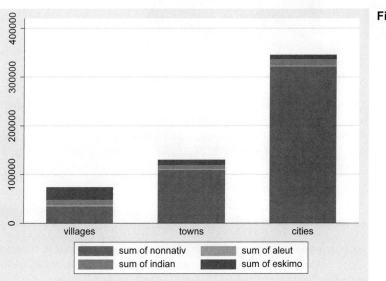

Figure 3.31

Figure 3.32 redraws this plot with better legend and axis labels. The **over** option now includes suboptions that relabel the community types so the horizontal axis is more informative. The **legend** option specifies four rows in the same vertical order as the bars themselves, and placed in the 11 o'clock position inside the plot space. It also improves legend labels. **ytitle**, **ylabel**, and **ytick** options format the vertical axis.

```
. graph bar (sum) nonnativ aleut indian eskimo,
      over(comtyp, relabel(1 "Villages <1,000" 2
        "Towns 1,000-10,000" 3 "Cities >10,000"))
      legend(rows(4) order(4 3 2 1) position(11) ring(0)
      label(1  "Non-native") label(2 "Aleut")
      label(3 "Indian") label(4 "Eskimo"))
      stack ytitle(Population)
      ylabel(0(100000)300000) ytick(50000(100000)350000)
```

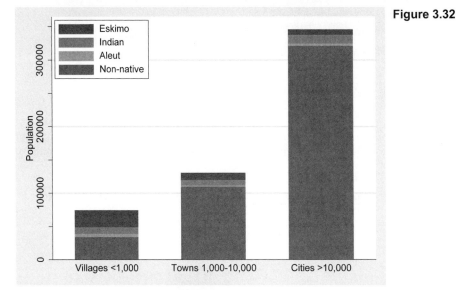

Figure 3.32

Figure 3.32 plots the same variables as the pie chart in Figure 3.27, but displays them quite differently. Whereas the pie charts show relative sizes (percentages) of ethnic groups within each community type, this bar chart shows their absolute sizes. Consequently, Figure 3.32 tells us something that Figure 3.27 could not: the majority of Alaska's Eskimo (Yupik and Inupiat) population lives in villages.

Dot Plots

Dot plots serve much the same purpose as bar charts: visually comparing statistical summaries of one or more measurement variables. The organization and Stata options for the two types of plot are broadly similar, including the choices of statistical summaries. To see a dot plot comparing the medians of variables x, y, z, and w, type

```
. graph dot (median) x y z w
```

For a dot plot comparing the mean of *y* across categories of *x*, type

```
. graph dot (mean) y, over(x)
```

Figure 3.33 shows a dot plot of male and female smoking rates by region, from *statehealth.dta*. The **over** option includes a suboption, **sort(*smokeM*)**, which calls for the regions to be sorted in order of their mean values of *smokeM* — that is, from lowest to highest smoking rates. We also specify a hollow triangle as the marker symbol for *smokeM*, and solid circle for *smokeF*.

```
. graph dot (mean) smokeM smokeF, over(region, sort(smokeM))
      marker(1, msymbol(Th)) marker(2, msymbol(O))
```

Figure 3.33

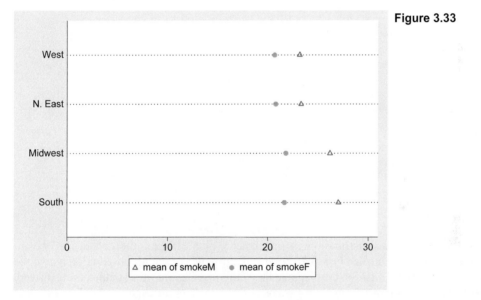

Although Figure 3.33 displays only eight means, it does so in a way that facilitates several comparisons. We see that smoking rates are generally higher for males; that among both sexes they are higher in the South and Midwest; and that regional variations are substantially greater for the male smoking rates. Bar charts could convey the same information, but one advantage of dot plots is their compactness. Dot plots (particularly when rows are sorted by the statistic of interest, as in Figure 3.33) remain easily readable even with a dozen or more rows.

Symmetry and Quantile Plots

Box plots, bar charts, and dot plots summarize measurement variable distributions, hiding individual data points to clarify overall patterns. Symmetry and quantile plots, on the other hand, include points for every observation in a distribution. They are harder to read than summary graphs, but convey more detailed information.

A histogram of per-capita energy consumption in the 50 U.S. states (from *states.dta*) appears in Figure 3.34. The distribution includes a handful of very high-consumption states, which happen to be oil producers. A superimposed normal (Gaussian) curve indicates that *energy* has a lighter-than-normal left tail, and a heavier-than-normal right tail — the definition of positive skew.

```
. histogram energy, start(100) width(100) xlabel(0(100)1000)
       frequency norm
```

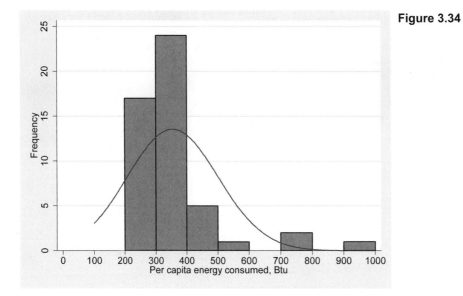

Figure 3.34

Figure 3.35 depicts this distribution as a symmetry plot. It plots the distance of the *i*th observation above the median (vertical) against the distance of the *i*th observation below the median. All points would lie on the diagonal line if this distribution were symmetrical. Instead, we see that distances above the median grow steadily larger than corresponding distances below the median, a symptom of positive skew.

. **symplot** *energy*

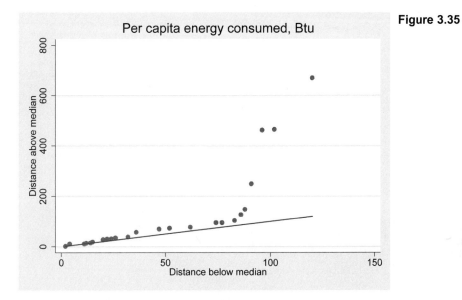

Figure 3.35

Quantiles are values below which a certain fraction of the data lie. For example, a .3 quantile is that value higher than 30% of the data (similar to the 30th percentile). If we sort *n* observations in ascending order, the *i*th value forms the (*i*–.5)/*n* quantile. Quantile plots automatically calculate what fraction of the observations lie below each data value, and display the results graphically as in Figure 3.36 on the following page. Quantile plots provide a graphic reference for someone who does not have the original data at hand. From well-labeled quantile plots, we can estimate order statistics such as median (.5 quantile) or quartiles (.25 and .75 quantiles). The IQR equals the rise between .25 and .75 quantiles. We could also read a quantile plot to estimate the fraction of observations falling below a given value.

. **quantile** *energy*

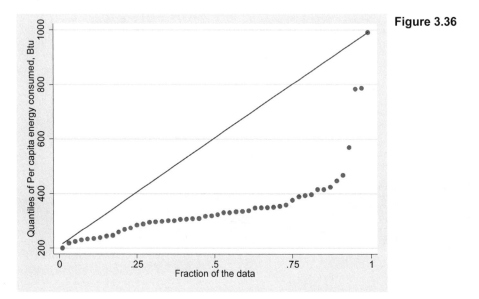

Figure 3.36

Quantile–normal plots, also called normal probability plots, compare quantiles of a variable's distribution with quantiles of a theoretical normal distribution having the same mean and standard deviation. They allow visual inspection for departures from normality in every part of a distribution, which can help guide decisions regarding normality assumptions and efforts to find a normalizing transformation. Figure 3.37 (following page), a quantile–normal plot of *energy*, confirms the severe positive skew that we had noticed earlier. The **grid** option calls for a set of lines marking the .05, .10, .25 (first quartile), .50 (median), .75 (third quartile), .90, and .95 quantiles of both distributions. The .05, .50, and .95 quantile values are printed along the top and right-hand axes.

`. qnorm energy, grid`

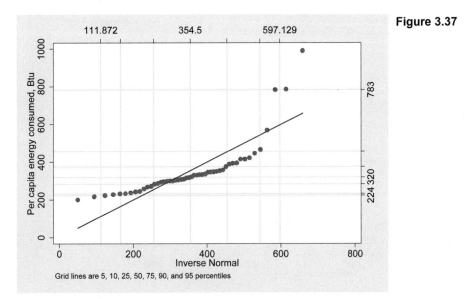

Figure 3.37

Quantile–quantile plots resemble quantile–normal plots, but compare quantiles (ordered data points) of two empirical distributions instead of comparing one empirical distribution with a theoretical normal distribution. Figure 3.38 shows a quantile–quantile plot of mean math and verbal SAT scores across U.S. states (*states.dta*). If the distributions were identical, points would lie on the diagonal line. Instead, they form a straight line roughly parallel to the diagonal, indicating that the two variables have different means but similar shapes and standard deviations.

. qqplot *msat vsat*

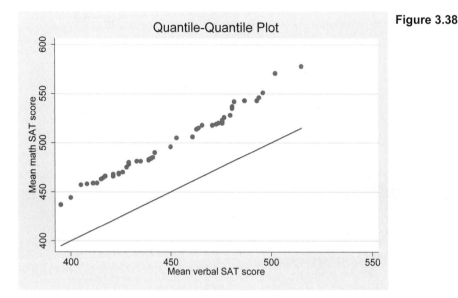

Figure 3.38

Regression with Graphics (Hamilton 1992a) includes an introduction to reading quantile-based plots. Chambers et al. (1983) provide more details. Related Stata commands include **pnorm** (standard normal probability plot), **pchi** (chi-squared probability plot), and **qchi** (quantile–chi-squared plot).

Adding Text to Graphs

Titles, captions, and notes can be added to make graphs more self-explanatory. The default versions of titles and subtitles appear above the plot space; notes (which might document the data source, for instance) and captions appear below. These defaults can be overridden, of course. Type **help title options** for more information about placement of titles, or **help textbox options** for details concerning their content. Figure 3.39 demonstrates the default versions of these four options in a scatterplot of the prevalence of smoking and college graduates among U.S. states, using *statehealth.dta*. Figure 3.39 also includes titles for both the left and right y axes, **yaxis(1 2)**, and top and bottom x axes, **xaxis(1 2)**. Subsequent **ytitle** and **xtitle** options refer to the second axes specifically, by including the **axis(2)** suboption. y axis 2 is not necessarily on the right, and x axis 2 is not necessarily on the top, as we will see later; but these are their default positions.

```
. graph twoway scatter smokeT college, yaxis(1 2) xaxis(1 2)
    title("This is the TITLE") subtitle("This is the SUBTITLE")
    caption("This is the CAPTION") note("This is the NOTE")
    ytitle("Percent adults smoking")
    ytitle("This is Y AXIS 2", axis(2))
    xtitle("Percent adults with Bachelor's degrees or higher")
    xtitle("This is X AXIS 2", axis(2))
```

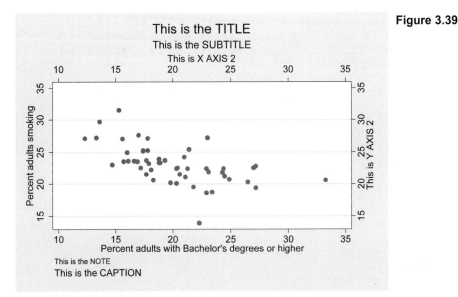

Figure 3.39

Titles add text boxes outside of the plot space. We can also add text boxes at specified coordinates within the plot space. Several outliers stand out in this scatterplot. Upon investigation, they turn out to be Washington DC (highest *college* value, at far right), Utah (lowest *smokeT* value, at bottom center), and Nevada (highest *smokeT* value, at upper left). Text boxes provide a way for us to identify these observations within our graph, as demonstrated in Figure 3.40. The option **text(15.5 22.5 "Utah")** places the word "Utah" at position $y = 15.5$, $x = 22.5$ in the scatterplot, directly above Utah's data point. Similarly, we place the word "Nevada" at $y = 33.5$, $x = 15$, and draw a box (with small margins; see **help marginstyle**) around that state's name. Three lines of left-justified text are placed next to Washington DC (each line specified in its own set of quotation marks). Any text box or title can have multiple lines in this fashion; we specify each line individually in its own set of quotations, then specify justification or other suboptions. The "Nevada" box uses a default shaded background, whereas for the "Washington DC" box we chose a white background color (see **help textbox options** and **help colorstyle**).

```
. graph twoway scatter smokeT college, yaxis(1 2) xaxis(1 2)
    title("This is the TITLE") subtitle("This is the SUBTITLE")
    caption("This is the CAPTION") note("This is the NOTE")
    ytitle("Percent adults smoking")
    ytitle("This is Y AXIS 2", axis(2))
    xtitle("Percent adults with Bachelor's degrees or higher")
    xtitle("This is X AXIS 2", axis(2))
    text(15.5 22.5 "Utah")
    text(33.5 15 "Nevada", box margin(small))
    text(24.5 32 "Washington DC" "is not actually" "a state",
        box justification(left) margin(small) bfcolor(white))
```

Figure 3.40

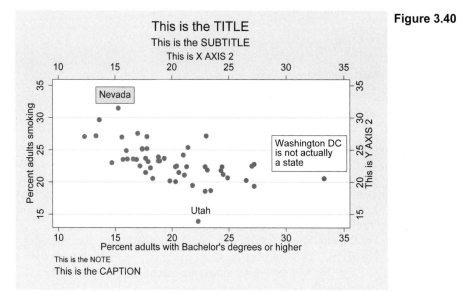

Overlaying Multiple Twoway Plots

Two or more plots from the versatile **graph twoway** family can be overlaid, one atop the other, to form a single unified image. Figure 1.1 in Chapter 1 gave a simple example. The **twoway** family includes several model-based types such as **lfit** (linear regression line), **qfit** (quadratic regression curve), and so forth. By themselves, such plots provide minimal information. For example, Figure 3.41 depicts the linear regression line, with 95% confidence bands for the conditional mean, from the regression of *smokeT* on *college* (from dataset *statehealth.dta*).

. `graph twoway lfitci` *smokeT college*

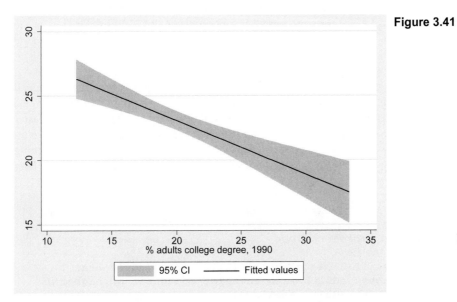

Figure 3.41

A more informative graph results when we overlay a scatterplot on top of the regression line plot, as seen in Figure 3.42. To do this, we essentially give two distinct graphing commands, separated by " || ".

. `graph twoway lfitci` *smokeT college*
    `||   scatter` *smokeT college*

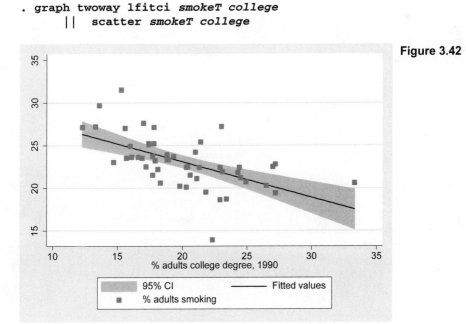

Figure 3.42

The second plot (scatterplot) overprints the first in Figure 3.42. This order has consequences for the default line style (solid, dashed, etc.) and also for the marker symbols (squares, circles, etc.) used by each sub-plot. More importantly, it superimposes the scatterplot points on the confidence bands so the points remain visible. Try reversing the order of the two plots in the command, to see how this works.

Figure 3.43 takes this idea a step further, improving the image through axis labeling and legend options. Because these options apply to the graph as a whole, not just to one of the subplots, the options are placed after a second ‖ separator, followed by a comma. To make them easier to read, in this book long **graph** commands with overlays are sometimes printed with extra spacing (as done below). The actual command would be typed interactively without hitting "Enter" until the end, or written as one logical line (for example, using **#delimit** as shown later in this chapter) within a do-file.

```
. graph twoway lfitci smokeT college

    || scatter smokeT college

    || , xlabel(12(2)34) ylabel(14(2)32, angle(horizontal))
    xtitle("Percent adults with Bachelor's degrees or higher")
    ytitle("Percent adults smoking")
    note("Data from CDC and US Census")
    legend(order(2 1) label(1 "95% c.i.")
        label(2 "regression line") rows(3) position(1) ring(0))
```

Figure 3.43

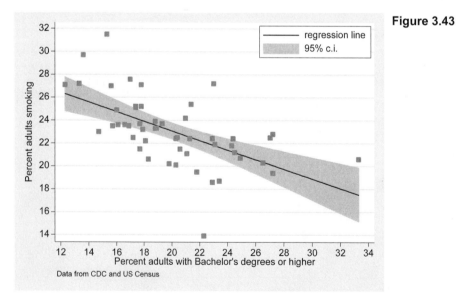

Most of the options for Figure 3.43 resemble those used in previous examples. The **order(2 1)** option here does something new: it omits one of the three legend items, so that only two of them (2, the regression line, followed by 1, the confidence interval) appear in the figure. Compare this legend with Figure 3.42 to see the difference. Although we list only two legend items in Figure 3.43, it still is necessary to specify a **rows(3)** legend format as if all three were retained.

The two separate plots (**lfitci** and **scatter**) overlaid in Figure 3.43 share the same y and x scales, so a single set of axes applies to both. When the variables of interest have different scales, we need independently scaled axes.

On the following page, Figure 3.44 illustrates this with an overlay of two line plots based on the Gulf of St. Lawrence environmental data in *gulf.dta*. This figure combines time series of the minimum mean temperature of the Gulf's cold intermediate layer waters (*cil*) in degrees Celsius, and maximum winter ice cover (*maxarea*) in thousands of square kilometers. The *cil* plot makes use of **yaxis(1)**, which by default is on the left. The *maxarea* plot makes use of **yaxis(2)** which, by default, is on the right. The various **ylabel, ytitle, yline,** and **yscale** options each include an **axis(1)** or **axis(2)** suboption, declaring which y axis they refer to. Extra spaces inside the quotation marks for **ytitle** provided a quick way to place the words of these titles where we want them, near the numerical labels. (For a different approach, see Figure 3.45 on page 120.) The text box containing "Northern Gulf fisheries decline and collapse" is drawn with medium-wide margins around the text; see **help marginstyle** for other choices. **yscale(range( ))** options give both y axes a range wider than their data, with specific values chosen after experimenting to find the best vertical separation between the two series.

```
. graph twoway line cil winter,  yaxis(1)
     yscale(range(-1,3) axis(1))
     ytitle("Degrees C                                ", axis(1))
     yline(0)  ylabel(-1(.5)1.5, axis(1) angle(horizontal) nogrid)
     text(1 1992 "Northern Gulf" "fisheries decline" "and collapse"
        , box margin(medium))

     ||  line maxarea winter,
     yaxis(2) ylabel(50(50)200, axis(2) angle(horizontal))
     yscale(range(-100,221) axis(2))
     ytitle("                                1000s of km^2", axis(2))
     yline(173.6, axis(2) lpattern(dot))

     ||  if winter > 1949,
     xtitle("")  xlabel(1950(10)2000) xtick(1950(2)2002)
     legend(position(11) ring(0) rows(2) order(2 1)
        label(1 "Max ice area") label(2 "Min CIL temp"))
     note("Source:  Hamilton, Haedrich and Duncan (2004); data from
        DFO (2003)")
```

Figure 3.44

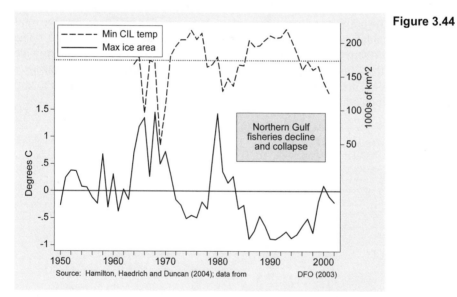

The text box on the right in Figure 3.44 marks the late-1980s and early-1990s period when key fisheries including the Northern Gulf cod declined or collapsed. As the graph shows, the fisheries declines coincided with the most sustained cold and ice conditions on record.

To place cod catches in the same graph with temperature and ice, we need three independent vertical scales. Figure 3.45 involves three overlaid plots:

connected *maxarea winter*
>   A connected-line plot of *maxarea* vs. *winter*, using *y* axis 3 (leftmost in our final graph). The *y* axis scale ranges from –300 to +220, with no grid of horizontal lines. Its title is "Ice area, 1000 km^2." This title is placed in the "northwest" position, **placement(nw)**.

line *cil winter*
>   A line plot of *cil* vs. *winter*, using *y* axis 2. *y* scale ranges from –4 to +3, with default labels.

connected *cod winter*
>   A connected-line plot of *cod* vs. *winter*, using *y* axis 1. The title placement is **sw**.

Bringing these three component plots together, the full command for Figure 3.45 appears on the next page. *y* ranges for each of the overlaid plots were chosen by experimenting to find the "right" amount of vertical separation among the three series. Options applied to the whole graph restrict the analysis to years since 1959, specify legend and *x* axis labeling, and request vertical grid lines.

```
. graph twoway connected maxarea winter, yaxis(3)
     yscale(range(-300,220) axis(3))
     ylabel(50(50)200, nogrid axis(3))
     ytitle("Ice area, 1000 km^2", axis(3) placement(nw))
     clpattern(dash)

     || line cil winter, yaxis(2) yscale(range(-4,3) axis(2))
     ylabel(, nogrid axis(2))
     ytitle("CIL temperature, degrees C", axis(2)) clpattern(solid)

     || connected cod winter, yaxis(1)
     yscale(range(0,200) axis(1))
     ylabel(, nogrid axis(1))
     ytitle("Cod catch, 1000 tons", axis(1) placement(sw))

     || if winter > 1959,
     legend(ring(0) position(7) label(1 "Max ice area")
        label(2 "Min CIL temp") label(3 "Cod catch") rows(3))
     xtitle("") xlabel(1960(5)2000, grid)
```

Figure 3.45

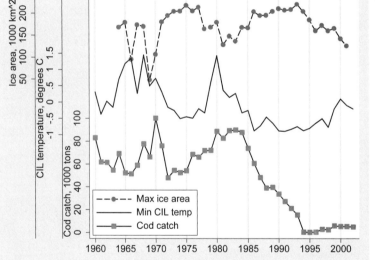

Graphing with Do-Files

Complicated graphics like Figure 3.45 require **graph** commands that are many physical lines long (although Stata views the whole command as one logical line). Do-files, introduced in Chapter 2, help in writing such multi-line commands. They also make it easy to save the command for future re-use, in case we later want to modify the graph or draw it again.

The following commands, typed into Stata's Do-file Editor and saved with the file name *fig03_45.do*, become a new do-file for drawing Figure 3.45.

```
#delimit ;
use gulf.dta, clear ;
graph twoway connected maxarea winter, yaxis(3)
    yscale(range(-300,220) axis(3)) ylabel(50(50)200, nogrid axis(3))
    ytitle("Ice area, 1000 km^2", axis(3) placement(nw))
    clpattern(dash)
    || line cil winter, yaxis(2) yscale(range(-4,3) axis(2))
    ylabel(, nogrid axis(2))
    ytitle("CIL temperature, degrees C", axis(2)) clpattern(solid)
    || connected cod winter, yaxis(1) yscale(range(0,200) axis(1))
    ylabel(, nogrid axis(1))
    ytitle("Cod catch, 1000 tons", axis(1) placement(sw))
    || if winter > 1959,
    legend(ring(0) position(7) label(1 "Max ice area")
        label(2 "Min CIL temp") label(3 "Cod catch") rows(3))
    xtitle("") xlabel(1960(5)2000, grid);
graph save Graph fig03_45.gph, replace;
graph export fig03_45.eps, as(eps) preview(on) replace;
#delimit cr
```

Once these commands are saved as *fig03_45.do*, simply typing the command

. do *fig03_45*

causes this do-file to execute, redrawing the graph and saving it in two formats.

The first line of this do-file, **#delimit ;** sets the semicolon (;) as end-of-line delimiter. Thereafter, Stata does not consider a line finished until it encounters a semicolon. The second line simply retrieves the dataset (*gulf.dta*) needed to draw Figure 3.45; note the semicolon that finishes this line. The long **graph twoway** command occupies the next 15 lines on this page, but Stata treats this all as one logical line that ends with the semicolon after the **xtitle( )** option.

The **graph save Graph** command saves the graph (by default, temporarily named "Graph" as seen in the Graph window) in Stata's .gph format. We can always specify our own temporary name for a graph, by adding to the **graph** command an option such as **name(*newgraph*)** or **name(*fig03_45*)**. Such temporary names for graphs become important when we have several graphs currently displayed, and want to save or print a particular one. Giving a temporary name to the graph as we create it does *not* save that graph to disk The temporary and saved-file names need not be the same. Type **help name option** for more discussion.

The do-file's **graph export** command creates a second version of the same graph in Encapsulated Postscript format, as indicated by the .eps suffix in the filename *fig04_45.eps*. Type **help graph export** to learn more about this command, which is particularly useful for writing programs or do-files that will create many graphs.

The do-file's final **#delimit cr** command re-sets a carriage return as the end-of-line delimiter, going back to Stata's usual mode. Although it is not visible on paper, the line **#delimit cr** must itself end with a carriage return (hit the Enter key), creating one last blank line at the end of the do-file.

Retrieving and Combining Graphs

Any graph saved in Stata's "live" .gph format can subsequently be retrieved into memory by the **graph use** command. For example, we could retrieve Figure 3.45 by typing

```
. graph use fig03_45
```

Once the graph is in memory, it is displayed onscreen and can be printed or saved again with a different name or format. From a graph saved earlier in .gph format, we could subsequently save versions in other formats such as Encapsulated Postscript (.eps), Portable Network Graphics (.png), or Enhanced Windows metafile (.emf). We also could change the color scheme, either through menus or directly in the **graph use** command. *fig03_45.gph* was saved in the s2 monochrome scheme, but we could see how it looks in the s1 color scheme by typing

```
. graph use fig03_45, scheme(s1color)
```

Graphs saved on disk can also be combined by the **graph combine** command. This provides a way to bring multiple plots into the same image. For illustration, we return to the Gulf of St. Lawrence data shown earlier in Figure 3.45. The following commands draw three simple time plots (not shown), saving them with the names *fig03_46a.gph*, *fig03_46b.gph*, and *fig03_46c.gph*. The **margin(medium)** suboptions specify the margin width for title boxes within each plot.

```
. graph twoway line maxarea winter if winter > 1964, xtitle("")
     xlabel(1965(5)2000, grid) ylabel(50(50)200, nogrid)
     title("Maximum winter ice area", position(4) ring(0) box
        margin(medium))
     ytitle("1000 km^2") saving(fig03_46a)

. graph twoway line cil winter if winter > 1964, xtitle("")
     xlabel(1965(5)2000, grid) ylabel(-1(.5)1.5, nogrid)
     title("Minimum CIL temperature", position(1) ring(0) box
        margin(medium))
     ytitle("Degrees C") saving(fig03_46b)

. graph twoway line cod winter if winter > 1964,  xtitle("")
     xlabel(1965(5)2000, grid) ylabel(0(20)100, nogrid)
     title("Northern Gulf cod catch", position(1) ring(0) box
        margin(medium))
     ytitle("1000 tons") saving(fig03_46c)
```

To combine these plots, we type the following command. Because the three plots have identical *x* scales, it makes sense to align the graphs vertically, in three rows. The **imargin** option specifies "very small" margins around the individual plots of Figure 3.46.

```
. graph combine fig03_46a.gph fig03_46b.gph fig03_46c.gph,
     imargin(vsmall) rows(3)
```

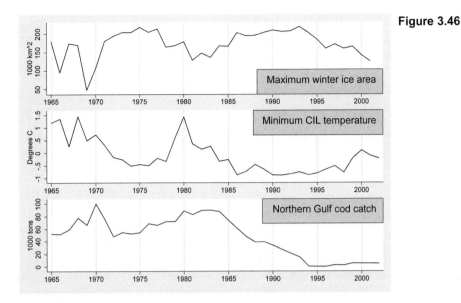

Figure 3.46

The story behind the cod and climate data in Figure 3.46, and the fisher folk who counted on the cod, is told in an article by Hamilton, Haedrich, and Duncan (2004). Figure 3.47 shows a more elaborate graphic that used **graph combine** to unify four plots with 13 time series including the human population of Newfoundland's Northern Peninsula, where life once centered around cod fishing.

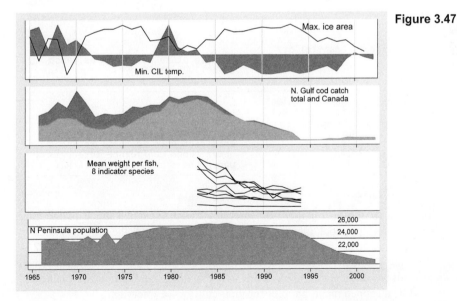

Figure 3.47

Type **help graph combine** for more information on this command. Options control details including the number of rows or columns, the size of text and markers (which otherwise become smaller as the number of plots increases), and the margins between individual plots. They can also specify whether *x* or *y* axes of twoway plots have common scales, or assign all components a common color scheme. Titles can be added to the combined graph, which can be printed, saved, retrieved, or for that matter combined again in the usual ways.

Graph Editor

With version 10, Stata added a Graph Editor to its toolkit. It is easier to learn about this useful feature by experimenting yourself, rather than reading from a book. A few examples showing the Graph Editor in practice should be enough to get you started.

Since 2001 the University of New Hampshire's Survey Research Center has been sampling public opinion about four times a year, with the Granite State Poll. One question repeated on polls from 2001 through 2007 asked whether respondents approved of the job George W. Bush was doing as president. Dataset *Granite1.dta* collects the approval ratings (percent saying they lean towards approval, approve somewhat, or strongly approve) from 23 of these polls. For each approval rating we have also calculated the lower and upper 95% confidence limits. Both approval ratings and confidence intervals reflect survey weighting, a statistical adjustment described in Chapter 14. Variable *date* is elapsed date (days since January 1, 1960), and formatted as %dM/Y so that its numerical values will have labels such as October/2001 (see **help dates and times**). This time series dataset has been **tsset date, daily** as described in Chapter 13.

```
. use "Granite1.dta", clear
(Granite State Poll -- Pres. Bush)

. describe

Contains data from C:\data\Granite1.dta
  obs:            23                          Granite State Poll -- Pres. Bush
  vars:            4                          3 Feb 2008 08:36
  size:          506 (99.9% of memory free)
```

variable name	storage type	display format	value label	variable label
date	int	%dM/Y		Month/year of poll
bushapp	float	%3.1f		President Bush approval rating
conlow	float	%3.1f		Lower bound confidence interval
conhigh	float	%3.1f		Upper bound confidence interval

```
Sorted by:  date
```

The 23 opinion polls (based on roughly 500 interviews each) spanned much of the Bush presidency. The first such poll took place in October 2001 shortly after the terrorist attacks of 9/11. The pollsters turned to questions about other politicians after September 2007, as campaigns for the next presidency began. From a high of 90% after the 9/11 attacks, approvals drifted down towards below 30% by 2007. Figure 3.48 visualizes these ratings as a connected-line plot, overlaid by a capped range-bar plot for their confidence intervals.

```
. graph twoway rcap conlow conhi date
      || connect bushapp date
      || , legend(off)
```

```
ytitle("President Bush approval rating")
xtitle("Month/year of poll")
```

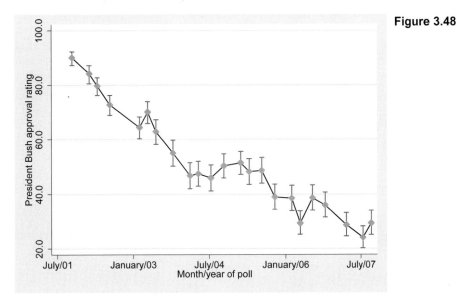

Figure 3.48

In the Graph window, select File > Start Graph Editor, or click the Graph Editor icon 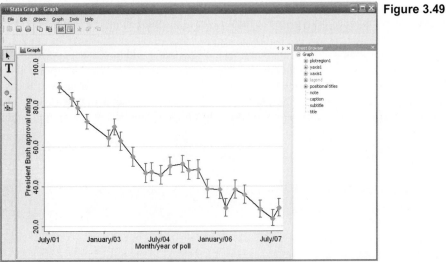. The view will change to include a Tools Toolbar at left and an Object Browser along the right. The Tools Toolbar contains a pointer tool for selecting parts of the graph, and other tools to add text, add lines, add markers, and edit the graph grid. The Object Browser presents a hierarchical list of graph contents. Some items are marked with a + sign, and clicking on this + will expand the list to show lower objects within it. We can select objects in the graph by clicking the pointer on their image, or by clicking names in the Object Browser (often easier in complex graphs).

Figure 3.49

In Figure 3.50, we have used the Object Browser to select plot2 within our plot region — the connected-line graph of *bushapp* vs. *year*. Plot2 is now highlighted both in the Object Browser, and in the graph itself. Selecting an object opens a Contextual Toolbar just above the graph, which gives information about the properties of that object. In this instance we see that plot2 is a connected plot with medium-size markers, the marker symbols are diamonds, and the line has medium width. If we click More... on the Contextual Toolbar we would see further options to control the properties of the line (such as solid, dashed, stairstep) or markers (such as symbol, color, size).

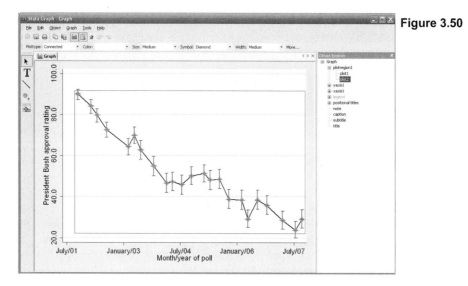

Figure 3.50

We might use the Graph Editor to add some historical notes to this graph, as seen in Figure 3.51. Doing so involved selecting the Add Line tool and choosing Arrowhead: Tail from the Contextual Toolbar above the graph. Then position the cursor where you want the arrowhead to be, click once, and drag the cursor to complete the arrow's tail. After drawing arrows, use the Add Text tool to position and type in labels, with the size and colors desired.

In general, the Graph Editor revises graph features in ways that could have been controlled by the original **graph** command. We cannot do things such as move an individual data point in the plot, although we can add or move new markers at any position. On the other hand, it is easy to change the properties of markers, lines, axis labels, or titles. We can also hide objects in a graph, making them invisible. Any changes made in the Graph Editor become permanent when we save the graph.

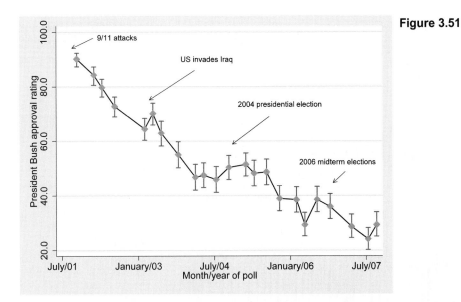

Figure 3.51

We can substantially alter a graph's appearance by selecting a plot and changing its plottype in the Contextual Toolbar. Figure 3.52 shows the result of changing plot1 from plottype rcap (capped range bars) to rarea (range area), colored a light gray (gray10). Plot2 has been changed to a medium-thick solid line.

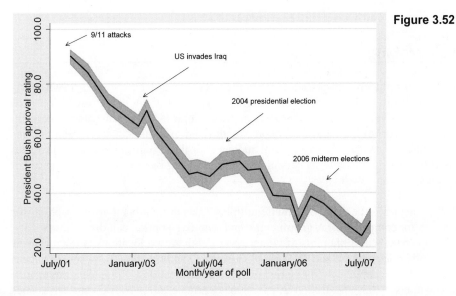

Figure 3.52

Figure 3.53 shows another variation, visually less attractive than Figure 3.52 but more clearly emphasizing the fact that our data come from 23 discrete surveys, not continuous measurements. For Figure 3.53, the confidence-limits (plot1) plottype has been changed to rbar (range bar), with black and medium-thick bars. The approval ratings (plot2) are now plottype bar, color "none" and width 50. Any of these alternative designs could have been specified in the original **graph** command, but the Graph Editor provides a fast and easy way to change graphs after they are saved, or to experiment with different looks.

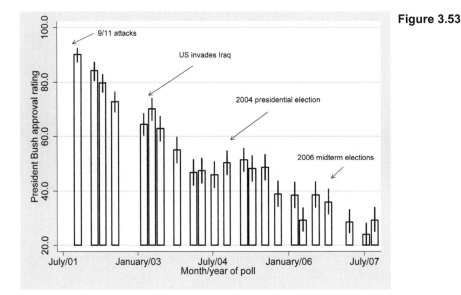

Figure 3.53

Creative Graphing

In his elegant and influential books about graphing data, Edward Tufte (1990, 1997, 2001) called for more effort at designing clear, information-packed graphics. Presenting a rich collection of impressively good or humorously awful examples, Tufte shows how successful graphics allow viewers to draw their own comparisons and examine details of relationships between variables. Stata users form a natural audience for these suggestions. Stata provides flexible tools for visualizing patterns in complex data, allowing basic plots to be enhanced or rearranged creatively in new images.

One of Tufte's themes is the value of "small multiples," sets of thumbnail-sized graphics that add dimensions for comparison. A **graph** command with **by( )** option can draw these nicely. Figure 3.54 illustrates with time plots of winter snow depth at two locations, a town in New Hampshire's White Mountains and the city of Boston, 225 kilometers to its south (dataset *whitemt1.dta*). Snow depth was measured daily at both locations; these data cover nine consecutive winters, 1997–98 through 2005–2006. Variable *dayseason* counts days since November 1 each winter season. *mtdepth* and *bosdepth* are snow depths in centimeters at the White Mountains location and in Boston, respectively. Variable *season* identifies the winter seasons, 1997–98 through 2005–06. The following command specifies a **twoway area** graph of *mtdepth* and *bosdepth* against *dayseason*, using lighter and darker gray (gs11 and gs5) for colors, in a 3×3 layout by *season*, and with a one-column legend positioned at 3 o'clock.

```
. graph twoway area mtdepth bosdepth dayseason
    if dayseason>29 & dayseason<160,
    bcolor(gs11 gs5) ytitle("Snow depth, cm")
    by(season, rows(3) note("") legend( position(3)))
    xlabel(30(30)150) ylabel(0(20)80)
    legend(cols(1) label(1 "White Mt") label(2 "Boston"))
```

Figure 3.54

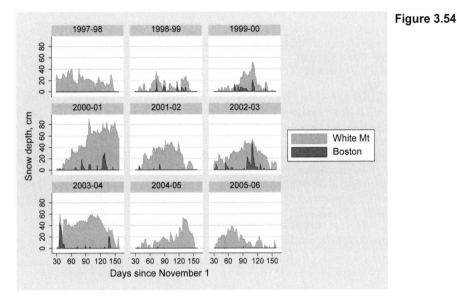

Figure 3.54 visualizes daily conditions through nine New England winters, showing how snow depth varies at two different places and on two different time scales. 2000–01 and 2003–04 stand out as heavy snow seasons in the mountains, with several significant storms in Boston. 1998–99 was much lighter in the mountains, with periods of no snow on the ground.

The data behind Figure 3.54 were assembled for research on how weather and climate affect attendance at ski areas (Hamilton, Brown, and Keim 2007). As New England's winter climate warmed in recent decades, low-snow winters became more common. That warming is troublesome from environmental and other perspectives, including that of winter recreation. Ski areas can feel not only the effects of local snow conditions, but also a "backyard effect" of snow conditions in distant cities such as Boston, where many skiers live. Figure 3.55 on the next page focuses on a single season, 1999–2000 (dataset *whitemt2.dta*), beginning with the same snow-depth shadow mountains as the top right plot in Figure 3.54.

Figure 3.55 overlays these shadow mountains (the **twoway area** plot) with a **line** plot showing the number of skier and snowboarder visits each day, at one ski area in the White Mountains close to where the snow-depth measurements were made. Both the observed visits (*visits*) and the number of visits predicted by a time series model (*model*) are graphed. The model, described in Chapter 13, predicts daily attendance as a function of weekly cyclical factors together with weather and snow conditions, both in the mountains and in Boston. The **graph** command creating Figure 3.55 assigns the left-hand *y* axis to snow depth in centimeters (*mtdepth* and *bosdepth*), and the right-hand *y* axis to observed and modeled number of visitors (*visits* and *model*).

Note that by carefully setting the **yscale(range( ))** and **ylabel( )** options for each of the two overlaid plots in Figure 3.55, we managed to align their scales so that the same horizontal grid lines work for both. This is not practical to do with all data, but can definitely improve the readability of graphs involving differently-scaled *y* variables.

```
. graph twoway area mtdepth bosdepth dayseason,  yaxis(1)
     ytitle("Snow depth, cm", axis(1)) bcolor(gs12 gs6)
     ylabel(0(10)60, axis(1))

     ||   line model visits dayseason, yaxis(2)
     lpattern(solid solid) lwidth(medthin medthick)
     ylabel(0(1)3, axis(2)) lcolor(gs1 gs0)

     ||   if dayseason>29 & dayseason<160,
     r2("Daily skier/snowboarder visits") xlabel(30(30)150)
     xtitle("Days since November 1")
     legend(rows(4) position(2) order(4 3 1 2) label(1 "White Mt")
     label(2 "Boston") label(3 "model") label(4 "attend"))
     yscale(range(0,51) axis(1)) ylabel(0(10)50, axis(1) grid)
     yscale(range(0,5100) axis(2)) ylabel(0(1000)5000, axis(2))
```

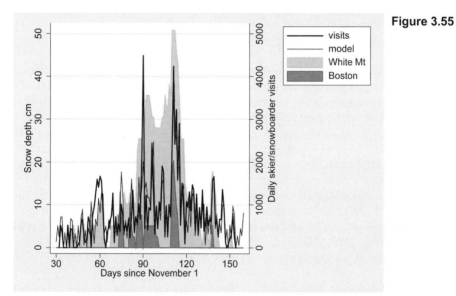

Figure 3.55

The two highest spikes in ski-area visits were school holiday periods that happened to coincide with snow in Boston. The significance of a "backyard effect" has been confirmed in more thorough analyses involving two ski areas and many seasons, as described by Hamilton, Brown, and Keim (2007). Graphically, it would be a simple step (not shown) from Figure 3.55 to a new set of small-multiples plots like Figure 3.54, visualizing the ski business together with the snow.

Population pyramids, widely used by demographers to represent the age-sex structure of populations, are not among Stata's plot types. They can, however, be constructed from horizontal bar charts, through slightly creative applications of **graph hbar**. There are several ways to do so. Figure 3.56 illustrates one approach, with a pyramid for the Greenland-born, predominantly Inuit, population of Greenland in 2006 (Hamilton and Rasmussen 2008). The number of females at each age is indicated by a bar to the right of center, and the number of males that same age by a bar to the left. The 90 one-year age groups seen here are too many to label individually, so they are marked off instead by gray bands every 20 years (0–19 years, 20–39 years, etc.). The graph indicates, for example, that in 2006 the Greenland-born population included almost 600 40-year-old males but fewer than 500 40-year-old females, reflecting sex differences in net outmigration. The central bulge in this pyramid marks a "baby boom" of adults now ages 35–49 (born in the 1950s and 60s), followed by much smaller cohorts of younger adults. We also see an "echo boom" of children, born in the 1980s and 90s to adults from the first baby boom. Ages 10–14 comprise the most numerous cohort among children.

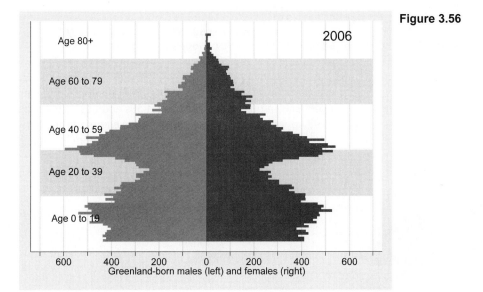

Figure 3.56

There are several tricks behind Figure 3.56. The raw data (*greenpop1.dta*) contain counts of the number of *males* and *females* at each *age*. In order to graph males on the left, we generate a new variable equal to the negative of the number of males,

```
. gen negmales = -males
```

A basic unlabeled pyramid could then be drawn by a command such as

```
. graph hbar (sum) negmales females if year==2006,
     over(age, descending gap(0) label(nolabel))
```

To place gray bands in the background, marking off 20-year groups, we define fake variables *maleGRAY* and *femGRAY* just to fill in the graph to plus or minus 700:

```
. gen maleGRAY = -(700-males) if (age>=20 & age<40)
      | (age>=60 & age<80)
. gen femGRAY = 700-females if (age>=20 & age<40)
      | (age>=60 & age<80)
```

Figure 3.56 now can be drawn by stacking *negmales*, *females*, *maleGRAY*, and *femGRAY* in a horizontal bar chart, with text to label the gray bands. We also apply labels such as "600" for –600 on the *y* axis so that the counts for males do not appear negative.

```
. graph hbar (sum) negmales females malGRAY femGRAY if year==2006,
      over(age, descending gap(0) label(nolabel))
      ylabel(-600 "600" -400 "400" -200 "200" 0 200 400 600)
      ytick(-700(100)700, grid) legend(off) stack
      ytitle("Greenland-born males (left) and females (right)")
      bar(1, color(gs8)) bar(2, color(gs3)) bar(3, color(gs14))
      bar(4, color(gs14)) text(550 97 "2006", size(large))
      text(-550 11 "Age 0 to 19")
      text(-550 33 "Age 20 to 39") text(-550 53 "Age 40 to 59")
      text(-550 76 "Age 60 to 79")  text(-550 95 "Age 80+")
```

Figure 3.57 takes this idea a step further by showing similar age pyramids for 1977, 1986, 1996, and 2006. In this sequence you can follow the rise of the large cohort born following improvements in Greenlanders' health and living standards in the 1950s and 60s. This baby boom shows up as teenagers in the 1977 pyramid. As the boom generation enters adulthood by the 1986 pyramid, we see the echo boom of their children. By 2006, this echo boom is waning.

Figure 3.57

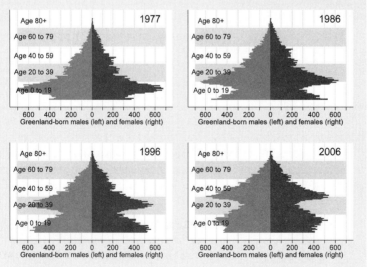

Although Figure 3.57 (constructed using separate images and **graph combine**) follows a small-multiples idea similar to Figure 3.54, these pyramids can be displayed in a more interesting way. For live presentations I drew a set of 30 annual pyramids, 1977 to 2006, using a do-file. These 30 Stata graphs (in .emf format) were then pasted onto one PowerPoint slide each, with automatic transitions at 1-second intervals, producing a 30-second animation of Greenland's demographic change. Another animation showed how the population of non-Greenlanders living in Greenland had changed over the same years, an interconnected but quite different demographic story (Hamilton and Rasmussen 2008).

Figure 3.58 is less dynamic, but combines five simple plots with text to form an image having some properties of both an illustration and a table. The resulting Stata graphic depicts population changes 1990–2000 among different ethnic groups living in rural counties of the U.S. South (based on U.S. Census data assembled by Voss et al. 2005; analysis from Hamilton 2006b). The left side of Figure 3.58 is a **twoway area** graph. To achieve the ramped effect showing population change, the variables graphed for each ethnic group (*popwbho*, *popwbh*, etc.) actually represent sums calculated as that group's population plus all the other populations that are graphically "below" it (dataset *southmig1.dta*). Important additional information, not evident from the area plot itself, is conveyed by two lines of labeling for each group in the legend. For example, readers can see from the legend that the Hispanic population of the rural South grew by 61% over this decade, from roughly 800,000 to 1.3 million people, and make their own visual or numerical comparisons with other populations.

```
. graph twoway area popwbho popwbh popwb popw year,
      legend(rows(4) position(3) symxsize(3)
      label(1 "Other, +50%" "0.3 to 0.5m")
      label(2 "Hispanic, +61%" "0.8 to 1.3m")
      label(3 "Black, +7%" "3.6 to 3.8m")
      label(4 "White, +6%" "14.9 to 15.7m"))
      xlabel(1990 2000) xtitle("")
      ylabel(0(5)20,angle(horizontal) grid)
      ytitle("Population in millions")
      title("Population growth 1990-2000")
```

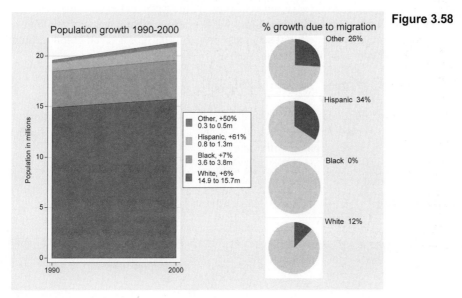

Figure 3.58

The right-hand part of Figure 3.58 consists of four pie charts showing the percentage of population growth due to net in-migration. Each pie chart was drawn separately using dataset *southmig2.dta*. For example, the bottom pie chart shows that 12% of the white population growth reflects net migration. Variables graphed are net migration (*netmig_w*, the total number of in-migrants minus out-migrants) and the remainder of population growth due to natural increase (*nonmig_w*, number of births minus deaths).

```
. graph pie nonmig_w netmig_w,
    legend(off)  pie(1, color(gs5)) pie(2, color(gs13))
    title("White  12%    ", position(2))
```

Each individual pie chart was saved with a file name such as *pie_white.gph*. After drawing and saving four such pie charts, they were brought together using **graph combine**.

```
. graph combine  pie_other.gph  pie_hisp.gph
    pie_black.gph  pie_white.gph,
    imargin(tiny) rows(4)
    title("% growth due to migration") fxsize(40)
```

An **fxsize(40)** option forces this four-pie-chart image to use only 40% of the width available. Consequently, when they are combined with the left-hand area plot to make Figure 3.58, the pie charts take up less than half of the total width.

Color can be a powerful element of graphical images, one not shown in this book. Graphs become more pleasant to look at, and often much easier to read, when information is color-coded. Although color printing remains too expensive for most books or research journals, even print journals increasingly also publish an online version in which color is free. Other online publications, printed reports, and PowerPoint-type presentations can make full use of color, so analysts may want to have both color and black-and-white versions of their best graphs. We can change from monochrome to color overall schemes (or vice versa) for a saved graph through a **graph use** command with **scheme( )** option, such as

```
. graph use fig03_47.gph, scheme(s2color)
```

Alternatively, the Graph Editor described in this chapter allows for more detailed alterations of color, shading, line properties, and so forth within a previously-saved image. Color further expands the scope for creative graphing with Stata.

4

Summary Statistics and Tables

The **summarize** command finds simple descriptive statistics such as medians, means, and standard deviations of measurement variables. More flexible arrangements of summary statistics are available through the command **tabstat**. For categorical or ordinal variables, **tabulate** obtains frequency distribution tables, cross-tabulations, assorted tests, and measures of association. **tabulate** can also construct one- or two-way tables of means and standard deviations across categories of other variables. A general table-making command, **table**, produces up to seven-way tables in which the cells contain statistics such as frequencies, sums, means, or medians. Finally, we review further one-variable procedures including normality tests, transformations, and displays for exploratory data analysis (EDA). Most of the analyses covered in this chapter can be accomplished either through the commands shown or through menu selections under Statistics > Summaries, tables & tests .

In addition to such general-purpose analyses, Stata provides many tables of particular interest to epidemiologists. These are not described in this chapter, but can be viewed by typing **help epitab**. Selvin (1996) introduces the topic.

Example Commands

. `summarize y1 y2 y3`

Calculates simple summary statistics (means, standard deviations, minimum and maximum values, and numbers of observations) for the variables listed.

. `summarize y1 y2 y3, detail`

Obtains detailed summary statistics including percentiles, median, mean, standard deviation, variance, skewness, and kurtosis.

. `summarize y1 if x1 > 3 & x2 < .`

Finds summary statistics for $y1$ using only those observations for which variable $x1$ is greater than 3, and $x2$ is not missing.

. `summarize y1 [fweight = w], detail`

Calculates detailed summary statistics for $y1$ using the frequency weights in variable w.

. `tabstat y1, stats(mean sd skewness kurtosis n)`

Calculates only the specified summary statistics for variable $y1$.

. **tabstat y1, stats(min p5 p25 p50 p75 p95 max) by(x1)**
Calculates the specified summary statistics (minimum, 5th percentile, 25th percentile, etc.) for measurement variable *y1*, within categories of *x1*.

. **tabulate x1**
Displays a frequency distribution table for all nonmissing values of variable *x1*.

. **tabulate x1, sort miss**
Displays a frequency distribution of *x1*, including the missing values. Rows (values) are sorted from most to least frequent.

. **tab1 x1 x2 x3 x4**
Displays a series of frequency distribution tables, one for each of the variables listed.

. **tabulate x1 x2**
Displays a two-variable cross-tabulation with *x1* as the row variable, and *x2* as the columns.

. **tabulate x1 x2, chi2 nof column**
Produces a cross-tabulation and Pearson χ^2 test of independence. Does not show cell frequencies, but instead gives the column percentages in each cell.

. **tabulate x1 x2, missing row all**
Produces a cross-tabulation that includes missing values in the table and in the calculation of percentages. Calculates "all" available statistics (Pearson and likelihood χ^2, Cramer's *V*, Goodman and Kruskal's gamma, and Kendall's τ_b).

. **tab2 x1 x2 x3 x4**
Performs all possible two-way cross-tabulations of the listed variables.

. **tabulate x1, summ(y)**
Produces a one-way table showing the mean, standard deviation, and frequency of *y* values within each category of *x1*.

. **tabulate x1 x2, summ(y) means**
Produces a two-way table showing the mean of *y* at each combination of *x1* and *x2* values.

. **by x3, sort: tabulate x1 x2, exact**
Creates a three-way cross-tabulation, with subtables for *x1* (row) by *x2* (column) at each value of *x3*. Calculates Fisher's exact test for each subtable. **by *varname*, sort:** works as a prefix for almost any Stata command where it makes sense. The **sort** option is unnecessary if the data already are sorted on *varname*.

. **table y x2 x3, by(x4 x5) contents(freq)**
Creates a five-way cross-tabulation, of *y* (row) by *x2* (column) by *x3* (supercolumn), by *x4* (superrow 1) by *x5* (superrow 2). Cells contain frequencies.

. **table x1 x2, contents(mean y1 median y2)**
Creates a two-way table of *x1* (row) by *x2* (column). Cells contain the mean of *y1* and the median of *y2*.

. **svy: tab** *y***, percent ci**
   Using survey-weighted data (as declared by **svyset**), obtains a one-way table of percentages for variable *y*, with 95% confidence intervals. Type **help svy tab** for more survey table options. Chapter 14 introduces survey data and analysis.

. **svy: tab** *y x***, column percent**
   Using survey-weighted data, obtains a two-way cross-tabulation of row variable *y* against column variable *x*, with adjusted χ^2 test of independence. Cells contain weighted column percentages.

Summary Statistics for Measurement Variables

Dataset *VTtown.dta* contains information from residents of a town in Vermont. A survey was conducted soon after routine state testing had detected trace amounts of toxic chemicals in the town's water supply. Higher concentrations were found in several private wells and near the public schools. Worried citizens held meetings to discuss possible solutions to this problem.

```
Contains data from c:\data\VTtown.dta
  obs:           153                          VT town survey (Hamilton 1985)
  vars:            7                          21 Jan 2008 09:32
  size:        2,295 (99.9% of memory free)

              storage  display    value
variable name   type   format     label      variable label

gender          byte   %8.0g      sexlbl     Respondent's gender
lived           byte   %8.0g                 Years lived in town
kids            byte   %8.0g      kidlbl     Have children <19 in town?
educ            byte   %8.0g                 Highest year school completed
meetings        byte   %8.0g      kidlbl     Attended meetings on pollution
contam          byte   %8.0g      contamlb   Believe own property/water
                                               contaminated
school          byte   %8.0g      close      School closing opinion

Sorted by:
```

To find the mean and standard deviation of the variable *lived* (years the respondent had lived in town), type

. **summarize** *lived*

Variable	Obs	Mean	Std. Dev.	Min	Max
lived	153	19.26797	16.95466	1	81

This table also gives the number of nonmissing observations and the variable's minimum and maximum values. If we had simply typed **summarize** with no variable list, we would obtain means and standard deviations for every numeric variable in the dataset.

To see more detailed summary statistics, type

```
. summarize lived, detail
```

```
                         Years lived in town

            Percentiles      Smallest
    1%           1               1
    5%           2               1
   10%           3               1          Obs               153
   25%           5               1          Sum of Wgt.       153

   50%          15                           Mean          19.26797
                              Largest        Std. Dev.     16.95466
   75%          29               65
   90%          42               65          Variance      287.4606
   95%          55               68          Skewness      1.208804
   99%          68               81          Kurtosis      4.025642
```

This **summarize, detail** output includes basic statistics plus the following:

Percentiles: Notably the first quartile (25th percentile), median (50th percentile), and third quartile (75th percentile). Because many samples do not divide evenly into quarters or other standard fractions, these percentiles are approximations.

Four smallest and four largest values, where outliers might show up.

Sum of weights: the **summarize** command permits frequency weights or **fweight**. For explanations see **help weight**.

Variance: Standard deviation squared (more properly, standard deviation equals the square root of variance).

Skewness: The direction and degree of asymmetry. A perfectly symmetrical distribution has skewness = 0. Positive skew (heavier right tail) results in skewness > 0; negative skew (heavier left tail) results in skewness < 0.

Kurtosis: Tail weight. A normal (Gaussian) distribution is symmetrical with kurtosis = 3. If a symmetrical distribution has heavier-than-normal tails (is sharply peaked), then kurtosis > 3. Kurtosis < 3 indicates lighter-than-normal tails.

The **tabstat** command provides a more flexible alternative to **summarize**. We can specify just which summary statistics we want to see. For example,

```
. tabstat lived, stats(mean range skewness)
```

variable	mean	range	skewness
lived	19.26797	80	1.208804

With a **by(*varname*)** option, **tabstat** constructs a table containing summary statistics for each value of *varname*. The following example gives means, standard deviations, medians, interquartile ranges, and number of nonmissing observations for *lived*, within each category of *gender*. The means and medians both indicate that, on average, the women in this sample had lived in town for fewer years than the men. Note that the median column is labeled "p50", meaning 50th percentile.

```
. tabstat lived, stats(mean sd median iqr n) by(gender)
Summary for variables: lived
      by categories of: gender (Respondent's gender)
```

gender	mean	sd	p50	iqr	N
male	23.48333	19.69125	19.5	28	60
female	16.54839	14.39468	13	19	93
Total	19.26797	16.95466	15	24	153

In addition to **mean**, **median**, or **iqr**, other statistics available for the **stats( )** option of **tabstat** include the same set listed earlier for **collapse** or **graph bar** (such as **count**, **sum**, **max**, **min**, **variance**, **sd**, and **p1** through **p99** for percentiles). Further **tabstat** options give control over the table layout and labeling. Type **help tabstat** to see a complete list.

The statistics produced by **summarize** or **tabstat** describe the sample at hand. We might also want to draw inferences about the population, for example, by constructing a 99% confidence interval for the mean of *lived*:

```
. ci lived, level(99)
```

Variable	Obs	Mean	Std. Err.	[99% Conf. Interval]	
lived	153	19.26797	1.370703	15.69241	22.84354

Based on this sample, we could be 99% confident that the population mean lies somewhere in the interval from 15.69 to 22.84 years. Here we used a **level( )** option to specify a 99% confidence interval. If we omit this option, **ci** defaults to a 95% confidence interval.

Other options allow **ci** to calculate exact confidence intervals for variables that follow binomial or Poisson distributions. A related command, **cii**, calculates normal, binomial, or Poisson confidence intervals directly from summary statistics, such as we might encounter in a published article. It does not require the raw data. Type **help ci** for details about both of these useful commands.

Exploratory Data Analysis

Statistician John Tukey invented a toolkit of methods for exploratory data analysis (EDA), which involves analyzing data in an exploratory and skeptical way without making unneeded assumptions (see Tukey 1977; also Hoaglin, Mosteller, and Tukey 1983, 1985). Box plots, introduced in Chapter 3, are one of Tukey's best-known innovations. Another is the stem-and-leaf display, a graphical arrangement of ordered data values in which initial digits form the "stems" and following digits for each observation make up the "leaves."

```
. stem lived
```

Stem-and-leaf plot for lived (Years lived in town)

```
   0*  111111122222333333344444444
   0.  5555555555556666666777889999
   1*  00000011222233333334
   1.  55555567788899
   2*  0000000111112224444
   2.  56778899
   3*  00000124
   3.  5555666789
   4*  0012
   4.  59
   5*  00134
   5.  556
   6*
   6.  5558
   7*
   7.
   8*  1
```

stem automatically chose a double-stem version here, in which 1* denotes first digits of 1 and second digits of 0–4 (that is, respondents who had lived in town 10–14 years). 1. denotes first digits of 1 and second digits of 5 to 9 (15–19 years). We can control the number of lines per initial digit with the **lines( )** option. For example, a five-stem version in which the 1* stem holds leaves of 0–1, 1t leaves of 2–3, 1f leaves of 4–5, 1s leaves of 6–7, and 1. leaves of 8–9 could be obtained by typing

```
. stem lived, lines(5)
```

Type **help stem** for information about other options.

Letter-value displays (**lv**) use order statistics to dissect a distribution.

```
. lv lived
```

#	153	Years lived in town			spread	pseudosigma
M	77		15			
F	39	5	17	29	24	17.9731
E	20	3	21	39	36	15.86391
D	10.5	2	27	52	50	16.62351
C	5.5	1	30.75	60.5	59.5	16.26523
B	3	1	33	65	64	15.15955
A	2	1	34.5	68	67	14.59762
Z	1.5	1	37.75	74.5	73.5	15.14113
	1	1	41	81	80	15.32737

				# below	# above
inner fence	-31		65	0	5
outer fence	-67		101	0	0

M denotes the median, and F the "fourths" (quartiles, using a different approximation than the quartile approximation used by **summarize, detail** and **tabsum**). E , D , C , . . . denote cutoff points such that roughly 1/8, 1/16, 1/32, . . . of the distribution remains outside in the tails. The second column of numbers gives the "depth" or distance from the nearest extreme, for each letter value. Within the center box, the middle column gives "midsummaries," which are averages of the two letter values. If midsummaries drift away from the median, as they do for *lived*, this tells us that the distribution becomes progressively more skewed as we move farther out into the tails. The "spreads" are differences between pairs of letter values. For instance, the spread between Fs equals the approximate interquartile range. Finally, "pseudosigmas" in the right-hand column estimate what the standard deviation should be if these letter values described a Gaussian population. The F pseudosigma, sometimes called a "pseudo standard deviation" (*PSD*), provides a simple and outlier-resistant check for approximate normality in symmetrical distributions:

1. Comparing mean with median diagnoses overall skew:
    mean > median   positive skew
    mean = median   symmetry
    mean < median   negative skew

2. *If* the mean and median are similar, indicating symmetry, then a comparison between standard deviation and *PSD* helps to evaluate tail normality:
    standard deviation > *PSD*   heavier-than-normal tails
    standard deviation = *PSD*   normal tails
    standard deviation < *PSD*   lighter-than-normal tails
    Let F_1 and F_3 denote 1st and 3rd fourths (approximate 25th and 75th percentiles). Then the interquartile range, *IQR*, equals $F_3 - F_1$, and $PSD = IQR / 1.349$.

lv also identifies mild and severe outliers.   We call an *x* value a "mild outlier" when it lies outside the inner fence, but not outside the outer fence:

$$F_1 - 3IQR \le x < F_1 - 1.5IQR \quad \text{or} \quad F_3 + 1.5IQR < x \le F_3 + 3IQR$$

The value of *x* is a "severe outlier" if it lies outside the outer fence:

$$x < F_1 - 3IQR \quad \text{or} \quad x > F_3 + 3IQR$$

lv gives these cutoffs and the number of outliers of each type. Severe outliers, values beyond the outer fences, occur sparsely (about two per million) in normal populations. Monte Carlo simulations suggest that the presence of any severe outliers in samples of $n = 15$ to about 20,000 should be sufficient evidence to reject a normality hypothesis at $\alpha = .05$ (Hamilton 1992b). Severe outliers create problems for many statistical techniques.

summarize, **stem** , and **lv** all confirm that *lived* has a positively skewed sample distribution, not at all resembling a theoretical normal curve. The next section introduces more formal normality tests, and transformations that can reduce a variable's skew.

Normality Tests and Transformations

Many statistical procedures work best when applied to variables that follow normal distributions. The preceding section described exploratory methods to check for approximate normality, extending the graphical tools (histograms, box plots, symmetry plots, and quantile–normal plots) presented in Chapter 3. A skewness–kurtosis test, making use of the skewness and kurtosis statistics shown by **summarize, detail**, can more formally evaluate the null hypothesis that the sample at hand came from a normally-distributed population.

```
. sktest lived
```

```
                     Skewness/Kurtosis tests for Normality
                                                      ──── joint ────
      Variable │  Pr(Skewness)   Pr(Kurtosis)   adj chi2(2)    Prob>chi2
    ──────────────────────────────────────────────────────────────────
         lived │     0.000          0.028          24.79        0.0000
```

sktest here rejects normality: *lived* appears significantly nonnormal in skewness ($P \approx .000$), kurtosis ($P = .028$), and in both statistics considered jointly ($P \approx .0000$). Stata rounds off displayed probabilities to three or four decimals; "0.0000" really means $P < .00005$.

Other normality tests include Shapiro–Wilk W (**swilk**) and Shapiro–Francia W' (**sfrancia**) methods. Type **help sktest** to see the options.

Nonlinear transformations such as square roots and logarithms are often employed to change distributions' shapes, with the aim of making skewed distributions more symmetrical and perhaps more nearly normal. Transformations might also help linearize relationships between variables (Chapter 8). Table 4.1 shows a progression called the "ladder of powers" (Tukey 1977) that provides guidance for choosing transformations to change distributional shape. The variable *lived* exhibits mild positive skew, so its square root might be more symmetrical. We could create a new variable equal to the square root of *lived* by typing

```
. generate srlived = lived ^.5
```

Instead of *lived* **^.5**, we could equally well have written **sqrt(*lived*)**.

Logarithms are another transformation that can reduce positive skew. To generate a new variable equal to the natural (base *e*) logarithm of *lived*, type

```
. generate loglived = ln(lived)
```

In the ladder of powers and related transformation schemes such as Box–Cox, logarithms take the place of a "0" power. Their effect on distribution shape is intermediate between .5 (square root) and –.5 (reciprocal root) transformations.

Table 4.1: Ladder of Powers

Transformation	Formula	Effect
cube	`new = old ^3`	reduce severe negative skew
square	`new = old ^2`	reduce mild negative skew
raw	`old`	no change (raw data)
square root	`new = old ^.5`	reduce mild positive skew
$\log_e$	`new = ln(old)`	reduce positive skew
(or $\log_{10}$)	`new = log10(old)`	
negative reciprocal root	`new = -(old ^-.5)`	reduce severe positive skew
negative reciprocal	`new = -(old ^-1)`	reduce very severe positive skew
negative reciprocal square	`new = -(old ^-2)`	"
negative reciprocal cube	`new = -(old ^-3)`	"

We take negatives of the result after raising to a power less than zero, to preserve the original order — the highest value of *old* becomes transformed into the highest value of *new*, and so forth. When *old* itself contains negative or zero values, it is necessary to add a constant before transformation. For example, if *arrests* measures the number of times a person has been arrested (0 for many people), then a suitable log transformation could be

. **generate** *logarrest* = ln(*arrests* + 1)

The **ladder** command combines the ladder of powers with **sktest** for normality. It tries each power on the ladder, and reports whether the result is significantly nonnormal. This can be illustrated using the skewed variable *energy*, per capita energy consumption, from *states.dta*.

. **ladder** *energy*

Transformation	formula	chi2(2)	P(chi2)
cubic	energy^3	53.74	0.000
square	energy^2	45.53	0.000
identity	energy	33.25	0.000
square root	sqrt(energy)	25.03	0.000
log	log(energy)	15.88	0.000
1/(square root)	1/sqrt(energy)	7.36	0.025
inverse	1/energy	1.32	0.517
1/square	1/(energy^2)	4.13	0.127
1/cubic	1/(energy^3)	11.56	0.003

It appears that the reciprocal (inverse) transformation, $1/energy$ (or $energy^{-1}$), best resembles a normal distribution. Most the other transformations (including the raw data) are significantly nonnormal. Figure 4.1, produced by the **gladder** command, visually supports this conclusion by comparing histograms of each transformation to normal curves.

. **gladder** *energy*

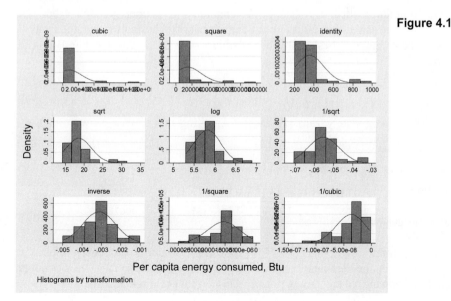

Figure 4.1

Histograms by transformation

Figure 4.2 shows a corresponding set of quantile–normal plots for these ladder of powers transformations, obtained by the "quantile ladder" command **qladder**. (Type **help ladder** for information about **ladder**, **gladder**, and **qladder**.) To make the tiny plots more readable in the example below, we scale the labels and marker symbols up by 25% with the **scale(1.25)** option. The axis labels (which would be unreadable and crowded) are suppressed by the options **ylabel(none) xlabel(none)**.

```
. qladder energy, scale(1.25) ylabel(none) xlabel(none)
```

Figure 4.2

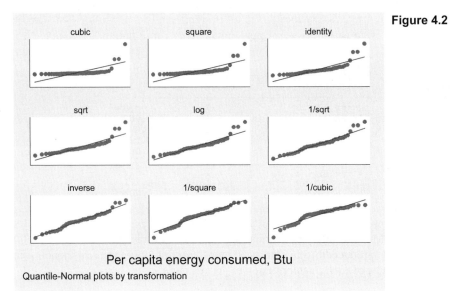

Per capita energy consumed, Btu

Quantile-Normal plots by transformation

An alternative technique called Box–Cox transformation offers finer gradations between transformations and automates the choice among them (easier for the analyst, but not always a good thing). The command **bcskew0** finds a value of λ (lambda) for the Box–Cox transformations

$$y^{(\lambda)} = \{y^{\lambda} - 1\} / \lambda \qquad\qquad \lambda > 0 \text{ or } \lambda < 0$$

or

$$y^{(\lambda)} = \ln(y) \qquad\qquad \lambda = 0$$

such that $y^{(\lambda)}$ has approximately zero skewness. Applying this to *energy*, we obtain the transformed variable *benergy*:

```
. bcskew0 benergy = energy, level(95)
```

Transform	L	[95% Conf. Interval]		Skewness
(energy^L-1)/L	-1.246052	-2.052503	-.6163383	.000281

(1 missing value generated)

That is, *benergy* = (*energy*$^{-1.246}$ − 1)/(−1.246) is the transformation that comes closest to symmetry (as defined by the skewness statistic). The Box–Cox parameter $\lambda = -1.246$ is not far from our ladder-of-powers choice, the −1 power. The confidence interval for λ,

$$-2.0525 < \lambda < -.6163$$

allows us to reject some other possibilities including logarithms ($\lambda = 0$) or square roots ($\lambda = .5$). Chapter 8 describes a Box–Cox approach to regression modeling.

Frequency Tables and Two-Way Cross-Tabulations

The methods described above apply to measurement variables. Categorical variables require other approaches, such as tabulation. Returning to the survey data in *VTtown.dta*, we could find the percentage of respondents who attended meetings concerning the pollution problem by tabulating the categorical variable *meetings*:

```
. tabulate meetings
```

Attended meetings on pollution	Freq.	Percent	Cum.
no	106	69.28	69.28
yes	47	30.72	100.00
Total	153	100.00	

tabulate can produce frequency distributions for variables that have thousands of values. To construct a manageable frequency distribution table for a variable with many values, however, you might first want to group those values by applying **generate** with its **recode** or **autocode** options (see Chapter 2 or **help generate**).

tabulate followed by two variable names creates a two-way cross-tabulation. For example, here is a cross-tabulation of *meetings* by *kids* (whether respondent has children under 19 living in town):

```
. tabulate meetings kids
```

Attended meetings on pollution	Have children <19 in town? no	yes	Total
no	52	54	106
yes	11	36	47
Total	63	90	153

The first-named variable forms the rows, and the second forms columns in the resulting table. We see that only 11 of these 153 people were non-parents who attended the meetings.

tabulate has a number of options that are useful with two-way tables, including:

cell   Shows total percentages for each cell.

chi2   Pearson χ^2 test of hypothesis that row and column variables are independent.

column   Shows column percentages for each cell.

exact Fisher's exact test of the independence hypothesis. Superior to **chi2** if the table contains thin cells with low expected frequencies. Often too slow to be practical in large tables, however.

expected Displays the expected frequency under the assumption of independence in each cell of a two-way table.

gamma Goodman and Kruskal's γ (gamma), with its asymptotic standard error (ASE). Measures association between ordinal variables, based on the number of concordant and discordant pairs (ignoring ties). $-1 \leq \gamma \leq 1$.

generate(*new*) Creates a set of dummy variables named *new1*, *new2*, and so on to represent the values of the tabulated variable.

lrchi2 Likelihood-ratio χ^2 test of independence hypothesis. Not obtainable if the table contains any empty cells.

missing Includes "missing" as one row and/or column of the table.

nofreq Does not show cell frequencies.

nolabel Shows numeric values rather than value labels of labeled numeric variables.

row Shows row percentages for each cell.

taub Kendall's τ_b (tau-b), with its asymptotic standard error (ASE). Measures association between ordinal variables. **taub** is similar to **gamma**, but uses a correction for ties. $-1 \leq \tau_b \leq 1$.

V Cramer's V (note capitalization), a measure of association for nominal variables. In 2×2 tables, $-1 \leq V \leq 1$. In larger tables, $0 \leq V \leq 1$.

tabulate can also save frequencies and variables names as matrices. Consult **help tabulate twoway** for a full listing of options and their syntax.

To obtain a table containing only column percentages (because the column variable, *kids*, is our independent variable here) and a χ^2 test for the cross-tabulation of *meetings* by *kids*, type

```
. tabulate meetings kids, column chi2
```

```
┌─────────────────────┐
│ Key                 │
├─────────────────────┤
│      frequency      │
│  column percentage  │
└─────────────────────┘
```

Attended meetings on pollution	Have children <19 in town? no	yes	Total
no	52 82.54	54 60.00	106 69.28
yes	11 17.46	36 40.00	47 30.72
Total	63 100.00	90 100.00	153 100.00

$$\text{Pearson chi2(1)} = 8.8464 \quad \text{Pr} = 0.003$$

Forty percent of the respondents with children attended meetings, compared with about 17% of the respondents without children. This association is statistically significant ($P = .003$).

Occasionally we might need to re-analyze a published table, without access to the original raw data. A special command, **tabi** ("immediate" tabulation), accomplishes this. Type the cell frequencies on the command line, with table rows separated by "\". For illustration, here is how **tabi** could reproduce the previous χ^2 analysis, directly from the four cell frequencies without need for any dataset:

```
. tabi 52 54 \ 11 36, column chi2
```

```
┌─────────────────────┐
│ Key                 │
├─────────────────────┤
│      frequency      │
│  column percentage  │
└─────────────────────┘
```

row	col 1	2	Total
1	52 82.54	54 60.00	106 69.28
2	11 17.46	36 40.00	47 30.72
Total	63 100.00	90 100.00	153 100.00

$$\text{Pearson chi2(1)} = 8.8464 \quad \text{Pr} = 0.003$$

Unlike **tabulate**, **tabi** does not require or refer to any data in memory. By adding the **replace** option, however, we can ask **tabi** to replace whatever data are in memory with the new cross-tabulation. Statistical options (**chi2**, **exact**, **nofreq**, and so forth) work the same for **tabi** as they do with **tabulate**; see **help tabulate twoway**.

Multiple Tables and Multi-Way Cross-Tabulations

With surveys and other large datasets, we sometimes need frequency distributions of many different variables. Instead of asking for each table separately, for example by typing **tabulate** *meetings*, then **tabulate** *gender*, and finally **tabulate** *kids*, we could simply use another specialized command, **tab1**:

. **tab1** *meetings gender kids*

Or, to produce one-way frequency tables for each variable from *gender* through *school* in this dataset (the maximum is 30 variables at one time), type

. **tab1** *gender-school*

Similarly, **tab2** creates multiple two-way tables. For example, the following command cross-tabulates every two-way combination of the listed variables:

. **tab2** *meetings gender kids*

tab1 and **tab2** offer the same options as **tabulate**.

To form multi-way contingency tables, it is possible to use **tabulate** with a **by** prefix. For example, below we have a three-way cross-tabulation of *meetings* by *kids* by *contam* (respondent believes his or her own property or water contaminated), with χ^2 tests for the independence of *meetings* and *kids* within each level of *contam*:

. **by** *contam*, **sort**: **tabulate** *meetings kids*, **nofreq col chi2**

```
-> contam = no

Attended |
meetings | Have children <19 in
     on  |        town?
pollution|    no         yes   |   Total
---------+---------------------+---------
      no |   91.30      68.75  |   78.18
     yes |    8.70      31.25  |   21.82
---------+---------------------+---------
   Total |  100.00     100.00  |  100.00

        Pearson chi2(1) =    7.9814   Pr = 0.005
```

```
-> contam = yes

Attended |
meetings | Have children <19 in
     on  |        town?
pollution|    no         yes   |   Total
---------+---------------------+---------
      no |   58.82      38.46  |   46.51
     yes |   41.18      61.54  |   53.49
---------+---------------------+---------
   Total |  100.00     100.00  |  100.00

        Pearson chi2(1) =    1.7131   Pr = 0.191
```

Parents were more likely to attend meetings, among both the contaminated and uncontaminated groups. Only among the larger uncontaminated group is this "parenthood effect" statistically significant, however. As multi-way tables separate the data into smaller subsamples, the size of these subsamples has noticeable effects on significance-test outcomes.

This approach can be extended to tabulations of greater complexity. To get a four-way cross-tabulation of *gender* by *contam* by *meetings* by *kids,* with χ^2 tests for each *meetings* by *kids* subtable (results not shown), type the command

. **by *gender contam*, sort: tabulate *meetings kids*, column chi2**

A better way to produce multi-way tables, if we do not need percentages or statistical tests, is through Stata's general table-making command, **table**. This versatile command has many options, only a few of which are illustrated here. To construct a simple frequency table of *meetings*, type

. **table *meetings*, contents(freq)**

Attended meetings on pollution	Freq.
no	106
yes	47

For a two-way frequency table or cross-tabulation, type

. **table *meetings kids*, contents(freq)**

Attended meetings on pollution	Have children <19 in town?	
	no	yes
no	52	54
yes	11	36

If we specify a third categorical variable, it forms the "supercolumns" of a three-way table:

. **table *meetings kids contam*, contents(freq)**

Attended meetings on pollution	Believe own property/water contaminated and Have children <19 in town?			
	— no —		— yes —	
	no	yes	no	yes
no	42	44	10	10
yes	4	20	7	16

More complicated tables require the **by( )** option, which allows up to four "supperrow" variables. **table** thus can produce up to seven-way tables: one row, one column, one supercolumn, and up to four superrows. Here is a four-way example:

```
. table meetings kids contam, contents(freq) by(gender)
```

Responden t's gender and Attended meetings on pollution	Believe own property/water contaminated and Have children <19 in town?			
	— no —		— yes —	
	no	yes	no	yes
male				
no	18	18	3	3
yes	2	7	3	6
female				
no	24	26	7	7
yes	2	13	4	10

The **contents( )** option of **table** specifies what statistics the table's cells contain:

contents(freq)	Frequency
contents(mean *varname*)	Mean of *varname*
contents(sd *varname*)	Standard deviation of *varname*
contents(sum *varname*)	Sum of *varname*
contents(rawsum *varname*)	Sums ignoring optionally specified weight
contents(count *varname*)	Count of nonmissing observations of *varname*
contents(n *varname*)	Same as **count**
contents(max *varname*)	Maximum of *varname*
contents(min *varname*)	Minimum of *varname*
contents(median *varname*)	Median of *varname*
contents(iqr *varname*)	Interquartile range (IQR) of *varname*
contents(p1 *varname*)	1st percentile of *varname* (so forth to **p99**)

The next section illustrates several more of these options.

Tables of Means, Medians, and Other Summary Statistics

tabulate produces tables of means and standard deviations within categories of the tabulated variable. For example, to tabulate means of *lived* within categories of *meetings*, type

```
. tabulate meetings, summ(lived)
```

Attended meetings on pollution	Summary of Years lived in town		
	Mean	Std. Dev.	Freq.
no	21.509434	17.743809	106
yes	14.212766	13.911109	47
Total	19.267974	16.954663	153

Meetings attenders appear to be relative newcomers, averaging 14.2 years in town, compared with 21.5 years for those who did not attend. We can also use **tabulate** to form a two-way table of means by typing

. **tabulate** *meetings kids,* **sum(***lived***) means**

```
                    Means of Years lived in town

 Attended │
 meetings │  Have children <19
       on │     in town?
pollution │      no        yes  │     Total
──────────┼──────────────────────┼────────────
       no │ 28.307692  14.962963 │ 21.509434
      yes │ 23.363636  11.416667 │ 14.212766
──────────┼──────────────────────┼────────────
    Total │ 27.444444  13.544444 │ 19.267974
```

Both parents and nonparents among the meeting attenders tend to have lived fewer years in town, so the newcomer/oldtimer division noticed in the previous table is not a spurious reflection of the fact that parents with young children were more likely to attend.

The **means** option used above called for a table containing only means. Otherwise we get a bulkier table with means, standard deviations, and frequencies in each cell. Chapter 5 describes statistical tests for hypotheses about subgroup means.

Although it performs no tests, **table** nicely builds up to seven-way tables containing means, standard deviations, sums, medians, or other statistics (see the option list in previous section). Here is a one-way table showing means of *lived* within categories of *meetings*:

. **table** *meetings,* **contents(mean** *lived***)**

```
Attended  │
meetings  │
on        │
pollution │  mean(lived)
──────────┼─────────────
       no │    21.5094
      yes │    14.2128
```

A two-way table of means is a straightforward extension:

. **table** *meetings kids,* **contents(mean** *lived***)**

```
Attended  │
meetings  │ Have children <19
on        │    in town?
pollution │    no       yes
──────────┼──────────────────
       no │ 28.3077   14.963
      yes │ 23.3636   11.4167
```

Table cells can contain more than one statistic. Suppose we want a two-way table with both means and medians of the variable *lived*:

```
. table meetings kids, contents(mean lived median lived)
```

Attended meetings on pollution	Have children <19 in town? no	yes
no	28.3077 27.5	14.963 12.5
yes	23.3636 21	11.4167 6

The medians in the table above confirm our earlier conclusion based on means: the meeting attenders, both parents and nonparents, tended to have lived fewer years in town than their non-attending counterparts. Medians within each cell are less than the means, reflecting the positive skew (means pulled up by a few long-time residents) of the variable *lived*. **table** could show means, medians, sums, or other summary statistics for two or more different variables.

Using Frequency Weights

summarize, **tabulate**, **table**, and related commands can be used with frequency weights that indicate the number of replicated observations. For example, file *sextab2.dta* contains results from a British survey of sexual behavior (Johnson et al. 1992). It apparently has 48 observations:

```
Contains data from c:\data\sextab2.dta
  obs:            48                        British sex survey (Johnson 92)
  vars:            4                        21 Jan 2008 12:44
  size:          624 (99.9% of memory free)
```

variable name	storage type	display format	value label	variable label
age	byte	%8.0g	age	Age
gender	byte	%8.0g	gender	Gender
lifepart	byte	%8.0g	partners	# heterosex partners lifetime
count	int	%8.0g		Number of individuals

```
Sorted by:  age  lifepart  gender
```

One variable, *count*, indicates the number of individuals with each combination of characteristics, so this small dataset actually contains information from over 18,000 respondents. For example, 405 respondents were male, ages 16 to 24, and reported having no heterosexual partners so far in their lives.

```
. list in 1/5
```

	age	gender	lifepart	count
1.	16-24	male	none	405
2.	16-24	female	none	465
3.	16-24	male	one	323
4.	16-24	female	one	606
5.	16-24	male	two	194

We use *count* as a frequency weight to create a cross-tabulation of *lifepart* by *gender*:

. tabulate *lifepart gender* [fweight = *count*]

# heterosex partners lifetime	Gender male	female	Total
none	544	586	1,130
one	1,734	4,146	5,880
two	887	1,777	2,664
3-4	1,542	1,908	3,450
5-9	1,630	1,364	2,994
10+	2,048	708	2,756
Total	8,385	10,489	18,874

The usual **tabulate** options work as expected with frequency weights. Below is the same table showing column percentages instead of frequencies:

. tabulate *lifepart gender* [fweight = *count*], column nof

# heterosex partners lifetime	Gender male	female	Total
none	6.49	5.59	5.99
one	20.68	39.53	31.15
two	10.58	16.94	14.11
3-4	18.39	18.19	18.28
5-9	19.44	13.00	15.86
10+	24.42	6.75	14.60
Total	100.00	100.00	100.00

Sampling or probability weights do not work with **tabulate**. Chapter 14 will show examples using the **svy: tabulate** command, which is designed for probability-weighted survey data.

A different application of frequency weights can be demonstrated with **summarize**. File *college1.dta* contains information on a random sample consisting of 11 U.S. colleges, drawn from *Barron's Compact Guide to Colleges* (1992).

```
Contains data from c:\data\college1.dta
  obs:            11                           Colleges sample 1 (Barron's 92)
  vars:            5                           11 Jul 2003 18:05
  size:          473  (99.9% of memory free)
```

variable name	storage type	display format	value label	variable label
school	str28	%28s		College or university
enroll	int	%8.0g		Full-time students 1991
pctmale	byte	%8.0g		Percent male 1991
msat	int	%8.0g		Average math SAT
vsat	int	%8.0g		Average verbal SAT

```
Sorted by:
```

The variables include *msat*, the mean math Scholastic Aptitude Test score at each of the 11 schools.

```
. list school enroll msat
```

		school	enroll	msat
1.	Brown University		5550	680
2.	U. Scranton		3821	554
3.	U. North Carolina/Asheville		2035	540
4.	Claremont College		849	660
5.	DePaul University		6197	547
6.	Thomas Aquinas College		201	570
7.	Davidson College		1543	640
8.	U. Michigan/Dearborn		3541	485
9.	Mass. College of Art		961	482
10.	Oberlin College		2765	640
11.	American University		5228	587

We can easily find the mean *msat* value among these 11 schools by typing

```
. summarize msat
```

Variable	Obs	Mean	Std. Dev.	Min	Max
msat	11	580.4545	67.63189	482	680

This summary table gives each school's mean math SAT score the same weight. DePaul University, however, has 30 times as many students as Thomas Aquinas College. To take the different enrollments into account we could weight by *enroll*,

```
. summarize msat [fweight = enroll]
```

Variable	Obs	Mean	Std. Dev.	Min	Max
msat	32691	583.064	63.10665	482	680

The enrollment-weighted mean, unlike the unweighted mean, is equivalent to the mean for the 32,691 students at these colleges (assuming they all took the SAT). Note, however, that we could not say the same thing about the standard deviation, minimum, or maximum. Apart from the mean, most individual-level statistics cannot be calculated simply by weighting data that already are aggregated. Thus, we need to use weights with caution. They might make sense in the context of one particular analysis, but seldom do for the dataset as a whole, when many different kinds of analyses are needed.

5

ANOVA and Other Comparison Methods

Analysis of variance (ANOVA) encompasses a set of methods for testing hypotheses about differences between means. Its applications range from simple analyses where we compare the means of y across categories of x, to more complicated situations with multiple categorical and measurement x variables. t tests for hypotheses regarding a single mean (one-sample) or a pair of means (two-sample) correspond to elementary forms of ANOVA.

Rank-based "nonparametric" tests, including sign, Mann–Whitney, and Kruskal–Wallis, take a different approach to comparing distributions. These tests make weaker assumptions about measurement, distribution shape, and spread. Consequently, they remain valid under a wider range of conditions than ANOVA and its "parametric" relatives. Careful analysts sometimes use parametric and nonparametric tests together, checking to see whether both point toward similar conclusions. Further troubleshooting is called for when parametric and nonparametric results disagree.

anova is the first of Stata's model-fitting commands to be introduced in this book. Like the others, it has considerable flexibility encompassing a wide variety of models. **anova** can fit one-way and N-way ANOVA or analysis of covariance (ANCOVA) for balanced and unbalanced designs, including designs with missing cells. It can also fit factorial, nested, mixed, or repeated-measures designs. One follow-up command, **predict**, calculates predicted values, several types of residuals, and assorted standard errors and diagnostic statistics after **anova**. Another followup command, **test**, obtains tests of user-specified null hypotheses. Both **test** and **predict** work in similar fashion with other Stata model-fitting commands, such as **regress** (Chapter 6).

The following menu choices give access to most operations described in this chapter:

Statistics > Summaries, tables, & tests > Classical tests of hypotheses

Statistics > Summaries, tables, & tests > Nonparametric tests of hypotheses

Statistics > Linear models and related > ANOVA/MANOVA

Statistics > Postestimation > Predictions, residuals, etc.

Graphics > Twoway graph (scatter, line, etc.)

Example Commands

. **anova** *y x1 x2*

Performs two-way ANOVA, testing for differences among the means of *y* across categories of *x1* and *x2*.

. **anova** *y x1 x2 x1*x2*

Performs a two-way factorial ANOVA, including both the main and interaction (*x1*x2*) effects of categorical variables *x1* and *x2*.

. **anova** *y x1 x2 x3 x1*x2 x1*x3 x2*x3 x1*x2*x3*

Performs a three-way factorial ANOVA, including the three-way interaction *x1*x2*x3*, as well as all two-way interactions and main effects.

. **anova** *reading curriculum / teacher|curriculum /*

Fits a nested model to test the effects of three types of curriculum on students' reading ability (*reading*). *teacher* is nested within *curriculum* (***teacher|curriculum***) because several different teachers were assigned to each curriculum. The *Base Reference Manual* provides other nested ANOVA examples, including a split-plot design.

. **anova** *headache subject medication, repeated(medication)*

Fits a repeated-measures ANOVA model to test the effects of three types of headache medication (*medication*) on the severity of subjects' headaches (*headache*). The sample consists of 20 subjects who report suffering from frequent headaches. Each subject tried each of the three medications at separate times during the study.

. **anova** *y x1 x2 x3 x4 x2*x3, continuous(x3 x4) regress*

Performs analysis of covariance (ANCOVA) with four independent variables, two of them (*x1* and *x2*) categorical and two of them (*x3* and *x4*) measurements. Includes the *x2*x3* interaction, and shows results in the form of a regression table instead of the default ANOVA table.

. **kwallis** *y, by(x)*

Performs a Kruskal–Wallis test of the null hypothesis that *y* has identical rank distributions across the *k* categories of *x* ($k > 2$).

. **oneway** *y x*

Performs a one-way analysis of variance (ANOVA), testing for differences among the means of *y* across categories of *x*. The same analysis, with a different output table, is produced by **anova** *y x*.

. **oneway** *y x, tabulate scheffe*

Performs one-way ANOVA, including a table of sample means and Scheffé multiple-comparison tests in the output.

. **ranksum y, by(x)**
Performs a Wilcoxon rank-sum test (also known as a Mann–Whitney U test) of the null hypothesis that y has identical rank distributions for both categories of dichotomous variable x. If we assume that both rank distributions possess the same shape, this amounts to a test for whether the two medians of y are equal.

. **serrbar ymean se x, scale(2)**
Constructs a standard-error-bar plot from a dataset of means. Variable *ymean* holds the group means of y; *se* the standard errors; and x the values of categorical variable x. **scale(2)** asks for bars extending to ± 2 standard errors around each mean (default is ± 1).

. **signrank y1 = y2**
Performs a Wilcoxon matched-pairs signed-rank test for the equality of the rank distributions of $y1$ and $y2$. We could test whether the median of $y1$ differs from a constant such as 23.4 by typing the command **signrank y1 = 23.4**.

. **signtest y1 = y2**
Tests the equality of the medians of $y1$ and $y2$ (assuming matched data; that is, both variables measured on the same sample of observations). Typing **signtest y1 = 5** would perform a sign test of the null hypothesis that the median of $y1$ equals 5.

. **ttest y = 5**
Performs a one-sample t test of the null hypothesis that the population mean of y equals 5.

. **ttest y1 = y2**
Performs a one-sample (paired difference) t test of the null hypothesis that the population mean of $y1$ equals that of $y2$. The default form of this command assumes that the data are paired. With unpaired data ($y1$ and $y2$ are measured from two independent samples), add the option **unpaired**.

. **ttest y, by(x) unequal**
Performs a two-sample t test of the null hypothesis that the population mean of y is the same for both categories of variable x. Does not assume that the populations have equal variances. (Without the **unequal** option, **ttest** does assume equal variances.)

One-Sample Tests

One-sample t tests have two seemingly different applications:

1. Testing whether a sample mean $\bar{y}$ differs significantly from an hypothesized value μ_0.

2. Testing whether the means of y_1 and y_2, two variables measured over the same set of observations, differ significantly from each other. This is equivalent to testing whether the mean of a "difference score" variable created by subtracting y_1 from y_2 equals zero.

We use essentially the same formulas for either application, although the second starts with information on two variables instead of one.

The data in *writing.dta* were collected to evaluate a college writing course based on word processing (Nash and Schwartz 1987). Measures such as the number of sentences completed in timed writing were collected both before and after students took the course. The researchers wanted to know whether the post-course measures showed improvement.

```
Contains data from c:\data\writing.dta
  obs:            24                          Nash and Schwartz (1987)
  vars:            9                          21 Jan 2008 14:17
  size:          408 (99.9% of memory free)
```

variable name	storage type	display format	value label	variable label
id	byte	%8.0g	slbl	Student ID
preS	byte	%8.0g		# of sentences (pre-test)
preP	byte	%8.0g		# of paragraphs (pre-test)
preC	byte	%8.0g		Coherence scale 0-2 (pre-test)
preE	byte	%8.0g		Evidence scale 0-6 (pre-test)
postS	byte	%8.0g		# of sentences (post-test)
postP	byte	%8.0g		# of paragraphs (post-test)
postC	byte	%8.0g		Coherence scale 0-2 (post-test)
postE	byte	%8.0g		Evidence scale 0-6 (post-test)

```
Sorted by:
```

Suppose that we knew that students in previous years were able to complete an average of 10 sentences. Before examining whether the students in *writing.dta* improved during the course, we might want to learn whether at the start of the course they were essentially like earlier students — in other words, whether their pre-test (*preS*) mean differs significantly from the mean of previous students (10). To see a one-sample *t* test of $H_0:\mu = 10$, type

```
. ttest preS = 10
```

One-sample t test

Variable	Obs	Mean	Std. Err.	Std. Dev.	[95% Conf. Interval]	
preS	24	10.79167	.9402034	4.606037	8.846708	12.73663

```
    mean = mean(preS)                                          t =    0.8420
Ho: mean = 10                                degrees of freedom =        23

   Ha: mean < 10              Ha: mean != 10                 Ha: mean > 10
 Pr(T < t) = 0.7958      Pr(|T| > |t|) = 0.4084          Pr(T > t) = 0.2042
```

The notation $\Pr(T < t)$ means "probability of a *t*-distribution value less than the observed *t*, if H_0 were true"— that is, the one-tail test probability. The two-tail probability of a greater absolute *t* appears as $\Pr(|T| > |t|) = 0.48084$. Because this probability is high, we have no reason to reject $H_0:\mu = 10$. Note that **ttest** automatically provides a 95% confidence interval for the mean, and this confidence interval includes the null-hypothesis value 10. We could find some other confidence interval, such as 90%, by adding a **level(90)** option to this command.

A nonparametric counterpart, the sign test, employs the binomial distribution to test hypotheses about single medians. For example, we could test whether the median of *preS* equals 10. **signtest** gives us no reason to reject that null hypothesis either.

`. signtest preS = 10`

Sign test

sign	observed	expected
positive	12	11
negative	10	11
zero	2	2
all	24	24

One-sided tests:
  Ho: median of preS - 10 = 0 vs.
  Ha: median of preS - 10 > 0
    Pr(#positive >= 12) =
      Binomial(n = 22, x >= 12, p = 0.5) =   0.4159

  Ho: median of preS - 10 = 0 vs.
  Ha: median of preS - 10 < 0
    Pr(#negative >= 10) =
      Binomial(n = 22, x >= 10, p = 0.5) =   0.7383

Two-sided test:
  Ho: median of preS - 10 = 0 vs.
  Ha: median of preS - 10 != 0
    Pr(#positive >= 12 or #negative >= 12) =
      min(1, 2*Binomial(n = 22, x >= 12, p = 0.5)) =   0.8318

Like **ttest**, **signtest** includes right-tail, left-tail, and two-tail probabilities. Unlike the symmetrical t distributions used by **ttest**, however, the binomial distributions used by **signtest** have different left- and right-tail probabilities. In this example, only the two-tail probability matters because we were testing whether the *writing.dta* students "differ" from the null-hypothesis median of 10.

Next, we can test for improvement during the course by testing the null hypothesis that the mean number of sentences completed before and after the course (that is, the means of *preS* and *postS*) are equal. The **ttest** command accomplishes this as well, finding a significant improvement.

`. ttest postS = preS`

Paired t test

Variable	Obs	Mean	Std. Err.	Std. Dev.	[95% Conf. Interval]	
postS	24	26.375	1.693779	8.297787	22.87115	29.87885
preS	24	10.79167	.9402034	4.606037	8.846708	12.73663
diff	24	15.58333	1.383019	6.775382	12.72234	18.44433

```
        mean(diff) = mean(postS - preS)                            t =   11.2676
Ho: mean(diff) = 0                             degrees of freedom =         23

Ha: mean(diff) < 0            Ha: mean(diff) != 0            Ha: mean(diff) > 0
Pr(T < t) = 1.0000       Pr(|T| > |t|) = 0.0000         Pr(T > t) = 0.0000
```

Because we expect "improvement," not just "difference" between the *preS* and *postS* means, a one-tail test is appropriate. The displayed right-tail probability rounds off to zero. Students' mean sentence completion does significantly improve. Based on this sample, we are 95% confident that it improves by between 12.7 and 18.4 sentences.

t tests ordinarily assume that variables are normally distributed around their group means. This assumption usually is not critical because the tests are moderately robust. When nonnormality involves severe outliers, however, or occurs in small samples, we might be safer turning to medians instead of means and employing a nonparametric test that does not assume normality. The Wilcoxon signed-rank test, for example, assumes only that the distributions are symmetrical and continuous. Applying a signed-rank test to these data yields essentially the same conclusion as **ttest**, that students' sentence completion significantly improved. Because both tests agree on this conclusion, we can state it with more assurance.

```
. signrank postS = preS
```

Wilcoxon signed-rank test

sign	obs	sum ranks	expected
positive	24	300	150
negative	0	0	150
zero	0	0	0
all	24	300	300

```
unadjusted variance      1225.00
adjustment for ties        -1.63
adjustment for zeros        0.00
                         _____
adjusted variance        1223.38

Ho: postS = preS
          z =    4.289
 Prob > |z| =    0.0000
```

Two-Sample Tests

The remainder of this chapter draws examples from a survey of college undergraduates by Ward and Ault (1990) (*student2.dta*).

```
Contains data from c:\data\student2.dta
  obs:           243                          Student survey (Ward 1990)
  vars:           19                          21 Jan 2008 15:51
  size:        7,533 (99.9% of memory free)
```

variable name	storage type	display format	value label	variable label
id	int	%8.0g		Student ID
year	byte	%9.0g	year	Year in college
age	byte	%8.0g		Age at last birthday
gender	byte	%9.0g	s	Gender (male)
major	byte	%8.0g		Student major
relig	byte	%8.0g	v4	Religious preference
drink	byte	%9.0g		33-point drinking scale
gpa	float	%9.0g		Grade Point Average
grades	byte	%8.0g	grades	Guessed grades this semester
belong	byte	%8.0g	belong	Belong to fraternity/sorority
live	byte	%8.0g	v10	Where do you live?
miles	byte	%8.0g		How many miles from campus?
study	byte	%8.0g		Avg. hours/week studying
athlete	byte	%8.0g	yes	Are you a varsity athlete?
employed	byte	%8.0g	yes	Are you employed?

allnight	byte	%8.0g	allnight	How often study all night?
ditch	byte	%8.0g	times	How many class/month ditched?
hsdrink	byte	%9.0g		High school drinking scale
aggress	byte	%9.0g		Aggressive behavior scale

Sorted by: **year**

About 19% of these students belong to a fraternity or sorority:

. **tabulate belong**

Belong to fraternity/ sorority	Freq.	Percent	Cum.
member	47	19.34	19.34
nonmember	196	80.66	100.00
Total	243	100.00	

Another variable, *drink*, measures how often and heavily a student drinks alcohol, on a 33-point scale. Campus rumors might lead one to suspect that fraternity/sorority members tend to differ from other students in their drinking behavior. Box plots comparing the median *drink* values of members and nonmembers, and a bar chart comparing their means, both appear consistent with these rumors. Figure 5.1 combines these two separate plot types in one image.

. **graph box** *drink***, over(***belong***) ylabel(0(5)35) saving(***fig05_01a***)**
. **graph bar (mean)** *drink***, over(***belong***) ylabel(0(5)35)**
       saving(*fig05_01b***)**
. **graph combine** *fig05_01a.gph fig05_01b.gph***, col(2) iscale(1.05)**

Figure 5.1

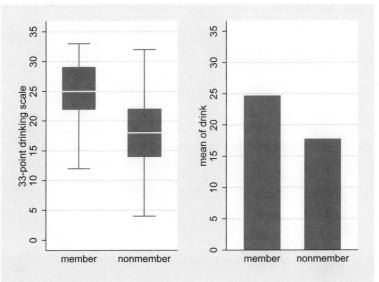

The **ttest** command, used earlier for one-sample and paired-difference tests, can perform two-sample tests as well. In this application its general syntax is **ttest** *measurement***, by(***categorical***)**. For example,

. ttest *drink*, by(*belong*)

Two-sample t test with equal variances

Group	Obs	Mean	Std. Err.	Std. Dev.	[95% Conf. Interval]	
member	47	24.7234	.7124518	4.884323	23.28931	26.1575
nonmembe	196	17.7602	.4575013	6.405018	16.85792	18.66249
combined	243	19.107	.431224	6.722117	18.25756	19.95643
diff		6.9632	.9978608		4.997558	8.928842

```
    diff = mean(member) - mean(nonmembe)                         t =   6.9781
Ho: diff = 0                                  degrees of freedom =       241

    Ha: diff < 0                 Ha: diff != 0                  Ha: diff > 0
Pr(T < t) = 1.0000         Pr(|T| > |t|) = 0.0000         Pr(T > t) = 0.0000
```

As the output notes, this *t* test rests on an equal-variances assumption. But the fraternity and sorority members' sample standard deviation appears somewhat lower — they are more alike than nonmembers in their reported drinking behavior. To perform a similar test without assuming equal variances, add the option **unequal**:

. ttest *drink*, by(*belong*) **unequal**

Two-sample t test with unequal variances

Group	Obs	Mean	Std. Err.	Std. Dev.	[95% Conf. Interval]	
member	47	24.7234	.7124518	4.884323	23.28931	26.1575
nonmembe	196	17.7602	.4575013	6.405018	16.85792	18.66249
combined	243	19.107	.431224	6.722117	18.25756	19.95643
diff		6.9632	.8466965		5.280627	8.645773

```
    diff = mean(member) - mean(nonmembe)                         t =   8.2240
Ho: diff = 0              Satterthwaite's degrees of freedom =       88.22

    Ha: diff < 0                 Ha: diff != 0                  Ha: diff > 0
Pr(T < t) = 1.0000         Pr(|T| > |t|) = 0.0000         Pr(T > t) = 0.0000
```

Adjusting for unequal variances does not alter our basic conclusion that members and nonmembers are significantly different. We can further check this conclusion by trying a nonparametric Mann–Whitney *U* test, also known as a Wilcoxon rank-sum test. Assuming that the rank distributions have similar shape, the rank-sum test on the following page indicates that we can reject the null hypothesis of equal population medians.

```
. ranksum drink, by(belong)
```

Two-sample Wilcoxon rank-sum (Mann-Whitney) test

belong	obs	rank sum	expected
member	47	8535	5734
nonmember	196	21111	23912
combined	243	29646	29646

```
unadjusted variance     187310.67
adjustment for ties       -472.30
                        _____
adjusted variance       186838.36

Ho: drink(belong==member) = drink(belong==nonmember)
             z =    6.480
    Prob > |z| =    0.0000
```

One-Way Analysis of Variance (ANOVA)

Analysis of variance (ANOVA) provides another way, more general than t tests, to test for differences among means. The simplest case, one-way ANOVA, tests whether the means of y differ across categories of x. One-way ANOVA can be performed by a **oneway** command with the general form **oneway** *measurement categorical*. For example,

```
. oneway drink belong, tabulate
```

Belong to fraternity/ sorority	Summary of 33-point drinking scale Mean	Std. Dev.	Freq.
member	24.723404	4.8843233	47
nonmember	17.760204	6.4050179	196
Total	19.106996	6.7221166	243

Source	Analysis of Variance SS	df	MS	F	Prob > F
Between groups	1838.08426	1	1838.08426	48.69	0.0000
Within groups	9097.13385	241	37.7474433		
Total	10935.2181	242	45.1868517		

Bartlett's test for equal variances: chi2(1) =   4.8378 Prob>chi2 = 0.028

The **tabulate** option produces a table of means and standard deviations in addition to the analysis of variance table itself. One-way ANOVA with a dichotomous x variable is equivalent to a two-sample t test, and its F statistic equals the corresponding t statistic squared. **oneway** offers more options and processes faster, but it lacks an **unequal** option for relaxing the equal-variances assumption.

oneway does, however, formally test the equal-variances assumption using Bartlett's χ^2. A low Bartlett's probability implies that an equal-variance assumption is implausible, in which case we should not trust the ANOVA F test results. In the **oneway** *drink belong* example above, Bartlett's $P = .028$ casts doubt on the ANOVA's validity.

One-way ANOVA's real value lies not in two-sample comparisons, but in comparisons of three or more means. For example, we could test whether mean drinking behavior varies by year in college:

```
. oneway drink year, tabulate scheffe
```

Year in college	Summary of 33-point drinking scale		
	Mean	Std. Dev.	Freq.
Freshman	18.975	6.9226033	40
Sophomore	21.169231	6.5444853	65
Junior	19.453333	6.2866081	75
Senior	16.650794	6.6409257	63
Total	19.106996	6.7221166	243

Analysis of Variance

Source	SS	df	MS	F	Prob > F
Between groups	666.200518	3	222.066839	5.17	0.0018
Within groups	10269.0176	239	42.9666008		
Total	10935.2181	242	45.1868517		

Bartlett's test for equal variances: chi2(3) = 0.5103 Prob>chi2 = 0.917

Comparison of 33-point drinking scale by Year in college
(Scheffe)

Row Mean- Col Mean	Freshman	Sophomor	Junior
Sophomor	2.19423 0.429		
Junior	.478333 0.987	-1.7159 0.498	
Senior	-2.32421 0.382	-4.51844 0.002	-2.80254 0.103

We can reject the hypothesis of equal means ($P = .0018$), but not the hypothesis of equal variances ($P = .917$). The latter is "good news" regarding the ANOVA's validity.

The box plots in Figure 5.2 support this conclusion, showing similar variation within each category. This figure, which combines separate box plots and dot plots, shows that differences among medians and among means follow similar patterns.

```
. graph hbox drink, over(year) ylabel(0(5)35) saving(fig05_02a)

. graph dot (mean) drink, over(year) ylabel(0(5)35, grid)
       marker(1, msymbol(Sh)) saving(fig05_02b)

. graph combine fig05_02a.gph fig05_02b.gph, row(2) iscale(1.05)
```

Figure 5.2

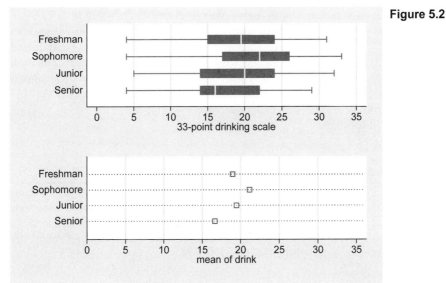

The **scheffe** option (Scheffé multiple-comparison test) produced a table showing the differences between each pair of means. The freshman mean equals 18.975 and the sophomore mean equals 21.16923, so the sophomore–freshman difference is 21.16923 – 18.975 = 2.19423, not statistically distinguishable from zero ($P = .429$). Of the six contrasts in this table, only the senior–sophomore difference, $16.6508 - 21.1692 = -4.5184$, is significant ($P = .002$). Thus, our overall conclusion that these four groups' means are not the same arises mainly from the contrast between seniors (the lightest drinkers) and sophomores (the heaviest).

oneway offers three multiple-comparison options: **scheffe**, **bonferroni**, and **sidak** (see *Base Reference Manual* for definitions) . The Scheffé test remains valid under a wider variety of conditions, although it is sometimes less sensitive.

The Kruskal–Wallis test (**kwallis**), a K-sample generalization of the two-sample rank-sum test, provides a nonparametric alternative to one-way ANOVA. It tests the null hypothesis of equal population medians.

```
. kwallis drink, by(year)
```

Kruskal-Wallis equality-of-populations rank test

year	Obs	Rank Sum
Freshman	40	4914.00
Sophomore	65	9341.50
Junior	75	9300.50
Senior	63	6090.00

```
chi-squared =      14.453 with 3 d.f.
probability =       0.0023

chi-squared with ties =       14.490 with 3 d.f.
probability =       0.0023
```

These **kwallis** results ($P = .0023$) agree with our **oneway** findings of significant differences in *drink* by year in college. Kruskal–Wallis is generally safer than ANOVA if we have reason to doubt ANOVA's equal-variances or normality assumptions, or if we suspect problems caused by outliers. **kwallis**, like **ranksum**, makes the weaker assumption of similar-shaped rank distributions within each group. In principle, **ranksum** and **kwallis** should produce similar results when applied to two-sample comparisons, but in practice this is true only if the data contain no ties. **ranksum** incorporates an exact method for dealing with ties, which makes it preferable for two-sample problems.

Two- and *N*-Way Analysis of Variance

One-way ANOVA examines how the means of measurement variable *y* vary across categories of one other variable *x*. *N*-way ANOVA generalizes this approach to deal with two or more categorical *x* variables. For example, we might consider how drinking behavior varies not only by fraternity or sorority membership, but also by gender. We start by examining a two-way table of means:

```
. table belong gender, contents(mean drink) row col
```

Belong to fraternity/sorority	Gender (male)		
	Female	Male	Total
member	22.44444	26.13793	24.7234
nonmember	16.51724	19.5625	17.7602
Total	17.31343	21.31193	19.107

It appears that in this sample, males drink more than females and members drink more than nonmembers. The member–nonmember difference appears similar among males and females. Stata's *N*-way ANOVA command, **anova**, can test for significant differences among these means attributable to belonging to a fraternity or sorority, gender, or the interaction of belonging and gender (written *belong*gender*).

```
. anova drink belong gender belong*gender
```

		Number of obs =	243	R-squared	= 0.2221
		Root MSE	= 5.96592	Adj R-squared =	0.2123

Source	Partial SS	df	MS	F	Prob > F
Model	2428.67237	3	809.557456	22.75	0.0000
belong	1406.2366	1	1406.2366	39.51	0.0000
gender	408.520097	1	408.520097	11.48	0.0008
belong*gender	3.78016612	1	3.78016612	0.11	0.7448
Residual	8506.54574	239	35.5922416		
Total	10935.2181	242	45.1868517		

In this example of "two-way factorial ANOVA," the output shows significant main effects for *belong* (P = .0000) and *gender* (P = .0008), but their interaction contributes little to the model (P = .7448). This interaction cannot be distinguished from zero, so we might prefer to fit a simpler model without the interaction term (results not shown):

. anova *drink belong gender*

To include any interaction term with **anova**, specify the variable names joined by *. Unless the number of observations with each combination of x values is the same (a condition called "balanced data"), it can be hard to interpret the main effects in a model that also includes interactions. This does not mean that the main effects in such models are unimportant, however. Regression analysis might help to make sense of complicated ANOVA results, as illustrated in the following section.

Analysis of Covariance (ANCOVA)

Analysis of Covariance (ANCOVA) extends N-way ANOVA to encompass a mix of categorical and continuous x variables. This is accomplished through the **anova** command if we specify which variables are continuous. For example, when we include *gpa* (college grade point average) among the independent variables, we find that it, too, is related to drinking behavior.

. anova *drink belong gender gpa*, continuous(*gpa*)

```
              Number of obs =      218   R-squared     =  0.2970
              Root MSE      = 5.68939   Adj R-squared =  0.2872

      Source |  Partial SS    df       MS            F     Prob > F
-------------+----------------------------------------------------
       Model |  2927.03087     3   975.676958       30.14    0.0000
             |
      belong |  1489.31999     1   1489.31999       46.01    0.0000
      gender |  405.137843     1   405.137843       12.52    0.0005
         gpa |    407.0089     1     407.0089       12.57    0.0005
-------------+----------------------------------------------------
    Residual |  6926.99206   214   32.3691218
-------------+----------------------------------------------------
       Total |  9854.02294   217   45.4102439
```

From this analysis we know that a significant relationship exists between *drink* and *gpa* when we control for *belong* and *gender*. Beyond their F tests for statistical significance, however, ANOVA or ANCOVA tables do not provide much descriptive information about how variables are related. Regression, with its explicit model and parameter estimates, does a better descriptive job. Because ANOVA and ANCOVA amount to special cases of regression, we could restate any of these analyses in regression form. Stata does so automatically if we add the **regress** option to **anova**. For instance, we might want to see regression output in order to understand results from the following ANCOVA.

```
. anova drink belong gender belong*gender gpa, continuous(gpa)
     regress
```

Source	SS	df	MS			
Model	2933.45823	4	733.364558			
Residual	6920.5647	213	32.4909141			
Total	9854.02294	217	45.4102439			

Number of obs =   218
F(4,   213) =   22.57
Prob > F   = 0.0000
R-squared   = 0.2977
Adj R-squared = 0.2845
Root MSE   = 5.7001

drink	Coef.	Std. Err.	t	P>|t|	[95% Conf. Interval]	
_cons	27.47676	2.439962	11.26	0.000	22.6672	32.28633
belong						
1	6.925384	1.286774	5.38	0.000	4.388942	9.461826
2	(dropped)					
gender						
1	-2.629057	.8917152	-2.95	0.004	-4.386774	-.8713407
2	(dropped)					
gpa	-3.054633	.8593498	-3.55	0.000	-4.748552	-1.360713
belong*gender						
1 1	-.8656158	1.946211	-0.44	0.657	-4.701916	2.970685
1 2	(dropped)					
2 1	(dropped)					
2 2	(dropped)					

With the **regress** option, we get the **anova** output formatted as a regression table. The top part gives the same overall F test and R^2 as a standard ANOVA table. The bottom part describes the following regression:

> We construct a separate dummy variable {0,1} representing each category of each x variable, except for the highest categories, which are dropped. Interaction terms (if specified in the variable list) are constructed from the products of every possible combination of these dummy variables. Regress y on all these dummy variables and interactions, and also on any continuous variables specified in the command line.

The previous example therefore corresponds to a regression of *drink* on four x variables:

1. a dummy coded 1 = fraternity/sorority member, 0 otherwise (highest category of *belong*, nonmember, gets dropped);
2. a dummy coded 1 = female, 0 otherwise (highest category of *gender*, male, gets dropped);
3. the continuous variable *gpa*;
4. an interaction term coded 1 = sorority female, 0 otherwise.

Interpret the individual dummy variables' regression coefficients as effects on predicted or mean y. For example, the coefficient on the first category of *gender* (female) equals –2.629057. This tells us that the mean drinking scale levels for females are about 2.63 points lower than those of males with the same grade point average and membership status. We also learn that among students of the same gender and membership status, mean drinking scale values decline by 3.054633 with each one-point increase in grades. Note that we have confidence intervals and individual t tests for each coefficient, too. There is much more information in the **anova, regress** output than in the ANOVA table alone.

Predicted Values and Error-Bar Charts

After **anova**, the followup command **predict** calculates predicted values, residuals, or standard errors and diagnostic statistics. One application for such statistics is in drawing graphical representations of the model's predictions, in the form of error-bar charts. For a simple illustration, we return to the one-way ANOVA of *drink* by *year*:

```
. anova drink year
```

| | | Number of obs = 243 | | R-squared = 0.0609 | |
| | | Root MSE = 6.55489 | | Adj R-squared = 0.0491 | |

Source	Partial SS	df	MS	F	Prob > F
Model	666.200518	3	222.066839	5.17	0.0018
year	666.200518	3	222.066839	5.17	0.0018
Residual	10269.0176	239	42.9666008		
Total	10935.2181	242	45.1868517		

To calculate predicted means from the recent **anova**, type **predict** followed by a new variable name:

```
. predict drinkmean
(option xb assumed; fitted values)
. label variable drinkmean "Mean drinking scale"
. predict SEdrink, stdp
```

With the **stdp** option, **predict** calculates standard errors of the predicted means. Using new variables *drinkmean* and *SEdrink*, we apply **serrbar** to draw an error-bar chart. The **scale(2)** option tells **serrbar** to draw error bars of plus and minus two standard errors, from

$$drinkmean - 2 \times SEdrink$$
to
$$drinkmean + 2 \times SEdrink$$

In a **serrbar** command, the first-listed variable should be the means or *y* variable; the second-listed, the standard error or standard deviation (depending on which you want to show); and the third-listed variable defines the *x* axis. The **addplot( )** option for **serrbar** can specify a second plot to overlay on the standard-error bars. In Figure 5.3, we overlay a line plot that connects the *drinkmean* values with solid line segments.

```
. serrbar drinkmean SEdrink year, scale(2)
     addplot(line drinkmean year, clpattern(solid)) legend(off)
```

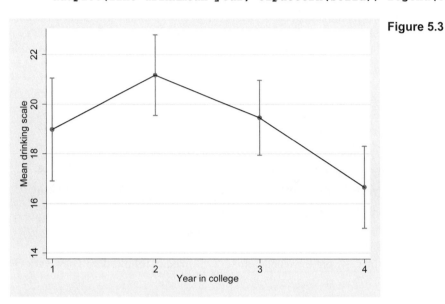

Figure 5.3

For a two-way factorial ANOVA, error-bar charts help us to visualize main and interaction effects. Although the usual error-bar command **serrbar** can, with effort, be adapted for this purpose, an alternative approach using the more flexible **graph twoway** family will be illustrated below. First, we perform ANOVA, obtain group means (predicted values) and their standard errors, then generate new variables equal to the group means plus or minus two standard errors. The example examines the relationship between students' aggressive behavior (*aggress*), gender, and year in college. Both the main effects of *gender* and *year*, and their interaction, are statistically significant.

```
. anova aggress gender year gender*year
```

	Number of obs =	243	R-squared	=	0.2503
	Root MSE	= 1.45652	Adj R-squared =		0.2280

Source	Partial SS	df	MS	F	Prob > F
Model	166.482503	7	23.7832147	11.21	0.0000
gender	94.3505972	1	94.3505972	44.47	0.0000
year	19.0404045	3	6.34680149	2.99	0.0317
gender*year	24.1029759	3	8.03432529	3.79	0.0111
Residual	498.538073	235	2.12143861		
Total	665.020576	242	2.74801891		

```
. predict aggmean
. label variable aggmean "Mean aggressive behavior scale"
. predict SEagg, stdp
. gen agghigh = aggmean + 2 * SEagg
. gen agglow = aggmean - 2 * SEagg

. graph twoway connected aggmean year
       || rcap agghigh agglow year
       || , by(gender, legend(off) note(""))
    ytitle("Mean aggressive behavior scale")
```

Figure 5.4

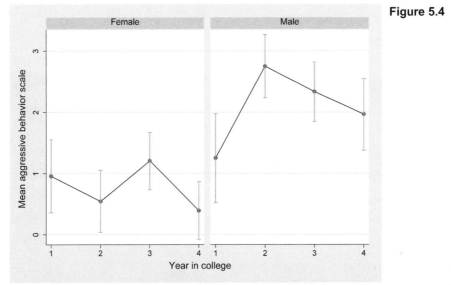

The error-bar charts in Figure 5.4 were created by overlaying two pairs of plots. The first pair are female and male connected-line plots, connecting group means of *aggress* (calculated using **predict**, and saved as variable *aggmean*). The second pair are female and male capped-spike range plots (**twoway rcap**) with vertical spikes connecting *agghigh* (group means of *aggress* plus two standard errors) and *agglow* (means minus two standard errors). The **by(*gender*)** option produced sub-plots for females and males. To suppress legends and notes in a graph that uses a **by( )** option, **legend(off)** and **note("")** must appear as suboptions within **by( )**.

The resulting error-bar chart shows female means on the aggressive-behavior scale fluctuating at comparatively low levels during the four years of college. Male means are higher throughout, with a sophomore-year peak that resembles the pattern seen earlier for drinking (Figures 5.2 and 5.3). Thus, the relationship between *aggress* and *year* is different for males and females. This graph helps us to understand and explain the significant interaction effect.

predict works the same way with regression analysis (**regress**) as it does with **anova** because the two share a common mathematical framework. A list of some other **predict** options appears in Chapter 6, and further examples using these options are given in Chapter 7. The options include residuals that can be used to check assumptions regarding error distributions, and also a suite of diagnostic statistics (such as leverage, Cook's *D*, and *DFBETA*) that measure the influence of individual observations on model results. Conditional effect plotting (Chapter 7) provides a graphical approach that can aid interpretation of more complicated regression, ANOVA, or ANCOVA models.

Linear Regression Analysis

Stata offers an exceptionally broad range of regression procedures. A partial list of the possibilities can be seen by typing **help regress**. This chapter introduces **regress** and related commands that perform simple and multiple ordinary least squares (OLS) regression. One followup command, **predict**, calculates predicted values, residuals, and diagnostic statistics such as leverage or Cook's D. Another followup command, **test**, performs tests of user-specified hypotheses. **regress** can accomplish other analyses including weighted least squares and two-stage least squares. Regression with dummy variables, interaction effects, polynomial terms, and stepwise variable selection are covered briefly in this chapter, along with a first look at residual analysis.

The following menus access most of the operations discussed:

Statistics > Linear models and related > Linear regression

Statistics > Linear models and related > Regression diagnostics

Statistics > Postestimation > Predictions, residuals, etc.

Graphics > Twoway graph (scatter, line, etc.)

Statistics > Longitudinal/panel data

Example Commands

. **regress** *y x*
    Performs ordinary least squares (OLS) regression of variable y on one predictor, x.

. **regress** *y x* if *ethnic* == 3 & *income* > 50 & *income* < .
    Regresses y on x using only that subset of the data for which variable *ethnic* equals 3 and *income* is greater than 50 (but not missing).

. **predict** *yhat*
    Generates a new variable (here arbitrarily named *yhat*) equal to the predicted values from the most recent regression.

. `predict e, resid`
Generates a new variable (here arbitrarily named *e*) equal to the residuals from the most recent regression.

. `graph twoway lfit y x || scatter y x`
Draws the simple regression line (**lfit** or linear fit) with a scatterplot of *y* vs. *x*.

. `graph twoway mspline yhat x || scatter y x`
Draws a simple regression line with a scatterplot of *y* vs. *x* by connecting (with a smooth cubic spline curve) the regression's predicted values (in this example named *yhat*).
    <u>Note:</u> There are many alternative ways to draw regression lines or curves in Stata. These alternatives include the **twoway** graph types **mspline** (illustrated above), **mband**, **line**, **lfit**, **lfitci**, **qfit**, and **qfitci**, each of which has its own advantages and options. Usually we combine (overlay) the regression line or curve with a scatterplot. If the scatterplot comes second in our **graph twoway** command, as in the example above, then scatterplot points will print on top of the regression line. Placing the scatterplot first in the command causes the line to print on top of the scatter. Examples throughout this and the following chapters illustrate some of these different possibilities.

. `rvfplot`
Draws a residual versus fitted (predicted values) plot, automatically based on the most recent regression.

. `graph twoway scatter e yhat, yline(0)`
Draws a residual versus predicted values plot using the variables *e* and *yhat*.

. `regress y x1 x2 x3`
Performs multiple regression of *y* on three predictor variables, *x1*, *x2*, and *x3*.

. `regress y x1 x2 x3, vce(robust)`
Calculates robust (Huber/White) estimates of standard errors. See the *User's Guide* for details. The **vce(robust)** option works with many other model fitting commands as well.

. `regress y x1 x2 x3, beta`
Performs multiple regression and includes standardized regression coefficients ("beta weights") in the output table.

. `correlate x1 x2 x3 y`
Displays a matrix of Pearson correlations, using only observations with no missing values on all of the variables specified. Adding the option **covariance** produces a variance–covariance matrix instead of correlations.

. `pwcorr x1 x2 x3 y, sig star(.05)`
Displays a matrix of Pearson correlations, using pairwise deletion of missing values and showing probabilities from *t* tests of null hypothesis $H_0:\rho = 0$, for each correlation. Statistically significant correlations ($P < .05$) are indicated by stars (*).

. **graph matrix** *x1 x2 x3 y*, **half**
Draws a scatterplot matrix. Because their variable lists are the same, this example yields a scatterplot matrix having the same organization as the correlation matrix produced by the preceding **pwcorr** command. Listing the dependent (*y*) variable last creates a matrix in which the bottom row forms a series of *y*-versus-*x* plots.

. **test** *x1 x2*
Performs an *F* test of the null hypothesis that coefficients on *x1* and *x2* both equal zero in the most recent regression model.

. **xi: regress** *y x1 x2* **i.**catvar*x2*
Performs "expanded interaction" regression of *y* on predictors *x1*, *x2*, a set of dummy variables created automatically to represent categories of *catvar*, and a set of interaction terms equal to those dummy variables times measurement variable *x2*. **help xi** obtains more details.

. **stepwise, pr(.05):   regress** *y x1 x2 x3*
Performs stepwise regression using backward elimination until all remaining predictors are significant at the .05 level. All listed predictors are entered on the first iteration. Thereafter, each iteration drops one predictor with the highest *P* value, until all predictors remaining have probabilities below the "probability to retain," pr(.05). Options permit forward or hierarchical selection. The **stepwise** prefix works with many other model-fitting commands as well; type **help stepwise** for a list.

. **regress** *y x1 x2 x3* **[aweight** = *w*]
Performs weighted least squares (WLS) regression of *y* on *x1*, *x2*, and *x3*. Variable *w* holds the analytical weights, which work as if we had multiplied each variable and the constant by the square root of *w*, and then performed an ordinary regression. Analytical weights are often employed to correct for heteroskedasticity when the *y* and *x* variables are means, rates, or proportions, and *w* is the number of individuals making up each aggregate observation (e.g., city or school) in the data. If the *y* and *x* variables are individual-level, and the weights indicate numbers of replicated observations, then use frequency weights **[fweight** = *w*] instead. See **help survey** if the weights reflect design factors such as disproportionate sampling (Chapter 14).

. **regress** *y1 y2 x* (*x z*)
. **regress** *y2 y1 z* (*x z*)
Estimates the reciprocal effects of *y1* and *y2*, using instrumental variables *x* and *z*. The first parts of these commands specify the structural equations:
$$y1 = \alpha_0 + \alpha_1 y2 + \alpha_2 x + \epsilon_1$$
$$y2 = \beta_0 + \beta_1 y1 + \beta_2 z + \epsilon_2$$
The parentheses in the commands enclose variables that are exogenous to all of the structural equations. **regress** uses two-stage least squares (2SLS) in this example.

The Regression Table

File *states.dta* contains educational data on the U.S. states and District of Columbia:

. **describe** *state csat expense percent income high college region*

```
                 storage  display    value
variable name    type     format     label    variable label

state            str20    %20s                 State
csat             int      %9.0g               Mean composite SAT score
expense          int      %9.0g               Per pupil expenditures prim&sec
percent          byte     %9.0g               % HS graduates taking SAT
income           double   %10.0g              Median household income, $1,000
high             float    %9.0g               % adults HS diploma
college          float    %9.0g               % adults college degree
region           byte     %9.0g      region   Geographical region
```

Political leaders occasionally use mean Scholastic Aptitude Test (SAT) scores to make pointed comparisons between the educational systems of different U.S. states. For example, some have raised the question of whether SAT scores are higher in states that spend more money on education. We might try to address this question by regressing mean composite SAT scores (*csat*) on per-pupil expenditures (*expense*). The appropriate Stata command has the form **regress** *y x*, where *y* is the predicted or dependent variable, and *x* the predictor or independent variable.

. **regress** *csat expense*

```
      Source |       SS           df       MS            Number of obs =      51
-------------+------------------------------            F( 1,    49) =   13.61
       Model |  48708.3001         1   48708.3001       Prob > F      =  0.0006
    Residual |   175306.21        49   3577.67775       R-squared     =  0.2174
-------------+------------------------------            Adj R-squared =  0.2015
       Total |   224014.51        50    4480.2902       Root MSE      =  59.814
```

```
        csat |      Coef.   Std. Err.      t    P>|t|     [95% Conf. Interval]
-------------+----------------------------------------------------------------
     expense |  -.0222756   .0060371    -3.69   0.001    -.0344077    -.0101436
       _cons |   1060.732    32.7009    32.44   0.000     995.0175     1126.447
```

This regression tells an unexpected story: the more money a state spends on education, the lower its students' mean SAT scores. A causal interpretation is premature at this point, but the regression table does convey information about the linear statistical relationship between *csat* and *expense*. At upper right it gives an overall F test, based on the sums of squares at the upper left. This F test evaluates the null hypothesis that coefficients on all x variables in the model (here there is only one x variable, *expense*) equal zero. The F statistic, 13.61 with 1 and 49 degrees of freedom, leads easily to rejection of this null hypothesis ($P = .0006$). Prob > F means "the probability of a greater F statistic" if we drew samples randomly from a population in which the null hypothesis is true.

At upper right, we also see the coefficient of determination, $R^2 = .2174$. Per-pupil expenditures explain about 22% of the variance in states' mean composite SAT scores. Adjusted R^2, $R^2_a = .2015$, takes into account the complexity of the model relative to the complexity of the data.

The lower half of the regression table gives the fitted model itself. We find coefficients (slope and *y*-intercept) in the first column, here yielding the prediction equation

$$\text{predicted } csat = 1060.732 - .0222756 expense$$

The second column lists estimated standard errors of the coefficients. These are used to calculate *t* tests (columns 3–4) and confidence intervals (columns 5–6) for each regression coefficient. The *t* statistics (coefficients divided by their standard errors) test null hypotheses that the corresponding population coefficients equal zero. At the $\alpha = .05$ significance level, we could reject this null hypothesis regarding both the coefficient on *expense* ($P = .001$) and the *y*-intercept ("$.000$", really meaning $P < .0005$). Stata's modeling commands print 95% confidence intervals routinely, but we can request other levels by specifying the **level( )** option, as shown in the following:

```
. regress csat expense, level(99)
```

After fitting a regression model, we could re-display the results just by typing **regress**, without arguments. Typing **regress, level(90)** would repeat the results but show 90% confidence intervals this time.

Because the data in this example do not represent a random sample from some larger population of U.S. states, hypothesis tests and confidence intervals lack their usual meanings. They are discussed in this chapter anyway for purposes of illustration.

The term _cons stands for the regression constant, usually set at one (so the coefficient on _cons equals the *y* intercept). Stata automatically includes a constant unless we tell it not to. A **nocons** option would cause Stata to suppress the constant, performing regression through the origin. For example,

```
. regress y x,   nocons
```

or

```
. regress y x1 x2 x3, nocons
```

In certain advanced applications, you might need to specify your own constant. If the "independent variables" include a user-supplied constant (named *c*, for example), employ the **hascons** option instead of **nocons**:

```
. regress y c x, hascons
```

Using **nocons** in this situation results in a misleading *F* test and R^2. Consult the *Base Reference Manual* or **help regress** for more about **hascons**.

Multiple Regression

Multiple regression allows us to estimate how *expense* predicts *csat*, while adjusting for a number of other possible predictor variables. We can incorporate other predictors of *csat* simply by listing these variables in the command

. regress *csat expense percent income high college*

Source	SS	df	MS		
Model	184663.309	5	36932.6617		
Residual	39351.2012	45	874.471137		
Total	224014.51	50	4480.2902		

Number of obs = 51
F(5, 45) = 42.23
Prob > F = 0.0000
R-squared = 0.8243
Adj R-squared = 0.8048
Root MSE = 29.571

csat	Coef.	Std. Err.	t	P>|t|	[95% Conf. Interval]	
expense	.0033528	.0044709	0.75	0.457	-.005652	.0123576
percent	-2.618177	.2538491	-10.31	0.000	-3.129455	-2.106898
income	.1055853	1.166094	0.09	0.928	-2.243048	2.454218
high	1.630841	.992247	1.64	0.107	-.367647	3.629329
college	2.030894	1.660118	1.22	0.228	-1.312756	5.374544
_cons	851.5649	59.29228	14.36	0.000	732.1441	970.9857

This yields the multiple regression equation

$$\text{predicted } csat = 851.56 + .00335 expense - 2.618 percent + .0001 income + 1.63 high + 2.03 college$$

Controlling for four other variables weakens the coefficient on *expense* from $-.0223$ to $.00335$, which is no longer statistically distinguishable from zero. The unexpected negative relationship between *expense* and *csat* found in our earlier simple regression evidently is spurious, and explained by other predictors.

Only the coefficient on *percent* (percentage of high school graduates taking the SAT) attains significance at the .05 level. We could interpret this "fourth-order partial regression coefficient" (so called because its calculation adjusts for four other predictors) as follows.

$b_2 = -2.618$: Predicted mean SAT scores decline by 2.618 points, with each one-point increase in the percentage of high school graduates taking the SAT — if *expense*, *income*, *high*, and *college* do not change.

Taken together, the five *x* variables in this model explain about 80% of the variance in states' mean composite SAT scores ($R^2_a = .8048$). In contrast, our earlier simple regression with *expense* as the only predictor explained only 20% of the variance in *csat*.

To obtain standardized regression coefficients ("beta weights") with any regression, add the **beta** option. Standardized coefficients are what we would see in a regression where all the variables had been transformed into standard scores (means 0, standard deviations 1).

```
. regress csat expense percent income high college, beta
```

Source	SS	df	MS
Model	184663.309	5	36932.6617
Residual	39351.2012	45	874.471137
Total	224014.51	50	4480.2902

```
Number of obs =      51
F( 5,    45) =   42.23
Prob > F     =  0.0000
R-squared    =  0.8243
Adj R-squared =  0.8048
Root MSE     =  29.571
```

csat	Coef.	Std. Err.	t	P>|t|	Beta
expense	.0033528	.0044709	0.75	0.457	.070185
percent	-2.618177	.2538491	-10.31	0.000	-1.024538
income	.1055853	1.166094	0.09	0.928	.0101321
high	1.630841	.992247	1.64	0.107	.1361672
college	2.030894	1.660118	1.22	0.228	.1263952
_cons	851.5649	59.29228	14.36	0.000	.

The standardized regression equation is

$$\text{predicted } csat^* = .07expense^* - 1.0245percent^* + .01income^* + .136high^* + .126college^*$$

where $csat^*$, $expense^*$, etc. denote these variables in standard-score form. We might interpret the standardized coefficient on *percent*, for example, as follows:

> $b_2^* = -1.0245$: Predicted mean SAT scores decline by 1.0245 standard deviations, with each one-standard-deviation increase in the percentage of high school graduates taking the SAT — if *expense*, *income*, *high*, and *college* do not change.

The F and t tests, R^2, and other aspects of the regression remain the same.

Predicted Values and Residuals

After any regression, the **predict** command can obtain predicted values, residuals, and other case statistics. Suppose we have just done a regression of composite SAT scores on their strongest single predictor:

```
. regress csat percent
```

Now, to create a new variable called *yhat* containing predicted *y* values from this regression, type

```
. predict yhat
. label variable yhat "Predicted mean SAT score"
```

Through the **resid** option, we can also create another new variable containing the residuals, here named *e*:

```
. predict e, resid
. label variable e "Residual"
```

We might instead have obtained the same predicted *y* and residuals through two **generate** commands:

```
. generate yhat0 = _b[_cons] + _b[percent]*percent
. generate e0 = csat - yhat0
```

Stata temporarily remembers coefficients and other details from the recent regression. Thus _b[*varname*] refers to the coefficient on independent variable *varname*. _b[_cons] refers to the coefficient on _cons (usually, the *y*-intercept). These stored values are useful in programming and some advanced applications, but for most purposes, **predict** saves us the trouble of generating *yhat0* and *e0* "by hand" in this fashion.

Residuals contain information about where the model fits poorly, and so are important for diagnostic or troubleshooting analysis. Such analysis might begin just by sorting and examining the residuals. Negative residuals occur when our model overpredicts the observed values. That is, in these states the mean SAT scores are lower than we would expect, based on what percentage of students took the test. To list the states with the five lowest residuals, type

```
. sort e
. list state percent csat yhat e in 1/5
```

	state	percent	csat	yhat	e
1.	South Carolina	58	832	894.3333	-62.3333
2.	West Virginia	17	926	986.0953	-60.09526
3.	North Carolina	57	844	896.5714	-52.5714
4.	Texas	44	874	925.6666	-51.66666
5.	Nevada	25	919	968.1905	-49.19049

The four lowest residuals belong to southern states, suggesting that we might be able to improve our model, or better understand variation in mean SAT scores, by somehow taking region into account.

Positive residuals occur when actual *y* values are higher than predicted. Because the data already have been sorted by *e*, to list the five highest residuals we add the qualifier **in -5/l**. "–5" in this qualifier means the 5th-from-last observation, and the letter "el" (note that this is not the number "1") stands for the last observations. The qualifiers **in –5/–1**, **in 47/l** or **in 47/51** each could accomplish the same thing.

```
. list state percent csat yhat e in -5/1
```

	state	percent	csat	yhat	e
47.	Massachusetts	79	896	847.3333	48.66673
48.	Connecticut	81	897	842.8571	54.14292
49.	North Dakota	6	1073	1010.714	62.28567
50.	New Hampshire	75	921	856.2856	64.71434
51.	Iowa	5	1093	1012.952	80.04758

predict also derives other statistics from the most recently-fitted model. Below are some **predict** options that can be used after **anova** or **regress**.

. **predict** *new*	Predicted values of *y*. **predict** *new*, **xb** means the same thing (referring to **Xb**, the vector of predicted *y* values).
. **predict** *new*, **cooksd**	Cook's *D* influence measures.
. **predict** *new*, **covratio**	*COVRATIO* influence measures; effect of each observation on the variance–covariance matrix of estimates.
. **predict** *DFx1*, **dfbeta(x1)**	*DFBETA*s measuring each observation's influence on the coefficient of predictor *x1*.
. **predict** *new*, **dfits**	*DFITS* influence measures.
. **predict** *new*, **hat**	Diagonal elements of hat matrix (**leverage** also works).
. **predict** *new*, **resid**	Residuals.
. **predict** *new*, **rstandard**	Standardized residuals.
. **predict** *new*, **rstudent**	Studentized (jackknifed) residuals.
. **predict** *new*, **stdf**	Standard errors of predicted individual *y*, sometimes called the standard errors of forecast or the standard errors of prediction.
. **predict** *new*, **stdp**	Standard errors of predicted mean *y*.
. **predict** *new*, **stdr**	Standard errors of residuals.
. **predict** *new*, **welsch**	Welsch's distance influence measures.

Further options obtain predicted probabilities and expected values; type **help regress** for a list. All **predict** options create case statistics, which are new variables (like predicted values and residuals) that have a value for each observation in the sample.

When using **predict**, substitute a new variable name of your choosing for *new* in the commands shown above. For example, to obtain Cook's *D* influence measures, type

. **predict** *D*, **cooksd**

Or you can find leverage (hat matrix diagonals) by typing

. **predict** *h*, **hat**

The names of variables created by **predict** (such as *yhat*, *e*, *D*, *h*) are arbitrary and are invented by the user. As with other elements of Stata commands, we could abbreviate the options to the minimum number of letters it takes to identify them uniquely. For example,

```
. predict e, resid
```

could be shortened to

```
. pre e, re
```

Basic Graphs for Regression

This section introduces some elementary graphs you can use to represent a regression model or examine its fit. Chapter 7 describes more specialized graphs that aid post-regression diagnostic work.

In simple regression, predicted values lie on the line defined by the regression equation. By plotting and connecting predicted values, we can make that line visible. The **lfit** (linear fit) command automatically draws a simple regression line.

```
. graph twoway lfit csat percent
```

Ordinarily, it is more interesting to overlay a scatterplot on the regression line, as in Figure 6.1.

```
. graph twoway lfit csat percent
    || scatter csat percent
    || , ytitle("Mean composite SAT score") legend(off)
```

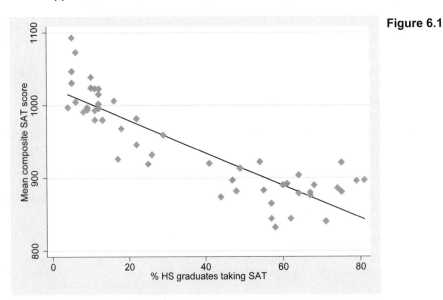

Figure 6.1

We could draw the same Figure 6.1 graph "by hand" using the predicted values (*yhat*) generated after the regression, and a command of the form

```
. graph twoway mspline yhat percent, bands(50)
      ||   scatter csat percent
      ||   , legend(off) ytitle("Mean composite SAT score")
```

The second approach is more work, but offers greater flexibility for advanced applications such as conditional effect plots or nonlinear regression. Working directly with the predicted values also keeps the analyst closer to the data, and to what a regression model is doing. **graph twoway mspline** (cubic spline curve fit to 50 cross-medians) simply draws a straight line when applied to linear predicted values, but will equally well draw a smooth curve in the case of nonlinear predicted values.

Residual-versus-predicted-values plots provide useful diagnostic tools (Figure 6.2). After any regression analysis (also after some other models, such as ANOVA) we can automatically draw a residual-versus-fitted (predicted values) plot just by typing

```
. rvfplot, yline(0)
```

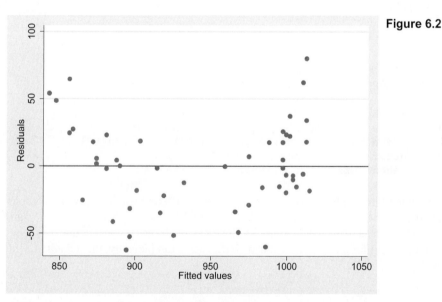

Figure 6.2

The "by-hand" alternative for drawing Figure 6.2 would be

```
. graph twoway scatter e yhat, yline(0)
```

Figure 6.2 reveals that our present model overlooks an obvious pattern in the data. The residuals or prediction errors appear to be mostly positive at first (due to too-high predictions), then mostly negative, followed by mostly positive residuals again. Later sections will seek a model that better fits these data.

predict can generate two kinds of standard errors for the predicted y values, which have different applications. These applications are sometimes distinguished by the names "confidence interval" and "prediction interval": A "confidence interval" in this context expresses our uncertainty in estimating the conditional mean of y at a given x value (or a given combination of x values, in multiple regression). Standard errors for this purpose are obtained through

```
. predict SE, stdp
```

Select an appropriate t value. With 49 degrees of freedom, for 95% confidence we should use $t = 2.01$, found by looking up the t distribution or simply by asking Stata:

```
. display invttail(49,.05/2)
2.0095752
```

Then the lower confidence limit is approximately

```
. generate low1 = yhat - 2.01*SE
```

and the upper confidence limit is

```
. generate high1 = yhat + 2.01*SE
```

Confidence bands in simple regression have an hourglass shape, narrowest at the mean of x. We could graph these using an overlaid **twoway** command such as the following.

```
. graph twoway mspline low1 percent, clpattern(dash) bands(50)
      ||   mspline high1 percent, clpattern(dash) bands(50)
      ||   mspline yhat percent, clpattern(solid) bands(50)
      ||   scatter csat percent
      ||   , legend(off) ytitle("Mean composite SAT score")
```

Shaded-area range plots (see **help twoway rarea**) offer a different way to draw such graphs, shading the range between *low1* and *high1*. Alternatively, **lfitci** can do this automatically, and take care of the confidence-band calculations, as illustrated on the next page in Figure 6.3. Note the **stdp** option, calling for a conditional-mean confidence band (actually, the default).

```
. graph twoway lfitci csat percent, stdp
      || scatter csat percent, msymbol(O)
      || , ytitle("Mean composite SAT score")  legend(off)
      title("Confidence bands for conditional means (stdp)")
```

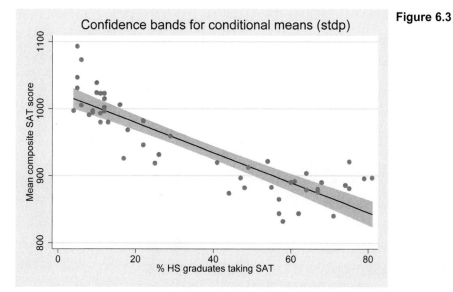

Figure 6.3

The second type of confidence interval for regression predictions is sometimes called a "prediction interval." This expresses our uncertainty in estimating the unknown value of y for an individual observation with known x value(s). Standard errors for this purpose are obtained by typing

```
. predict SEyhat, stdf
```

Figure 6.4 (next page) graphs this prediction band using **lfitci** with the **stdf** option. Predicting the y values of individual observations as done in Figure 6.4 inherently involves greater uncertainty, and hence wider bands, than does predicting the conditional mean of y (Figure 6.3). In both instances, the bands are narrowest at the mean of x.

```
. graph twoway lfitci csat percent, stdf
      ||   scatter csat percent, msymbol(O)
      ||   , ytitle("Mean composite SAT score")  legend(off)
      title("Confidence bands for individual-case predictions
         (stdf)")
```

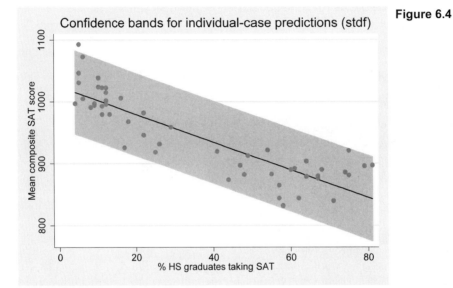

Figure 6.4

As with other confidence intervals and hypothesis tests in OLS regression, the standard errors and bands just described depend on the assumption of independent and identically distributed errors. Figure 6.2 has cast doubt on this assumption, so the results in Figures 6.3 and 6.4 could be misleading.

Correlations

correlate obtains Pearson product-moment correlations between variables.

```
. correlate csat expense percent income high college
(obs=51)
```

	csat	expense	percent	income	high	college
csat	1.0000					
expense	-0.4663	1.0000				
percent	-0.8758	0.6509	1.0000			
income	-0.4713	0.6784	0.6733	1.0000		
high	0.0858	0.3133	0.1413	0.5099	1.0000	
college	-0.3729	0.6400	0.6091	0.7234	0.5319	1.0000

correlate uses only a subset of the data that has no missing values on any of the variables listed. (With these particular variables, that does not matter because no observations have missing values.) In this respect, the **correlate** command resembles **regress** , and given the same variable list, they will use the same subset of the data. Analysts not employing regression or other multivariable techniques, however, might prefer to find correlations based upon all of the observations available for each variable pair. The command **pwcorr** (pairwise correlation) accomplishes this, and can also furnish t-test probabilities for the null hypotheses that each individual correlation equals zero. In this example, the **star(.05)** option requests stars (*) marking correlations individually significant at the $\alpha = .05$ level.

```
. pwcorr csat expense percent income high college, sig star(.05)
```

	csat	expense	percent	income	high	college
csat	1.0000					
expense	-0.4663* 0.0006	1.0000				
percent	-0.8758* 0.0000	0.6509* 0.0000	1.0000			
income	-0.4713* 0.0005	0.6784* 0.0000	0.6733* 0.0000	1.0000		
high	0.0858 0.5495	0.3133* 0.0252	0.1413 0.3226	0.5099* 0.0001	1.0000	
college	-0.3729* 0.0070	0.6400* 0.0000	0.6091* 0.0000	0.7234* 0.0000	0.5319* 0.0001	1.0000

It is worth recalling, however, that if we drew many random samples from a population in which all variables really had 0 correlations, about 5% of the sample correlations would nonetheless test "statistically significant" at the .05 level. Analysts who review many individual hypothesis tests, such as those in a **pwcorr** matrix, to identify the handful that are significant at the .05 level, therefore run a much higher than .05 risk of making a Type I error. This problem is called the "multiple comparison fallacy." **pwcorr** offers two methods, Bonferroni and Šidák, for adjusting significance levels to take multiple comparisons into account. Of these, the Šidák method is more accurate. Significance-test probabilities are adjusted for the number of comparisons made.

```
. pwcorr csat expense percent income high college, sidak sig star(.05)
```

	csat	expense	percent	income	high	college
csat	1.0000					
expense	-0.4663* 0.0084	1.0000				
percent	-0.8758* 0.0000	0.6509* 0.0000	1.0000			
income	-0.4713* 0.0072	0.6784* 0.0000	0.6733* 0.0000	1.0000		
high	0.0858 1.0000	0.3133 0.3180	0.1413 0.9971	0.5099* 0.0020	1.0000	
college	-0.3729 0.1004	0.6400* 0.0000	0.6091* 0.0000	0.7234* 0.0000	0.5319* 0.0009	1.0000

Comparing the test probabilities in the table above with those of the previous **pwcorr** provides some idea of how much adjustment occurs. In general, the more variables we correlate, the more the adjusted probabilities will exceed their unadjusted counterparts. See the *Base Reference Manual*'s discussion of **oneway** for the formulas involved.

correlate itself offers several important options. Adding the **covariance** option produces a matrix of variances and covariances instead of correlations.

```
. correlate w x y z, covariance
```

Typing the following after a regression analysis displays the matrix of correlations between estimated coefficients, sometimes used to diagnose multicollinearity (see Chapter 7).

```
. correlate, _coef
```

The following command will display the estimated coefficients' variance–covariance matrix, from which standard errors are derived.

```
. correlate, _coef covariance
```

A Pearson correlation coefficient measures how well an OLS regression line fits a pair of variables. Such correlations consequently share the assumptions and weaknesses of OLS, and like OLS, should generally not be interpreted without a look at the corresponding scatterplots. Scatterplot matrices provide a quick way to do this, using the same organization as the correlation matrix. Figure 6.5 shows a scatterplot matrix corresponding to the **pwcorr** matrix given earlier. Only the lower-triangular half of the matrix is drawn, and plus signs are used as plotting symbols. We suppress *y* and *x*-axis labeling here to keep the graph uncluttered.

```
. graph matrix csat expense percent income high college,
    half msymbol(+) maxis(ylabel(none) xlabel(none))
```

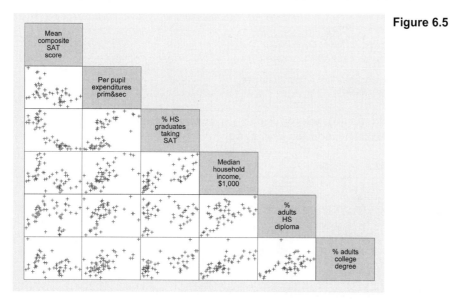

Figure 6.5

To obtain a scatterplot matrix corresponding to a **correlate** correlation matrix, from which all observations having missing values have been dropped, we would need to qualify the command. If all of the variables had some missing values, we could have excluded these with a "not missing" option (**!missing**):

```
. graph matrix csat expense percent income high college if
    !missing(csat, expense, income, high, college)
```

To reduce the likelihood of confusion and mistakes, it might make sense to create a new dataset keeping only those observations that have no missing values:

```
. keep if !missing(csat, expense, income, high, college)
. save nmvstate
```

In the example above, we immediately saved the reduced dataset with a new name, so as to avoid inadvertently writing over and losing the information in the old, more complete dataset. An alternative way to eliminate missing values uses **drop** instead of **keep**. The **missing** option evaluates to 1 if any of the listed variables' values are missing for an observation, and 0 otherwise.

```
. drop if missing(csat, expense, income, high, college) == 1
. save nmvstate
```

In addition to Pearson correlations, Stata can also calculate several rank-based correlations. These can be employed to measure associations between ordinal variables, or as an outlier-resistant alternative to Pearson correlation for measurement variables. To obtain the Spearman rank correlation between *csat* and *expense*, equivalent to the Pearson correlation if these variables were transformed into ranks, type

```
. spearman csat expense
```

```
 Number of obs =        51
Spearman's rho =    -0.4282
```

```
Test of Ho: csat and expense are independent
    Prob > |t| =        0.0017
```

Kendall's τ_a (tau-a) and τ_b (tau-b) rank correlations can be found easily for these data, although with larger datasets their calculation becomes slow:

```
. ktau csat expense
```

```
  Number of obs =        51
Kendall's tau-a =    -0.2925
Kendall's tau-b =    -0.2932
Kendall's score =      -373
    SE of score =    123.095   (corrected for ties)
```

```
Test of Ho: csat and expense are independent
     Prob > |z| =        0.0025   (continuity corrected)
```

For comparison, here is the Pearson correlation with its (unadjusted) *P*-value:

```
. pwcorr csat expense, sig
```

	csat	expense
csat	1.0000	
expense	-0.4663	1.0000
	0.0006	

In this example, both **spearman** (–.4282) and **pwcorr** (–.4663) yield higher correlations than **ktau** (–.2925 or –.2932). All three agree that null hypotheses of no association can be rejected.

Hypothesis Tests

Two types of hypothesis tests appear in **regress** output tables. As with other common hypothesis tests, they begin from the assumption that observations in the sample at hand were drawn randomly and independently from an infinitely large population.

1. Overall *F* test: The *F* statistic at the upper right in the regression table evaluates the null hypothesis that in the population, coefficients on all the model's *x* variables equal zero.

2. Individual *t* tests: The third and fourth columns of the regression table contain *t* tests for each individual regression coefficient. These evaluate the null hypotheses that in the population, the coefficient on each particular *x* variable equals zero.

The *t* test probabilities are two-sided. For one-sided tests, divide these *P*-values in half.

In addition to these standard F and t tests, Stata can perform F tests of user-specified hypotheses. The **test** command refers back to the most recently fitted model such as **anova** or **regress**. For example, individual t tests from the following regression report that neither the percent of adults with at least high school diplomas (*high*) nor the percent with college degrees (*college*) has a statistically significant individual effect on composite SAT scores.

```
. regress csat expense percent income high college
```

Conceptually, however, both predictors reflect the level of education attained by a state's population, and for some purposes we might want to test the null hypothesis that *both* have zero effect. To do this, we begin by repeating the multiple regression **quietly**, because we do not need to see its full output again. Then use the **test** command:

```
. quietly regress csat expense percent income high college
. test high college
```

```
 ( 1)   high = 0
 ( 2)   college = 0

       F(  2,    45) =    3.32
           Prob > F =    0.0451
```

Unlike the individual null hypotheses, the joint hypothesis that coefficients on *high* and *college* both equal zero can reasonably be rejected ($P = .0451$). Such tests on subsets of coefficients are useful when we have several conceptually related predictors or when individual coefficient estimates appear unreliable due to multicollinearity (Chapter 7).

test could duplicate the overall F test:

```
. test expense percent income high college
```

test could also duplicate the individual-coefficient tests:

```
. test expense
. test percent
. test income
```

and so forth. Applications of **test** more useful in advanced work include the following.

1. Test whether a coefficient equals a specified constant. For example, to test the null hypothesis that the coefficient on *income* equals 1 ($H_0: \beta_3 = 1$), instead of testing the usual null hypothesis that it equals 0 ($H_0: \beta_3 = 0$), type

    ```
    . test income = 1
    ```

2. Test whether two coefficients are equal. For example, the following command evaluates the null hypothesis $H_0: \beta_4 = \beta_5$

    ```
    . test high = college
    ```

3. Finally, **test** understands some algebraic expressions. We could request something like the following, which would test $H_0: \beta_3 = (\beta_4 + \beta_5) / 100$

```
. test income = (high + college)/100
```

Consult **help test** for more information and examples.

Dummy Variables

Categorical variables can become predictors in a regression when they are expressed as one or more {0,1} dichotomies called "dummy variables." For example, we have reason to suspect that regional differences exist in states' mean SAT scores. The **tabulate** command generates one dummy variable for each category of the tabulated variable if we include a **gen** (generate) option. Below, we create four dummy variables from the four-category variable *region*. The dummies are named *reg1*, *reg2*, *reg3* and *reg4*. *reg1* equals 1 for Western states and 0 for others; *reg2* equals 1 for Northeastern states and 0 for others; and so forth.

```
. tabulate region, gen(reg)
```

Geographical region	Freq.	Percent	Cum.
West	13	26.00	26.00
N. East	9	18.00	44.00
South	16	32.00	76.00
Midwest	12	24.00	100.00
Total	50	100.00	

```
. describe reg1-reg4
```

variable name	storage type	display format	value label	variable label
reg1	byte	%8.0g		region==West
reg2	byte	%8.0g		region==N. East
reg3	byte	%8.0g		region==South
reg4	byte	%8.0g		region==Midwest

```
. tabulate reg1
```

region==West	Freq.	Percent	Cum.
0	37	74.00	74.00
1	13	26.00	100.00
Total	50	100.00	

```
. tabulate reg2
```

region==N. East	Freq.	Percent	Cum.
0	41	82.00	82.00
1	9	18.00	100.00
Total	50	100.00	

Regressing *csat* on one dummy variable, *reg2* (Northeast), is equivalent to performing a two-sample *t* test of whether mean *csat* is the same across categories of *reg2*. That is, is the mean *csat* the same in the Northeast as in other U.S. states?

. **regress csat reg2**

Source	SS	df	MS		
Model	35191.4017	1	35191.4017		
Residual	177769.978	48	3703.54121		
Total	212961.38	49	4346.15061		

Number of obs =	50
F(1, 48) =	9.50
Prob > F =	0.0034
R-squared =	0.1652
Adj R-squared =	0.1479
Root MSE =	60.857

csat	Coef.	Std. Err.	t	P>|t|	[95% Conf. Interval]	
reg2	-69.0542	22.40167	-3.08	0.003	-114.0958	-24.01262
_cons	958.6098	9.504224	100.86	0.000	939.5002	977.7193

The dummy variable coefficient's *t* statistic ($t = -3.08$, $P = .003$) indicates a significant difference. According to this regression, mean SAT scores are 69.0542 points lower (because $b = -69.0542$) among Northeastern states. We get exactly the same result ($t = 3.08$, $P = .003$) from a simple *t* test, which also shows the means as 889.5556 (Northeast) and 958.6098 (other states), a difference of 69.0542.

. **ttest csat, by(reg2)**

Two-sample t test with equal variances

Group	Obs	Mean	Std. Err.	Std. Dev.	[95% Conf. Interval]	
0	41	958.6098	10.36563	66.37239	937.66	979.5595
1	9	889.5556	4.652094	13.95628	878.8278	900.2833
combined	50	946.18	9.323251	65.92534	927.4442	964.9158
diff		69.0542	22.40167		24.01262	114.0958

```
     diff = mean(0) - mean(1)                                  t =    3.0825
Ho: diff = 0                              degrees of freedom =        48

   Ha: diff < 0                  Ha: diff != 0                  Ha: diff > 0
Pr(T < t) = 0.9983        Pr(|T| > |t|) = 0.0034        Pr(T > t) = 0.0017
```

This conclusion proves spurious, however, once we control for the percentage of students taking the test. We do so with a multiple regression of *csat* on both *reg2* and *percent*.

```
. regress csat reg2 percent
```

Source	SS	df	MS
Model	174664.983	2	87332.4916
Residual	38296.3969	47	814.816955
Total	212961.38	49	4346.15061

```
Number of obs =      50
F( 2,    47) =  107.18
Prob > F      =  0.0000
R-squared     =  0.8202
Adj R-squared =  0.8125
Root MSE      =  28.545
```

csat	Coef.	Std. Err.	t	P>|t|	[95% Conf. Interval]	
reg2	57.52437	14.28326	4.03	0.000	28.79015	86.25858
percent	-2.793009	.2134796	-13.08	0.000	-3.222475	-2.363544
_cons	1033.749	7.270285	142.19	0.000	1019.123	1048.374

The Northeastern region variable *reg2* now has a statistically significant *positive* coefficient (b = 57.52437, $P < .0005$). The earlier negative relationship was misleading. Although mean SAT scores among Northeastern states really are lower, they are lower *because higher percentages of students take this test in the Northeast*. A smaller, more "elite" group of students, often less than 20% of high school seniors, take the SAT in many non-Northeast states where an alternative test, the ACT, is more prevalent. In all Northeastern states, however, large majorities (64% to 81%) take the SAT. Once we adjust for differences in the percentages taking the test, SAT scores actually tend to be higher in the Northeast.

To understand dummy variable regression results, it can help to write out the regression equation, substituting zeroes and ones. For Northeastern states, the equation is approximately

predicted *csat* = 1033.7 + 57.5*reg2* – 2.8*percent*
= 1033.7 + 57.5 × 1 – 2.8*percent*
= 1091.2 – 2.8*percent*

For other states, the predicted *csat* is 57.5 points lower at any given level of *percent*:

predicted *csat* = 1033.7 + 57.5 × 0 – 2.8*percent*
= 1033.7 – 2.8*percent*

Dummy variables in models such as this are termed "intercept dummy variables" because they describe a shift in the *y*-intercept or constant.

From a categorical variable with k categories we can define k dummy variables, but one of these will be redundant. Once we know a state's values on the West, Northeast, and Midwest dummy variables, for example, we can already guess its value on the South variable. For this reason, no more than $k - 1$ of the dummy variables (three, in the case of *region*) can be included in a regression. If we try to include all the possible dummies, Stata will automatically drop one because multicollinearity otherwise makes a unique answer impossible.

```
. regress csat reg1 reg2 reg3 reg4 percent
```

Source	SS	df	MS
Model	181378.099	4	45344.5247
Residual	31583.2811	45	701.850691
Total	212961.38	49	4346.15061

```
                                 Number of obs =       50
                                 F( 4,    45) =    64.61
                                 Prob > F     =   0.0000
                                 R-squared    =   0.8517
                                 Adj R-squared =  0.8385
                                 Root MSE     =   26.492
```

csat	Coef.	Std. Err.	t	P>|t|	[95% Conf. Interval]	
reg1	-23.77315	11.12578	-2.14	0.038	-46.18162	-1.364676
reg2	25.79985	16.96365	1.52	0.135	-8.366694	59.96639
reg3	-33.29951	10.85443	-3.07	0.004	-55.16146	-11.43757
reg4	(dropped)					
percent	-2.546058	.2140196	-11.90	0.000	-2.977116	-2.115001
_cons	1047.638	8.273625	126.62	0.000	1030.974	1064.302

The model's fit — including R^2, F tests, predictions, and residuals — remains essentially the same regardless of which dummy variable we (or Stata) choose to omit. Interpretation of the coefficients, however, occurs with reference to that omitted category. In this example, the Midwest dummy variable (*reg4*) was omitted. The regression coefficients on *reg1*, *reg2*, and *reg3* tell us that, at any given level of *percent*, the predicted mean SAT scores are approximately as follows:

23.8 points lower in the West (*reg1* = 1) than in the Midwest;

25.8 points higher in the Northeast (*reg2* = 1) than in the Midwest; and

33.3 points lower in the South (*reg3* = 1) than in the Midwest.

The West and South both differ significantly from the Midwest in this respect, but the Northeast does not.

An alternative command, **areg**, fits the same model without going through dummy variable creation. Instead, it "absorbs" the effect of a *k*-category variable such as *region*. The model's fit, F test on the absorbed variable, and other key aspects of the results are the same as those we could obtain through explicit dummy variables. Note, however, that **areg** does not provide estimates of the coefficients on individual dummy variables.

```
. areg csat percent, absorb(region)
```

Linear regression, absorbing indicators

```
                                 Number of obs =       50
                                 F( 1,    45) =   141.52
                                 Prob > F     =   0.0000
                                 R-squared    =   0.8517
                                 Adj R-squared =  0.8385
                                 Root MSE     =   26.492
```

csat	Coef.	Std. Err.	t	P>|t|	[95% Conf. Interval]	
percent	-2.546058	.2140196	-11.90	0.000	-2.977116	-2.115001
_cons	1035.445	8.38689	123.46	0.000	1018.553	1052.337

region	F(3, 45) =	9.465	0.000	(4 categories)

Although its output is less informative than regression with explicit dummy variables, **areg** does have two advantages. It speeds up exploratory work, providing quick feedback about whether a dummy variable approach is worthwhile. Secondly, when the variable of interest has many values, creating dummies for each of them could lead to too many variables or too large a model for our particular Stata configuration. **areg** thus works around the usual limitations on dataset and matrix size.

Explicit dummy variables have other advantages, however, including ways to model interaction effects. Interaction terms called "slope dummy variables" can be formed by multiplying a dummy times a measurement variable. For example, to model an interaction between Northeast/other region and *percent*, we create a slope dummy variable called *reg2perc*.

```
. generate reg2perc = reg2 * percent
(1 missing value generated)
. save states_new.dta
```

The last **save** command preserves our modified version of these data, under a new name, so the generated variables can be used again later in this chapter. The variable, *reg2perc*, equals *percent* for Northeastern states and zero for all other states. We can include this interaction term among the regression predictors:

```
. regress csat reg2 percent reg2perc
```

Source	SS	df	MS		Number of obs =	50
					F(3, 46) =	82.27
Model	179506.19	3	59835.3968		Prob > F =	0.0000
Residual	33455.1897	46	727.286733		R-squared =	0.8429
					Adj R-squared =	0.8327
Total	212961.38	49	4346.15061		Root MSE =	26.968

csat	Coef.	Std. Err.	t	P>|t|	[95% Conf. Interval]	
reg2	-241.3574	116.6278	-2.07	0.044	-476.1171	-6.597818
percent	-2.858829	.2032947	-14.06	0.000	-3.26804	-2.449618
reg2perc	4.179666	1.620009	2.58	0.013	.9187559	7.440576
_cons	1035.519	6.902898	150.01	0.000	1021.624	1049.414

The interaction is statistically significant ($t = 2.58$, $P = .013$). Because this analysis includes both intercept (*reg2*) and slope (*reg2perc*) dummy variables, it is worthwhile to write out the equations. The regression equation for Northeastern states is approximately

$$\text{predicted } csat = 1035.5 - 241.4reg2 - 2.9percent + 4.2reg2perc$$
$$= 1035.5 - 241.4 \times 1 - 2.9percent + 4.2 \times 1 \times percent$$
$$= 794.1 + 1.3percent$$

For other states it is

$$\text{predicted } csat = 1035.5 - 241.4 \times 0 - 2.9percent + 4.2 \times 0 \times percent$$
$$= 1035.5 - 2.9percent$$

An interaction implies that the effect of one variable changes depending on the values of some other variable. From this regression, it appears that *percent* has a relatively weak and positive effect among Northeastern states, whereas its effect is stronger and negative among the rest.

To visualize the results from a slope-and-intercept dummy variable regression, we have several graphing possibilities. Without even fitting the model, we could ask **lfit** to do the work as follows, with the results seen in Figure 6.6.

```
. label define reg2 0 "other regions" 1 "Northeast"

. label values reg2 reg2

. graph twoway lfit csat percent
      ||   scatter csat percent
      ||   , by(reg2, legend(off) note(""))
      ytitle("Mean composite SAT score")
```

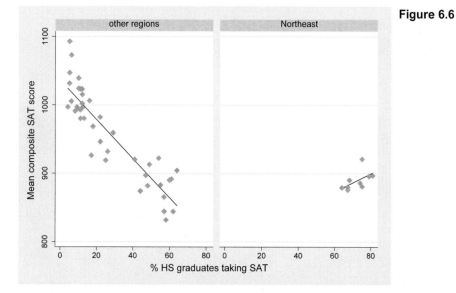

Figure 6.6

Alternatively, we could fit the regression model, calculate predicted values, and use those to draw a more refined plot such as Figure 6.7, on the following page. The **bands(50)** options with both **mspline** commands specify median splines based on 50 vertical bands, which is more than enough to cover the range of the data.

```
. quietly regress csat reg2 percent reg2perc

. predict yhat1
```

```
. graph twoway scatter csat percent if reg2 == 0

    || mspline yhat1 percent if reg2 == 0, clpattern(solid)
    bands(50)

    || scatter csat percent if reg2 == 1, msymbol(Sh)

    || mspline yhat1 percent if reg2 == 1, clpattern(solid)
    bands(50)

    || , ytitle("Composite mean SAT score")
    legend(order(1 3) label(1 "other regions")
       label(3 "Northeast states") position(12) ring(0))
```

Figure 6.7

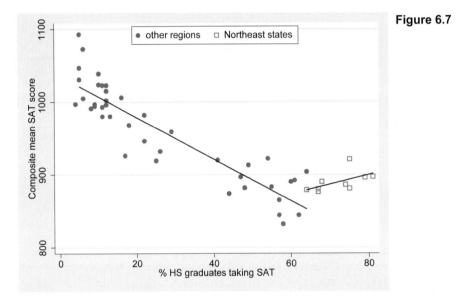

Figure 6.7 involves four overlays: two scatterplots (*csat* vs. *percent* for Northeast and other states) and two median-spline plots (connecting predicted values, *yhat1*, graphed against *percent* for Northeast and others). The Northeast states are plotted as hollow squares, **msymbol(Sh)**. **ytitle** and **legend** options simplify the *y*-axis title and the legend; in their default form, both would be crowded and unclear.

Figures 6.6 and 6.7 both show the striking difference, captured by our interaction effect, between Northeastern and other states. This raises the question of what other regional differences exist. Figure 6.8 explores this question by drawing a *csat–percent* scatterplot with different symbols for each of the four regions. In this plot, the Midwestern states, with one exception (Indiana), seem to have their own steeply negative regional pattern at the left side of the graph. Southern states are the most heterogeneous group.

```
. graph twoway scatter csat percent if reg1 ==1

    ||   scatter csat percent if reg2 ==1, msymbol(Sh)

    ||   scatter csat percent if reg3 == 1, msymbol(T)

    ||   scatter csat percent if reg4 == 1, msymbol(+)

    ||  , legend(position(1) ring(0) label(1 "West")
       label(2 "Northeast") label(3 "South") label(4 "Midwest"))
```

Figure 6.8

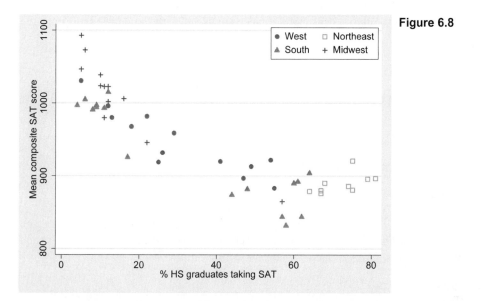

Automatic Categorical-Variable Indicators and Interactions

The **xi** (**ex**pand **i**nteractions) command simplifies the jobs of expanding multiple-category variables into sets of dummy and interaction variables, and including these as predictors in regression or other models. For example, in dataset *student2.dta* (introduced in Chapter 5) there is a four-category variable *year*, representing a student's year in college (freshman, sophomore, etc.). We could automatically create a set of three dummy variables by typing

```
. xi, prefix(ind) i.year
```

The three new dummy variables will be named *indyear_2*, *indyear_3*, and *indyear_4*. The **prefix( )** option specified the prefix used for their names. If we typed simply

```
. xi i.year
```

giving no **prefix( )** option, the names *_Iyear_2*, *_Iyear_3*, and *_Iyear_4* would be assigned (and any previously calculated variables with those names would be overwritten). Typing

```
. drop _I*
```

employs the wildcard * notation to drop all variables that have names beginning with _I.

By default, **xi** omits the lowest value of the categorical variable when creating dummies, but this can be controlled. Typing the command

```
. char _dta[omit] prevalent
```

will cause subsequent **xi** commands to automatically omit the most prevalent category (note the use of square brackets). **char _dta[ ]** preferences are saved with the data. To restore the default, type

```
. char _dta[omit]
```

Typing

```
. char year[omit] 3
```

would omit *year* 3. To restore the default, type

```
. char year[omit]
```

xi can also create interaction terms involving two categorical variables, or one categorical and one measurement variable. For example, we could create a set of interaction terms for *year* and *gender* by typing

```
. xi i.year*i.gender
```

From the four categories of *year* and the two categories of *gender*, this **xi** command creates seven new variables: four dummy variables and three interactions. Because their names all begin with _I , we can use the wildcard notation _I* to **describe** these variables.

```
. describe _I*
```

variable name	storage type	display format	value label	variable label
_Iyear_2	byte	%8.0g		year==2
_Iyear_3	byte	%8.0g		year==3
_Iyear_4	byte	%8.0g		year==4
_Igender_1	byte	%8.0g		gender==1
_IyeaXgen_2_1	byte	%8.0g		year==2 & gender==1
_IyeaXgen_3_1	byte	%8.0g		year==3 & gender==1
_IyeaXgen_4_1	byte	%8.0g		year==4 & gender==1

To create interaction terms for categorical variable *year* and measurement variable *drink* (33-point drinking behavior scale), type

```
. xi i.year*drink
```

Six new variables result: three dummy variables for *year*, and three interaction terms representing each of the *year* dummies times *drink*. For example, for a sophomore student *_Iyear2* = 1 and *_IyeaXdrink_2* = 1×*drink* = *drink*. For a junior student, *_Iyear2* = 0 and *_IyearXdrink_2* = 0×*drink* =0; also *_Iyear3* = 1 and *_IyeaXdrink_3* = 1×*drink* = *drink*, and so forth.

```
. describe _Iyea*

                storage  display   value
variable name   type     format    label     variable label
───────────────────────────────────────────────────────────────
_Iyear_2        byte     %8.0g               year==2
_Iyear_3        byte     %8.0g               year==3
_Iyear_4        byte     %8.0g               year==4
_IyeaXdrink_2   byte     %8.0g               (year==2)*drink
_IyeaXdrink_3   byte     %8.0g               (year==3)*drink
_IyeaXdrink_4   byte     %8.0g               (year==4)*drink
```

The real convenience of **xi** comes from its ability to generate dummy variables and interactions automatically within a regression or other model-fitting command, so that explicit generation of indicators as shown above is not needed. For example, to regress variable *gpa* (student's college grade point average) on *drink* and a set of dummy variables for *year*, simply type

```
. xi:  regress gpa drink i.year
```

This command automatically creates the necessary dummy variables, following the same rules described above. Similarly, to regress *gpa* on *drink*, *year*, and the interaction of *drink* and *year* (which will automatically include their main effects as well), type

```
. xi:  regress gpa i.year*drink

i.year          _Iyear_1-4        (naturally coded; _Iyear_1 omitted)
i.year*drink    _IyeaXdrink_#     (coded as above)

      Source │      SS        df       MS              Number of obs =     218
─────────────┼──────────────────────────────          F(  7,   210) =    3.75
       Model │ 5.08865901      7   .726951288          Prob > F      =  0.0007
    Residual │ 40.6630801    210   .193633715          R-squared     =  0.1112
─────────────┼──────────────────────────────          Adj R-squared =  0.0816
       Total │ 45.7517391    217   .210837507          Root MSE      =  .44004

─────────────┬───────────────────────────────────────────────────────────────
         gpa │     Coef.    Std. Err.      t     P>|t|     [95% Conf. Interval]
─────────────┼───────────────────────────────────────────────────────────────
    _Iyear_2 │ -.5839268    .314782     -1.86    0.065    -1.204464    .0366107
    _Iyear_3 │ -.2859424   .3044178     -0.94    0.349    -.8860487    .3141639
    _Iyear_4 │ -.2203783   .2939595     -0.75    0.454     -.799868    .3591114
       drink │ -.0285369   .0140402     -2.03    0.043    -.0562146   -.0008591
 _IyeaXdrin~2 │  .0199977   .0164436      1.22    0.225    -.0124179    .0524133
 _IyeaXdrin~3 │  .0108977    .016348      0.67    0.506    -.0213297     .043125
 _IyeaXdrin~4 │  .0104239    .016369      0.64    0.525    -.0218446    .0426925
       _cons │  3.432132   .2523984     13.60    0.000     2.934572    3.929691
─────────────┴───────────────────────────────────────────────────────────────
```

The **xi:** command can be applied in the same way before many other model-fitting procedures such as **logistic** (Chapter 10). In general, it allows us to include predictor (right-hand-side) variables such as the following, without first creating the actual dummy variable or interaction terms.

`i.catvar`	Creates j–1 dummy variables representing the j categories of *catvar*.
`i.catvar1*i.catvar2`	Creates j–1 dummy variables representing the j categories of *catvar1*; k–1 dummy variables from the k categories of *catvar2*; and $(j–1)(k–1)$ interaction variables (dummy × dummy).
`i.catvar*measvar`	Creates j–1 dummy variables representing the j categories of *catvar*, and j–1 variables representing interactions with the measurement variable (dummy × *measvar*).

After any **xi** command, the new variables remain in the dataset.

Stepwise Regression

With the regional dummy variable terms we added earlier to the state-level data in *states.dta*, (and you may have saved as *states_new.dta*) we have many possible predictors of *csat*. This results in an overly complicated model with several coefficients statistically indistinguishable from zero.

```
. regress csat expense percent income college high reg1 reg2
    reg2perc reg3
```

Source	SS	df	MS		Number of obs =	50
					F(9, 40) =	49.51
Model	195420.517	9	21713.3908		Prob > F =	0.0000
Residual	17540.863	40	438.521576		R-squared =	0.9176
					Adj R-squared =	0.8991
Total	212961.38	49	4346.15061		Root MSE =	20.941

csat	Coef.	Std. Err.	t	P>|t|	[95% Conf. Interval]	
expense	-.0022508	.0041333	-0.54	0.589	-.0106046	.006103
percent	-2.93786	.2302596	-12.76	0.000	-3.403232	-2.472488
income	-.4919133	1.025469	-0.48	0.634	-2.564464	1.580638
college	3.900087	1.719409	2.27	0.029	.4250318	7.375142
high	2.175542	1.171767	1.86	0.071	-.192688	4.543771
reg1	-33.78456	9.302983	-3.63	0.001	-52.58659	-14.98253
reg2	-143.5149	101.1244	-1.42	0.164	-347.8949	60.86509
reg2perc	2.506616	1.404483	1.78	0.082	-.3319506	5.345183
reg3	-8.799205	12.54658	-0.70	0.487	-34.15679	16.55838
_cons	839.2209	76.35942	10.99	0.000	684.8927	993.549

We might now try to simplify this model, dropping first that predictor with the highest *t* probability (*income*, $P = .634$), then refitting the model and deciding whether to drop something else. Through this process of backward elimination, we seek a more parsimonious model; one that is simpler but fits almost as well. Ideally, this strategy is pursued with attention to both the statistical results and to the substantive or theoretical implications of keeping or discarding certain variables.

For analysts in a hurry, stepwise methods provide ways to automate the process of model selection. They work either by subtracting predictors from a complicated model, or by adding predictors to a simpler one according to some pre-set statistical criteria. Stepwise methods cannot consider the substantive or theoretical implications of their choices, nor can they do much troubleshooting to evaluate possible weaknesses in the models produced at each step. They produce badly biased models in many instances due to overfitting. Despite their well-known limitations, stepwise methods meet some practical needs and have been widely used.

For automatic backward elimination, we type our **regress** command with a **stepwise** prefix. The example below includes all of our possible predictor variables, and a maximum *P* value required to retain them. Setting the *P*-to-retain criteria as **pr(.05)** ensures that only predictors having coefficients that are significantly different from zero at the .05 level will be kept in the model.

```
. stepwise, pr(.05): regress csat expense percent income college
        high reg1 reg2 reg2perc reg3
```

```
                            begin with full model
p = 0.6341 >= 0.0500  removing income
p = 0.5273 >= 0.0500  removing reg3
p = 0.4215 >= 0.0500  removing expense
p = 0.2107 >= 0.0500  removing reg2
```

Source	SS	df	MS		
Model	194185.761	5	38837.1521	Number of obs =	50
Residual	18775.6194	44	426.718624	F(5, 44) =	91.01
				Prob > F =	0.0000
				R-squared =	0.9118
				Adj R-squared =	0.9018
Total	212961.38	49	4346.15061	Root MSE =	20.657

csat	Coef.	Std. Err.	t	P>|t|	[95% Conf. Interval]	
reg1	-30.59218	8.479395	-3.61	0.001	-47.68128	-13.50308
percent	-3.119155	.1804553	-17.28	0.000	-3.482839	-2.755471
reg2perc	.5833272	.1545969	3.77	0.000	.2717577	.8948967
college	3.995495	1.359331	2.94	0.005	1.255944	6.735046
high	2.231294	.8178968	2.73	0.009	.5829312	3.879657
_cons	806.672	49.98744	16.14	0.000	705.9289	907.4151

stepwise dropped first *income*, then *reg3*, *expense*, and finally *reg2* before settling on the final model. Although it has four fewer coefficients, this final model has almost the same R^2 (.9118 versus .9176) and a higher R^2_a (.9018 versus .8991) compared with the earlier version.

If, instead of a *P*-to-retain, **pr(.05)**, we specify a *P*-to-enter value such as **pe(.05)**, then **stepwise** performs forward inclusion (starting with an "empty" or constant-only model) instead of backward elimination. Other stepwise options include hierarchical selection and locking certain predictors into the model. For example, the following command specifies that the first term (*x1*) should be locked into the model and not subject to possible removal:

```
. stepwise, pr(.05) lockterm1: regress y x1 x2 x3
```

The following command calls for forward inclusion of any predictors found significant at the .10 level, but with variables *x4*, *x5*, and *x6* treated as one unit, either entered or left out together:

```
. stepwise, pe(.10):  regress y x1 x2 x3 (x4 x5 x6)
```

The following command invokes hierarchical backward elimination with a *P* = .20 criterion:

```
. stepwise, pr(.20) hier:   regress y x1 x2 x3 (x4 x5 x6) x7
```

The **hier** option specifies that the terms are ordered; consider dropping the last term (*x7*) first, and stop if it is not dropped. If *x7* is dropped, next consider the second-to-last term *(x4 x5 x6)*, and so forth.

The **stepwise** prefix works in a similar way with many other modeling commands, including **glm**, **logit**, **nbreg**, and **qreg**. Type **help stepwise** for a complete list, and other details about its options and logic.

Polynomial Regression

Earlier in this chapter, Figures 6.1 and 6.2 revealed an apparently curvilinear relationship between mean composite SAT scores (*csat*) and the percentage of high school seniors taking the test (*percent*). Figure 6.6 illustrated one way to model the upturn in SAT scores at high *percent* values: as a phenomenon peculiar to the Northeastern states. That interaction model fit reasonably well ($R^2_a = .8327$). But Figure 6.9, a residuals versus predicted values plot for the interaction model, still exhibits signs of trouble. Residuals appear to trend upwards at both high and low predicted values.

```
. quietly regress csat reg2 percent reg2perc
```

```
. rvfplot, yline(0)
```

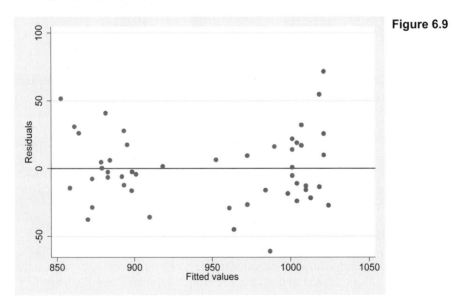

Figure 6.9

Chapter 8 presents a variety of techniques for curvilinear and nonlinear regression. "Curvilinear regression" here refers to intrinsically linear OLS regressions (for example, **regress**) that include nonlinear transformations of the original y or x variables. Although curvilinear regression fits a curved model with respect to the original data, this model remains linear in the transformed variables. (Nonlinear regression, also discussed in Chapter 8, applies non-OLS methods to fit models that cannot be linearized through transformation.)

One simple type of curvilinear regression, called polynomial regression, often succeeds in fitting U or inverted-U shaped curves. It includes as predictors both an independent variable and its square (and possibly higher powers if necessary). Because the *csat–percent* relationship appears somewhat U-shaped, we generate a new variable equal to *percent* squared, then include *percent* and *percent*^{2} as predictors of *csat*. Figure 6.10 graphs the resulting curve.

```
. generate percent2 = percent^2
```

```
. regress csat percent percent2
```

Source	SS	df	MS		
Model	193721.829	2	96860.9146		
Residual	30292.6806	48	631.097513		
Total	224014.51	50	4480.2902		

	Number of obs = 51
	F(2, 48) = 153.48
	Prob > F = 0.0000
	R-squared = 0.8648
	Adj R-squared = 0.8591
	Root MSE = 25.122

csat	Coef.	Std. Err.	t	P>|t|	[95% Conf. Interval]	
percent	-6.111993	.6715406	-9.10	0.000	-7.462216	-4.76177
percent2	.0495819	.0084179	5.89	0.000	.0326566	.0665072
_cons	1065.921	9.285379	114.80	0.000	1047.252	1084.591

```
. predict yhat2
(option xb assumed; fitted values)
```

```
. graph twoway mspline yhat2 percent, bands(50)
        ||   scatter csat percent
        ||   , legend(off) ytitle("Mean composite SAT score")
```

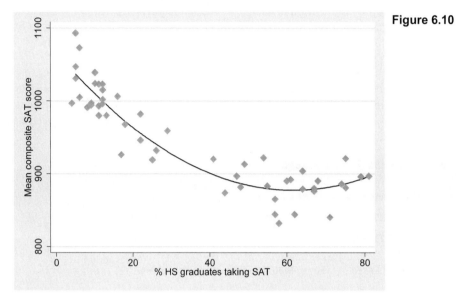

Figure 6.10

If we only wanted to see the graph, and did not need the regression analysis, there is a quicker way to achieve this. The command **graph twoway qfit** fits a quadratic (second-order polynomial) regression model; **qfitci** draws confidence bands as well. For example, a curve similar to Figure 6.10 could have been obtained by typing

```
. graph twoway qfit csat percent
        ||   scatter csat percent
```

The polynomial model in Figure 6.10 matches the data slightly better than our interaction model in Figure 6.6 (R^2_a = .8591 versus .8327). Because the curvilinear pattern is now less striking in a residual versus predicted values plot (Figure 6.11), the usual assumption of independent, identically distributed errors also appears more plausible with respect to this polynomial model.

```
. quietly regress csat percent percent2

. rvfplot, yline(0)
```

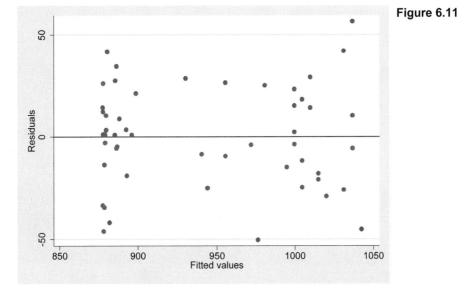

Figure 6.11

In Figures 6.7 and 6.10, we have two alternative models for the observed upturn in SAT scores at high levels of student participation. Statistical evidence seems to lean towards the polynomial model at this point. For serious research, however, we ought to choose between similar-fitting alternative models on substantive as well as statistical grounds. Which model seems more useful, or makes more sense? Which, if either, regression model suggests or corresponds to a good real-world explanation for the upturn in test scores at high levels of student participation?

Although it can closely fit sample data, polynomial regression also has important statistical weaknesses. The different powers of x might be highly correlated with each other, giving rise to multicollinearity. Furthermore, polynomial regression tends to track observations that have unusually large positive or negative x values, so a few data points can exert disproportionate influence on the results. For both reasons, polynomial regression results can sometimes be sample-specific, fitting one dataset well but generalizing poorly to other data. Chapter 7 takes a second look at this example, using tools that check for potential problems.

7

Regression Diagnostics

Do the data give us any reason to distrust our regression results? Can we find better ways to specify the model, or to estimate its parameters? Careful diagnostic work, checking for potential problems and evaluating the plausibility of key assumptions, forms a crucial step in modern data analysis. We fit an initial model, but then look closely at our results for signs of trouble or ways in which the model needs improvement. Many of the general methods introduced in earlier chapters, such as scatterplots, box plots, normality tests, or just sorting and listing the data, prove useful for troubleshooting. Stata also provides a toolkit of specialized diagnostic techniques designed for this purpose.

Autocorrelation, a complication that often affects regression with time series data, is not covered in this chapter. Chapter 13, Time Series Analysis, introduces Stata's library of time series procedures including Durbin–Watson tests, autocorrelation graphs, lag operators, and time-series regression modeling.

Regression diagnostic procedures are grouped under the following menus:

Statistics > Linear models and related > Regression diagnostics

Statistics > Postestimation

Graphics > Regression diagnostic plots

Example Commands

The commands illustrated in this section all assume that we have just fit a model using either **anova** or **regress**. The commands' results refer back to that model. These followup commands are of three basic types. Type **help regress postestimation** or **help diagnostic plots** for complete lists, and links to more detailed descriptions of particular commands.

1. **predict** options that generate new variables containing case statistics such as predicted values, residuals, standard errors, and influence statistics.
2. Diagnostic tests for statistical problems such as heteroskedasticity, multicollinearity, or specification errors.
3. Diagnostic plots such as added-variable or leverage plots, residual-versus-fitted plots, residual-versus-predictor plots, and component-versus-residual plots.

predict Options

. predict *new*, cooksd

Generates a new variable equal to Cook's distance D, summarizing how much each observation influences the fitted model.

. predict *new*, covratio

Generates a new variable equal to Belsley, Kuh, and Welsch's *COVRATIO* statistic. *COVRATIO* measures the ith case's influence upon the variance–covariance matrix of the estimated coefficients.

. predict *DFx1*, dfbeta(*x1*)

Generates *DFBETA* case statistics measuring how much each observation affects the coefficient on predictor $x1$. To create a complete set of *DFBETA*s for all predictors in the model, simply type the command **dfbeta** without arguments.

. predict *new*, dfits

Generates *DFITS* case statistics, summarizing the influence of each observation on the fitted model (similar in purpose to Cook's D and Welsch's W).

Diagnostic Tests

. estat hettest

Performs Cook and Weisberg's test for heteroskedasticity. If we have reason to suspect that heteroskedasticity is a function of a particular predictor $x1$, we could focus on that predictor by typing **estat hettest** *x1*. Type **help regress postestimation** for a complete list of options available after **regress**. Different postestimation choices exist for other modeling methods.

. estat ovtest, rhs

Performs the Ramsey regression specification error test (*RESET*) for omitted variables. The option **rhs** calls for using powers of the right-hand-side variables, instead of powers of predicted y (default).

. estat vif

Calculates variance inflation factors to check for multicollinearity.

. estat dwatson

Calculates the Durbin–Watson test for first-order autocorrelation in time series (**tsset**) data. Chapter 13 gives examples of this and other time series procedures.

Diagnostic Plots

. acprplot *x1*, mspline msopts(bands(7))

Constructs an augmented component-plus-residual plot (also known as an augmented partial residual plot), often better than **cprplot** in screening for nonlinearities. The options **mspline msopts(bands(7))** call for connecting with line segments the cross-medians of seven vertical bands. Alternatively, we might ask for a lowess-smoothed curve with bandwidth 0.5 by specifying the options **lowess lsopts(bwidth(.5))**.

. **avplot** *x1*

Constructs an added-variable plot (also called a partial-regression or leverage plot) showing the relationship between *y* and *x1*, both adjusted for other *x* variables. Such plots help to notice outliers and influence points.

. **avplots**

Draws and combines in one image all the added-variable plots from the recent **anova** or **regress**.

. **cprplot** *x1*

Constructs a component-plus-residual plot (also known as a partial-residual plot) showing the adjusted relationship between *y* and predictor *x1*. Such plots help detect nonlinearities in the data.

. **lvr2plot**

Constructs a leverage-versus-squared-residual plot (also known as an L–R plot).

. **rvfplot**

Graphs the residuals versus the fitted (predicted) values of *y*.

. **rvpplot** *x1*

Graphs the residuals against values of predictor *x1*.

SAT Score Regression, Revisited

Diagnostic techniques have been described as tools for "regression criticism" because they help us examine our regression models for possible flaws and for ways that the models could be improved. In this spirit, we return now to the state Scholastic Aptitude Test regressions of Chapter 6 (*states.dta*). A three-predictor model explains about 92% of the variance in mean state SAT scores. The predictors are *percent* (percent of high school graduates taking the test), *percent2* (*percent* squared), and *high* (percent of adults with a high school diploma).

. **generate** *percent2 = percent^2*

. **regress** *csat percent percent2 high*

Source	SS	df	MS		Number of obs =	51
					F(3, 47) =	193.37
Model	207225.103	3	69075.0343		Prob > F =	0.0000
Residual	16789.4069	47	357.221424		R-squared =	0.9251
					Adj R-squared =	0.9203
Total	224014.51	50	4480.2902		Root MSE =	18.9

csat	Coef.	Std. Err.	t	P>|t|	[95% Conf. Interval]	
percent	-6.520312	.5095805	-12.80	0.000	-7.545455	-5.495168
percent2	.0536555	.0063678	8.43	0.000	.0408452	.0664659
high	2.986509	.4857502	6.15	0.000	2.009305	3.963712
_cons	844.8207	36.63387	23.06	0.000	771.1228	918.5185

The regression equation is

$$\text{predicted } csat = 844.82 - 6.52 percent + .05 percent2 + 2.99 high$$

The scatterplot matrix in Figure 7.1 depicts interrelations among these four variables. As noted in Chapter 6, the squared term *percent2* allows our regression model to fit the visibly curvilinear relationship between *csat* and *percent*.

```
. graph matrix percent percent2 high csat, half msymbol(+)
```

Figure 7.1

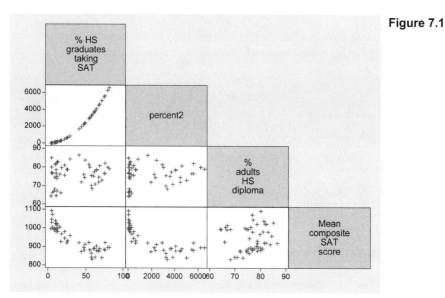

Several post-regression hypothesis tests perform checks on the model specification. The omitted-variables test **estat ovtest** essentially regresses y on the x variables, and also the second, third, and fourth powers of predicted y (after standardizing $\hat{y}$ to have mean 0 and variance 1). It then performs an F test of the null hypothesis that all three coefficients on those powers of $\hat{y}$ equal zero. If we reject this null hypothesis, further polynomial terms would improve the model. With the *csat* regression, we need not reject the null hypothesis.

```
. estat ovtest
```

```
Ramsey RESET test using powers of the fitted values of csat
      Ho:  model has no omitted variables
                 F(3, 44) =      1.48
                 Prob > F =      0.2319
```

A heteroskedasticity test, **estat hettest**, tests the assumption of constant error variance by examining whether squared standardized residuals are linearly related to $\hat{y}$ (see Cook and Weisberg 1994 for discussion and example). Results from the *csat* regression suggest that in this instance we should reject the null hypothesis of constant variance.

```
. estat hettest
```

```
Breusch-Pagan / Cook-Weisberg test for heteroskedasticity
        Ho: Constant variance
        Variables: fitted values of csat

        chi2(1)      =      4.86
        Prob > chi2  =    0.0274
```

"Significant" heteroskedasticity implies that our standard errors and hypothesis tests might be invalid. Figure 7.2, in the next section, shows why this result occurs.

Diagnostic Plots

Chapter 6 demonstrated how **predict** can create new variables holding residual and predicted values after a **regress** command. To obtain these values from our regression of *csat* on *percent*, *percent2*, and *high*, we type the two commands:

```
. predict yhat3
. predict e3, resid
```

The new variables named *e3* (residuals) and *yhat3* (predicted values) could be displayed in a residual-versus-predicted graph by typing **graph twoway scatter** *e3 yhat*, **yline(0)**. The **rvfplot** (residual-versus-fitted) command obtains such graphs in a single step. The version in Figure 7.2 includes a horizontal line at 0 (the residual mean), which helps in reading such plots.

```
. rvfplot, yline(0)
```

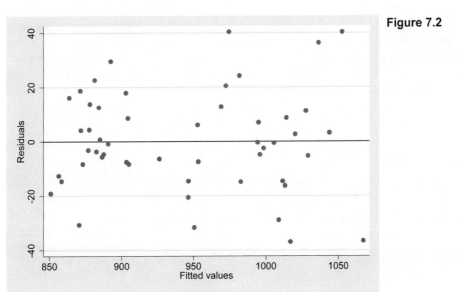

Figure 7.2

Figure 7.2 shows residuals symmetrically distributed around 0 (symmetry is consistent with the normal-errors assumption), and with no evidence of outliers or curvilinearity. The dispersion of the residuals appears somewhat greater for above-average predicted values of *y*, however, which is why **estat hettest** earlier rejected the constant-variance hypothesis.

Residual-versus-fitted plots provide a one-graph overview of the regression residuals. For more detailed study, we can plot residuals against each predictor variable separately through a series of "residual-versus-predictor" commands. To graph the residuals against predictor *high* (not shown), type

```
. rvpplot high
```

The one-variable graphs described in Chapter 3 can also be employed for residual analysis. For example, we could use box plots to check the residuals for outliers or skew, or quantile–normal plots to evaluate the assumption of normal errors.

Added-variable plots are valuable diagnostic tools known by different, names including partial-regression leverage plots, adjusted partial residual plots, or adjusted variable plots. They depict the relationship between *y* and one *x* variable, adjusting for the effects of other *x* variables. If we regressed *y* on *x2* and *x3*, and likewise regressed *x1* on *x2* and *x3*, then took the residuals from each regression and graphed these residuals in a scatterplot, we would obtain an added-variable plot for the relationship between *y* and *x1*, adjusted for *x2* and *x3*. An **avplot** command performs the necessary calculations automatically. We can draw the adjusted-variable plot for predictor *high*, for example, just by typing

```
. avplot high
```

Speeding the process further, we could type **avplots** to obtain a complete set of tiny added-variable plots with each of the predictor variables in the preceding regression. Figure 7.3 shows the results from the regression of *csat* on *percent*, *percent2*, and *high*. The lines drawn in added-variable plots have slopes equal to the corresponding partial regression coefficients. For example, the slope of the line at lower left in Figure 7.3 equals 2.99, which is the coefficient on *high*.

`. avplots`

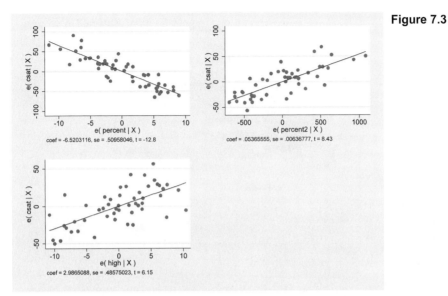

Figure 7.3

Added-variable plots help to uncover observations exerting a disproportionate influence on the regression model. In simple regression with one x variable, ordinary scatterplots suffice for this purpose. In multiple regression, however, the signs of influence become more subtle. An observation with an unusual combination of values on several x variables might have high leverage, or potential to influence the regression, even though none of its individual x values is unusual by itself. High-leverage observations show up in added-variable plots as points horizontally distant from the rest of the data. We see no such problems in Figure 7.3, however.

If outliers appear, we might identify which observations these are by including observation labels for the markers in an added-variable plot. This is done using the **mlabel( )** option, just as with scatterplots. Figure 7.4 on the next page illustrates using state names (values of the string variable *state*) as labels. Although such labels tend to overprint each other where the data are dense, individual outliers remain more readable.

```
. avplot high, mlabel(state)
```

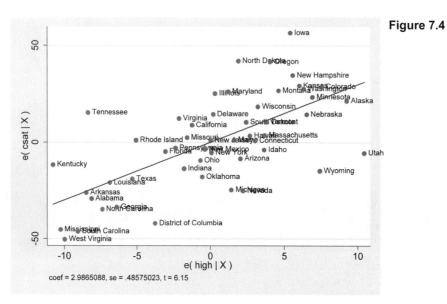

Figure 7.4

Component-plus-residual plots, produced by commands of the form **cprplot** *x1*, take a different approach to graphing multiple regression. The component-plus residual plot for variable *x1* graphs each observation's residual plus its component predicted from *x1*

$$e_i + b_1 x1_i$$

against values of *x1*. Such plots might help diagnose nonlinearities and suggest alternative functional forms. An augmented component-plus-residual plot (Mallows 1986) works somewhat better, although both types often seem inconclusive. Figure 7.5 shows an augmented component-plus-residual plot from the regression of *csat* on *percent*, *percent2*, and *high*.

`. acprplot `*`high`*`, lowess`

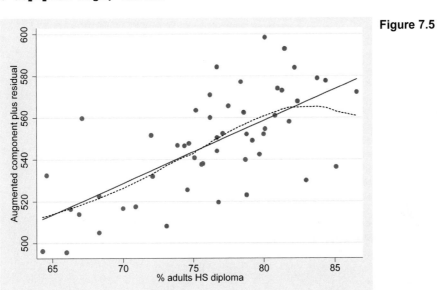

Figure 7.5

The straight line in Figure 7.5 corresponds to the regression model. The curved line reflects lowess smoothing based on the default bandwidth of .5, or half the data. The curve's downturn at far right can be disregarded as a lowess artifact because only a few cases determine its location toward the extremes (see Chapter 8). If more central parts of the lowess curve showed a systematically curved pattern, departing from the linear regression model, we would have reason to doubt the model's adequacy. In Figure 7.5, however, the component-plus-residuals medians closely follow the regression model. This plot reinforces the conclusion we reached earlier from Figure 7.2, that the present regression model adequately accounts for all nonlinearity visible in the raw data (Figure 7.1), leaving none apparent in its residuals.

As its name implies, a leverage-versus-squared-residuals plot graphs leverage (hat matrix diagonals) against the residuals squared. Figure 7.6 shows such a plot for the *csat* regression. To identify individual outliers, we label the markers with the values of *state*. The option **mlabsize(medsmall)** calls for "medium small" marker labels, somewhat larger than the default size of "small." (See **help testsizestyle** for a list of other choices.) Most of the state names form a jumble at lower left in Figure 7.6, but a few outliers stand out.

. `lvr2plot, mlabel(`*`state`*`) mlabsize(medsmall)`

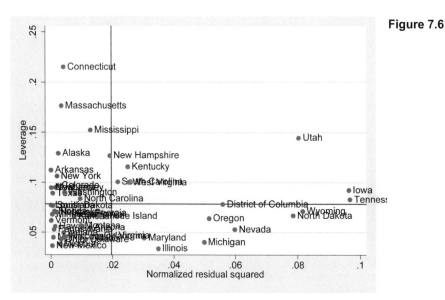

Figure 7.6

Lines in a leverage-versus-squared-residuals plot mark the means of leverage (horizontal line) and squared residuals (vertical line). Leverage tells us how much potential for influencing the regression an observation has, based on its particular combination of *x* values. Extreme *x* values or unusual combinations give an observation high leverage. A large squared residual indicates an observation with *y* value much different from that predicted by the regression model. Connecticut, Massachusetts, and Mississippi have the greatest potential leverage, but the model fits them relatively well. (This is not necessarily good. Sometimes, although not here, high-leverage observations exert so much influence that they control the regression, and it *must* fit them well.) Iowa and Tennessee are poorly fit, but have less potential influence. Utah stands out as one observation that is both ill fit and potentially influential. We can read its values by listing just this state. Because *state* is a string variable, we enclose the value "Utah" in double quotes.

. `list `*`csat yhat3 percent high e3`*` if `*`state`*` == "Utah"`

	csat	yhat3	percent	high	e3
45.	1031	1067.712	5	85.1	-36.71239

Only 5% of Utah students took the SAT, and 85.1% of the state's adults graduated from high school. This unusual combination of near-extreme values on both *x* variables is the source of the state's leverage, and leads our model to predict mean SAT scores 36.7 points higher than what Utah students actually achieved. To see exactly how much difference this one observation makes, we could repeat the regression using Stata's "not equal to" qualifier != to set Utah aside.

```
. regress csat percent percent2 high if state != "Utah"
```

Source	SS	df	MS
Model	201097.423	3	67032.4744
Residual	15214.0968	46	330.741235
Total	216311.52	49	4414.52082

Number of obs =	50
F(3, 46) =	202.67
Prob > F =	0.0000
R-squared =	0.9297
Adj R-squared =	0.9251
Root MSE =	18.186

csat	Coef.	Std. Err.	t	P>|t|	[95% Conf. Interval]	
percent	-6.778706	.5044217	-13.44	0.000	-7.794054	-5.763357
percent2	.0563562	.0062509	9.02	0.000	.0437738	.0689387
high	3.281765	.4865854	6.74	0.000	2.302319	4.26121
_cons	827.1159	36.17138	22.87	0.000	754.3067	899.9252

In the $n = 50$ (instead of $n = 51$) regression, all three coefficients strengthened a bit because we deleted an ill-fit observation. The general conclusions remain unchanged, however.

Chambers et al. (1983) and Cook and Weisberg (1994) provide more detailed examples and explanations of diagnostic plots and other graphical methods for data analysis.

Diagnostic Case Statistics

After using **regress** or **anova**, we can obtain a variety of diagnostic statistics through the **predict** command, as listed in Chapter 6 or in the dialog box for
    Statistics > Postestimation > Predictions, residuals, etc.
The variables created by **predict** are case statistics, meaning that they have values for each observation in the data. Diagnostic work often begins by calculating the predicted values and residuals.

There is some overlap in purpose among other **predict** statistics. Many attempt to measure how much each observation influences regression results. "Influencing regression results," however, could refer to several different things: effects on the y-intercept, on a particular slope coefficient, on all the slope coefficients, or on the estimated standard errors, for example. Consequently, we have a variety of alternative case statistics designed to measure influence.

Standardized and studentized residuals (**rstandard** and **rstudent**) help to identify outliers among the residuals — observations that particularly contradict the regression model. Studentized residuals have the most straightforward interpretation. They correspond to the t statistic we would obtain by including in the regression a dummy predictor coded 1 for that observation and 0 for all others. Thus, they test whether a particular observation significantly shifts the y-intercept.

Hat matrix diagonals (**hat**) measure leverage, meaning the potential to influence regression coefficients. Observations possess high leverage when their x values (or their combination of x values) are unusual.

Several other statistics measure actual influence on coefficients. *DFBETA*s indicate by how many standard errors the coefficient on $x1$ would change if observation i were dropped from the regression. These can be obtained for a single predictor, *x1*, in either of two ways: through the **predict** option **dfbeta(*x1*)** or through the command **dfbeta**.

Cook's *D* (**cooksd**), Welsch's distance (**welsch**), and *DFITS* (**dfits**), unlike *DFBETA*, all summarize how much observation *i* influences the regression model as a whole — or equivalently, how much observation *i* influences the set of predicted values. *COVRATIO* measures the influence of the *i*th observation on the estimated standard errors. Below we generate a full set of diagnostic statistics including *DFBETA*s for all three predictors. Note that **predict** supplies variable labels automatically for the variables it creates, but **dfbeta** does not. We begin by repeating our original regression to ensure that these post-regression diagnostics refer to the proper (*n* = 51) model.

```
. quietly regress csat percent percent2 high
. predict standard, rstandard
. predict student, rstudent
. predict h, hat
. predict D, cooksd
. predict DFITS, dfits
. predict W, welsch
. predict COVRATIO, covratio
. dfbeta
             DFpercent:   DFbeta(percent)
            DFpercent2:   DFbeta(percent2)
               DFhigh:   DFbeta(high)

. describe standard - DFhigh
```

variable name	storage type	display format	value label	variable label
standard	float	%9.0g		Standardized residuals
student	float	%9.0g		Studentized residuals
h	float	%9.0g		Leverage
D	float	%9.0g		Cook's D
DFITS	float	%9.0g		Dfits
W	float	%9.0g		Welsch distance
COVRATIO	float	%9.0g		Covratio
DFpercent	float	%9.0g		Dfbeta percent
DFpercent2	float	%9.0g		Dfbeta percent2
DFhigh	float	%9.0g		Dfbeta high

```
. summarize standard - DFhigh
```

Variable	Obs	Mean	Std. Dev.	Min	Max
standard	51	-.0031359	1.010579	-2.099976	2.233379
student	51	-.00162	1.032723	-2.182423	2.336977
h	51	.0784314	.0373011	.0336437	.2151227
D	51	.0219941	.0364003	.0000135	.1860992
DFITS	51	-.0107348	.3064762	-.896658	.7444486
W	51	-.089723	2.278704	-6.854601	5.52468
COVRATIO	51	1.092452	.1316834	.7607449	1.360136
DFpercent	51	.000938	.1498813	-.5067295	.5269799
DFpercent2	51	-.0010659	.1370372	-.440771	.4253958
DFhigh	51	-.0012204	.1747835	-.6316988	.3414851

summarize shows us the minimum and maximum values of each statistic, so we can quickly check whether any are large enough to cause concern. For example, special tables could be used to determine whether the observation with the largest absolute studentized residual (*student*) constitutes a significant outlier. A simple but conservative alternative would be to apply the Bonferroni inequality and *t* distribution table: max| *student* | is significant at level α if | *t* | is significant at α/n. In this example, we have max| *student* | = 2.337 (Iowa) and $n = 51$. For Iowa to be a significant outlier (cause a significant shift in intercept) at $\alpha = .05$, $t = 2.337$ must be significant at .05/51

```
. display .05/51
.00098039
```

Stata's **ttail( )** function can approximate the probability of | *t* | > 2.337, given $df = n - K - 1 = 51 - 3 - 1 = 47$

```
. display 2*ttail(47, 2.337)
.02375138
```

The obtained *P*-value ($P = .0238$) is not below $\alpha/n = .00098$, so Iowa is not a significant outlier at $\alpha = .05$

Studentized residuals measure the *i*th observation's influence on the *y*-intercept. Cook's *D*, *DFITS*, and Welsch's distance all measure the *i*th observation's influence on all coefficients in the model (or, equivalently, on all *n* predicted *y* values). To list the 5 most influential observations as measured by Cook's *D*, type

```
. sort D
. list state yhat3 e3 D DFITS W in -5/1
```

	state	yhat3	e3	D	DFITS	W
47.	North Dakota	1036.696	36.30439	.0705921	.5493086	4.020527
48.	Wyoming	1017.005	-37.00463	.0789454	-.5820746	-4.270465
49.	Tennessee	974.6981	40.30194	.111718	.6992343	5.162398
50.	Iowa	1052.78	40.22015	.1265392	.7444486	5.52468
51.	Utah	1067.712	-36.71239	.1860992	-.896658	-6.854601

The **in -5/l** qualifier tells Stata to list only the fifth-from-last (–5) through last (lowercase letter "l") observations. Figure 7.7 shows one way to display influence graphically: symbols in a residual-versus-predicted plot are given sizes proportional to values of Cook's *D*, through the "analytical weight" option **[aweight = *D*]**. Five influential observations stand out, with large positive or negative residuals and high predicted *csat* values.

```
. graph twoway scatter e3 yhat4 [aweight = D], msymbol(oh) yline(0)
```

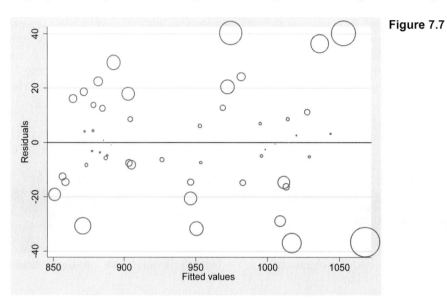

Figure 7.7

Although they have different statistical rationales, Cook's *D*, Welsch's distance, and *DFITS* are closely related. In practice they tend to flag the same observations as influential. Figure 7.8 shows their similarity in the example at hand.

```
. graph matrix D W DFITS, half
```

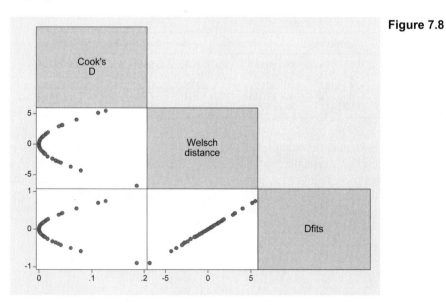

Figure 7.8

*DFBETA*s indicate how much each observation influences each regression coefficient. Typing **dfbeta** after a regression automatically generates *DFBETA*s for each predictor. In this example, they received the names *DFpercent* (*DFBETA* for predictor *percent*), *DFpercent2*, and *DFhigh*. Figure 7.9 graphs their distributions as box plots, with values of *state* applied to label the outlier marker symbols.

```
. graph box DFpercent DFpercent2 DFhigh, legend(cols(3))
      marker(1, mlabel(state)) marker(2, mlabel(state))
      marker(3, mlabel(state))
```

Figure 7.9

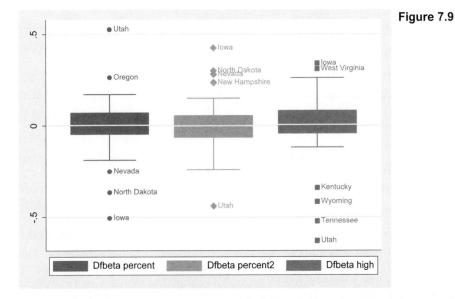

From left to right, Figure 7.9 shows the distributions of *DFBETA*s for *percent*, *percent2*, and *high*. (We could more easily distinguish them in color.) The extreme values in each plot belong to Iowa and Utah, which also have the two highest Cook's *D* values. For example, Utah's *DFhigh* = −.63. This tells us that Utah causes the coefficient on *high* to be .63 standard errors lower than it would be if Utah were set aside. Similarly, *DFpercent* = .53 indicates that with Utah present, the coefficient on *percent* is .53 standard errors higher (because the *percent* regression coefficient is negative, "higher" means closer to 0) than it otherwise would be. Thus, Utah weakens the apparent effects of both *high* and *percent*.

The most direct way to learn how particular observations affect a regression is to repeat the regression with those observations set aside. For example, we could set aside all states that move any coefficient by half a standard error (that is, have absolute *DFBETA*s of .5 or more):

```
. regress csat percent percent2 high if abs(DFpercent) < .5 &
     abs(DFpercent2) < .5 & abs(DFhigh) < .5
```

Source	SS	df	MS
Model	175366.782	3	58455.5939
Residual	11937.1351	44	271.298525
Total	187303.917	47	3985.18972

```
Number of obs =       48
F(  3,    44) =   215.47
Prob > F      =   0.0000
R-squared     =   0.9363
Adj R-squared =   0.9319
Root MSE      =   16.471
```

csat	Coef.	Std. Err.	t	P>|t|	[95% Conf. Interval]	
percent	-6.510868	.4700719	-13.85	0.000	-7.458235	-5.5635
percent2	.0538131	.005779	9.31	0.000	.0421664	.0654599
high	3.35664	.4577103	7.33	0.000	2.434186	4.279095
_cons	815.0279	33.93199	24.02	0.000	746.6424	883.4133

Careful inspection will reveal the details in which this regression table (based on $n = 48$) differs from its $n = 51$ or $n = 50$ counterparts seen earlier. Our central conclusion — that mean state SAT scores are well predicted by the percent of adults with high school diplomas and, curvilinearly, by the percent of students taking the test — remains unchanged, however.

Diagnostic statistics draw attention to influential observations, but they do not say whether we should set those observations aside. That requires a substantive decision based on careful evaluation of the data and research context. In this example, we have no substantive reason to discard any states, and even the most influential do not fundamentally change our conclusions.

Using any fixed definition of what constitutes an "outlier," we are liable to see more of them in larger samples. For this reason, sample-size-adjusted cutoffs are sometimes recommended for identifying unusual observations. After fitting a regression model with K coefficients (including the constant) based on n observations, we might look more closely at those observations for which any of the following are true:

leverage $h > 2K/n$
Cook's $D > 4/n$
$DFITS > 2\sqrt{K/n}$
Welsch's $W > 3\sqrt{K}$
$DFBETA > 2/\sqrt{n}$
$|COVRATIO - 1| \geq 3K/n$

The reasoning behind these cutoffs, and the diagnostic statistics more generally, can be found in Cook and Weisberg (1982, 1994); Belsley, Kuh, and Welsch (1980); or Fox (1991).

Multicollinearity

If perfect multicollinearity exists among the predictors, regression equations lack unique solutions. Stata warns us and then drops one of the offending predictors. High but not perfect multicollinearity causes more subtle problems. When we add a new predictor that is strongly related to predictors already in the model, symptoms of possible trouble include the following:

Substantially higher standard errors, with correspondingly lower t statistics.
Unexpected changes in coefficient magnitudes or signs.
Nonsignificant coefficients despite a high R^2.

Multiple regression attempts to estimate the independent effects of each x variable. There is little information for doing so, however, if one or more of the x variables does not have much independent variation. The symptoms listed above warn that coefficient estimates have become unreliable, and might shift drastically with small changes in the sample or model. Further troubleshooting is needed to determine whether multicollinearity really is at fault and, if so, what should be done about it.

Multicollinearity cannot necessarily be detected, or ruled out, by examining a matrix of correlations between variables. A more definitive assessment could be obtained by regressing each x on all of the other x variables, and then calculating $1 - R^2$ from that regression to see what fraction of the first x variable's variance is independent of the others. After **regress**, Stata saves the R^2 result as a scalar named e(r2) — type **ereturn list** to see a complete list of the saved estimation results. Using e(r2), we could accomplish the calculation as follows:

```
. quietly regress high percent percent2
. display 1 - e(r2)
.96942331
```

We see that almost 97% of *high*'s variance is independent of *percent* and *percent2*. A similar calculation finds that only 4% of *percent*'s variance is independent of the other two predictor variables, however.

```
. quietly regress percent high percent2
. display 1 - e(r2)
.04010307
```

This finding about *percent* and *percent2* is not surprising. In polynomial regression or regression with interaction terms, some x variables are calculated directly from other x variables. Although strictly speaking their relationship is nonlinear, it often is close enough to linear to raise problems of multicollinearity.

The post-regression command **estat vif**, for **v**ariance **i**nflation **f**actor, performs similar calculations automatically. This gives a quick and straightforward check for multicollinearity.

```
. quietly regress csat percent percent2 high
. estat vif
```

Variable	VIF	1/VIF
percent	24.94	0.040103
percent2	24.78	0.040354
high	1.03	0.969423
Mean VIF	16.92	

The 1/VIF column at right in an **estat vif** table gives values equal to $1 - R^2$ from the regression of each x on the other x variables, as can be seen by comparing the values for *high* (.969423) or *percent* (.040103) with our earlier **display** calculations. That is, 1/VIF (or $1 - R^2$) tells us what proportion of an x variable's variance is independent of all the other x variables. A low proportion, such as the .04 (4% independent variation) of *percent* and *percent2*, indicates potential trouble. Some analysts set a minimum level, called *tolerance*, for the 1/VIF value, and automatically exclude predictors that fall below their tolerance criterion.

The VIF column at center in the table reflects the degree to which other coefficients' variances (and standard errors) are increased due to the inclusion of that predictor. We see that *high* has virtually no impact on other variances, but *percent* and *percent2* affect the variances substantially. VIF values provide guidance but not direct measurements of the increase in coefficient variances. The following commands show the impact directly by displaying standard error estimates for the coefficient on *percent*, when *percent2* is and is not included in the model.

```
. quietly regress csat percent percent2 high
. display _se[percent]
.50958046
```

```
. quietly regress csat percent high
. display _se[percent]
.16162193
```

With *percent2* included in the model, the standard error for *percent* is three times higher:
.50958046 /.16162193 = 3.1529166
This corresponds to a tenfold increase in the coefficient's variance.

How much variance inflation is too much? Chatterjee, Hadi, and Price (2000) suggest the following as guidelines for the presence of multicollinearity:
The largest VIF is greater than 10; or
the mean VIF is considerably larger than 1.
With our largest VIFs close to 25, and the mean almost 17, the *csat* regression clearly meets both criteria. How troublesome the problem is, and what, if anything, should be done about it, are the next questions to consider.

Because *percent* and *percent2* are closely related, we cannot estimate their separate effects with nearly as much precision as we could the effect of either predictor alone. That is why the standard error for the coefficient on *percent* increases threefold when we compare the regression of *csat* on *percent* and *high* to a polynomial regression of *csat* on *percent*, *percent2*, and *high*. Despite this loss of precision, however, we can still distinguish all the coefficients from zero. Moreover, the polynomial regression obtains a better prediction model. For these reasons, the multicollinearity in this regression does not necessarily pose a great problem or require a solution. We could simply live with it as one feature of an otherwise acceptable model.

When solutions are needed, a simple trick called "centering" often succeeds in reducing multicollinearity in polynomial or interaction-effect models. Centering involves subtracting the mean from x variable values before generating polynomial or product terms. Subtracting the mean creates a new variable centered on zero and much less correlated with its own squared values. The resulting regression has the same fit (the same R^2, overall F test, predictions, and so forth) as an uncentered version. By reducing multicollinearity, centering often (*but not always*) yields more precise coefficient estimates with lower standard errors. The commands on the next page generate a centered version of *percent* named *Cpercent*, and then obtain squared values of *Cpercent* named *Cpercent2*.

```
. summarize percent
```

Variable	Obs	Mean	Std. Dev.	Min	Max
percent	51	35.76471	26.19281	4	81

```
. generate Cpercent = percent - r(mean)
. generate Cpercent2 = Cpercent ^2
. correlate Cpercent Cpercent2 percent percent2 high csat
(obs=51)
```

	Cpercent	Cperce~2	percent	percent2	high	csat
Cpercent	1.0000					
Cpercent2	0.3791	1.0000				
percent	1.0000	0.3791	1.0000			
percent2	0.9794	0.5582	0.9794	1.0000		
high	0.1413	-0.0417	0.1413	0.1176	1.0000	
csat	-0.8758	-0.0428	-0.8758	-0.7946	0.0858	1.0000

Whereas *percent* and *percent2* have a near-perfect correlation with each other ($r = .9794$), the centered versions *Cpercent* and *Cpercent2* are just moderately correlated ($r = .3791$). Otherwise, correlations involving *percent* and *Cpercent* are identical because centering is a linear transformation. Correlations involving *Cpercent2* are different from those with *percent2*, however. Figure 7.10 shows scatterplots that help to visualize these correlations, and the transformation's effects.

```
. graph matrix Cpercent Cpercent2 percent percent2 high csat,
       half msymbol(+)
```

Figure 7.10

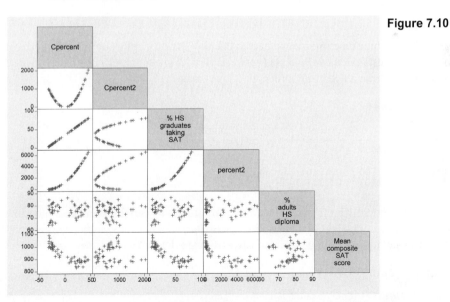

The R^2, overall F test, predictions, and many other aspects of a model should be unchanged after centering. Differences will be most noticeable in the centered variable's coefficient and standard error.

. regress *csat Cpercent Cpercent2 high*

Source	SS	df	MS
Model	207225.103	3	69075.0343
Residual	16789.407	47	357.221426
Total	224014.51	50	4480.2902

Number of obs = 51
F(3, 47) = 193.37
Prob > F = 0.0000
R-squared = 0.9251
Adj R-squared = 0.9203
Root MSE = 18.9

csat	Coef.	Std. Err.	t	P>|t|	[95% Conf. Interval]
Cpercent	-2.682362	.1119085	-23.97	0.000	-2.907493 -2.457231
Cpercent2	.0536555	.0063678	8.43	0.000	.0408452 .0664659
high	2.986509	.4857502	6.15	0.000	2.009305 3.963712
_cons	680.2552	37.82329	17.99	0.000	604.1646 756.3458

In this example, the standard error of the coefficient on *Cpercent* is actually lower (.11 compared with .16) when *Cpercent2* is included in the model. The *t* statistic is correspondingly larger. Thus, it appears that centering did improve that coefficient estimate's precision. The VIF table now gives less cause for concern: each of the three predictors has more than 80% independent variation, compared with 4% for *percent* and *percent2* in the uncentered regression.

. estat vif

Variable	VIF	1/VIF
Cpercent	1.20	0.831528
Cpercent2	1.18	0.846991
high	1.03	0.969423
Mean VIF	1.14	

Another diagnostic table sometimes consulted to check for multicollinearity is the matrix of correlations between estimated coefficients (*not* variables). This matrix can be displayed after **regress**, **anova**, or other model-fitting procedures by typing

. correlate, _coef

	Cpercent	Cperce~2	high	_cons
Cpercent	1.0000			
Cpercent2	-0.3893	1.0000		
high	-0.1700	0.1040	1.0000	
_cons	0.2105	-0.2151	-0.9912	1.0000

High correlations between pairs of coefficients on predictor variables indicate possible collinearity problems. By adding the option **covariance**, we can see the coefficients' variance–covariance matrix, from which standard errors are derived.

. correlate, _coef covariance

	Cpercent	Cperce~2	high	_cons
Cpercent	.012524			
Cpercent2	-.000277	.000041		
high	-.009239	.000322	.235953	
_cons	.891126	-.051817	-18.2105	1430.6

8

Fitting Curves

Basic regression and correlation methods assume linear relationships. Linear models provide reasonable and simple approximations for many real phenomena, over a limited range of values. But analysts also encounter phenomena where linear approximations are too simple; these call for nonlinear alternatives. This chapter describes three broad approaches to modeling nonlinear or curvilinear relationships:

Nonparametric methods, including band regression and lowess smoothing.

Linear regression with transformed variables ("curvilinear regression"), including Box–Cox methods.

Nonlinear regression.

Nonparametric regression serves as an exploratory tool because it can summarize data patterns visually without requiring the analyst to specify a particular model in advance. Transformed variables extend the usefulness of linear parametric methods, such as OLS regression (**regress**), to encompass curvilinear relationships as well. Nonlinear regression, on the other hand, requires a different class of methods that can estimate parameters of intrinsically nonlinear models.

The following menu groups cover many of the operations discussed in this chapter. The final topic, nonlinear regression, requires a command-based approach.

Graphics > Twoway graph (scatter, line, etc.)

Statistics > Nonparametric analysis > Lowess smoothing

Data > Create or change variables > Create new variable

Statistics > Linear models and related

Example Commands

. **boxcox** *y x1 x2 x3*, **model(lhs)**

Finds maximum-likelihood estimates of the parameter λ (lambda) for a Box–Cox transformation of *y*, assuming that $y^{(\lambda)}$ is a linear function of *x1*, *x2*, and *x3* plus Gaussian constant-variance errors. The **model(lhs)** option restricts transformation to the left-hand-side variable *y*. Other options could transform right-hand-side (*x*) variables by the same or different parameters, and control further details of the model. Type **help boxcox** for the syntax and a complete list of options. The *Base Reference Manual* gives technical details.

. **graph twoway mband** *y x*, **bands(10)** || **scatter** *y x*
Produces a *y* versus *x* scatterplot with line segments connecting the cross-medians (median *x*, median *y* points) within 10 equal-width vertical bands. This is one form of "band regression." Typing **mspline** in place of **mband** in this command would result in the cross-medians being connected by a smooth cubic spline curve instead of by line segments.

. **graph twoway lowess** *y x*, **bwidth(.4)** || **scatter** *y x*
Draws a lowess-smoothed curve with a scatterplot of *y* versus *x*. Lowess calculations use a bandwidth of .4 (40% of the data). In order to calculate and keep the smoothed values as a new variable, use the related command **lowess** .

. **lowess** *y x*, **bwidth(.3) gen(*newvar*)**
Draws a lowess-smoothed curve on a scatterplot of *y* versus *x*, using a bandwidth of .3 (30% of the data). Predicted values for this curve are saved as a variable named *newvar*. The **lowess** command offers more options than **graph twoway lowess**, including fitting methods and the ability to save predicted values. See **help lowess** for details.

. **nl (y1 = {b1=1}*{b2=1}^x)**
Uses iterative nonlinear least squares to fit a 2-parameter exponential growth model, $y = b_1 b_2{}^x$. Two parameters to be estimated, b1 and b2, are bound in {braces}, together with their suggested starting values (1). Instead of writing out our model in the command line, we might save time by calling one of the common models supplied with Stata, or write a new program to define our own model. The 2-parameter exponential happens to be one of these common models, defined by an existing program named **exp2**. Consequently, we could accomplish the same thing as the above command simply by typing **nl exp2:** *y x*, **init(b1 1 b2 1)**. After **nl**, use **predict** to generate predicted values or residuals.

. **nl log4:** *y x*, **init(b0 5 b1 25 b2 .1 b3 50)**
Fits a 4-parameter logistic growth model (**log4**) of the form
$$y = b_0 + b_1/(1 + \exp(-b_2(x - b_3)))$$
Sets initial parameter values for the iterative estimation process at b0 = 5, b1 = 25, b2 = .1, and b3 = 50. **log4**, like **exp2**, is one of the nonlinear models supplied with Stata.

. **regress** *lny x1 sqrtx2 invx3*
Performs curvilinear regression using the variables *lny*, *x1*, *sqrtx2*, and *invx3*. These variables were previously generated by nonlinear transformations of the raw variables *y*, *x2*, and *x3* through commands such as the following:
. **generate** *lny* = **ln(*y*)**
. **generate** *sqrtx2* = **sqrt(*x2*)**
. **generate** *invx3* = **1/*x3***
When, as in this example, the *y* variable was transformed, the predicted values generated by **predict** *yhat*, or residuals generated by **predict** *e*, **resid**, will be also in transformed units. For graphing or other purposes, we might want to return predicted values or residuals to raw-data units, using inverse transformations such as
. **replace** *yhat* = **exp(*yhat*)**

Band Regression

Nonparametric regression methods generally do not yield an explicit regression equation. They are primarily graphic tools for displaying the relationship, possibly nonlinear, between y and x. Stata can draw a simple kind of nonparametric regression, band regression, onto any scatterplot or scatterplot matrix. For illustration, consider these sobering Cold War data (*missile.dta*) from MacKenzie (1990). The observations are 48 types of long-range nuclear missiles, deployed by the U.S. and Soviet Union during their arms race, 1958 to 1990:

```
Contains data from c:\data\missile.dta
  obs:            48                          Missiles (MacKenzie 1990)
  vars:            6                          18 May 2008 18:39
  size:        1,584 (99.9% of memory free)
```

variable name	storage type	display format	value label	variable label
missile	str15	%15s		Missile
country	byte	%8.0g	soviet	US or Soviet missile?
year	int	%8.0g		Year of first deployment
type	byte	%8.0g	type	ICBM or submarine-launched?
range	int	%8.0g		Range in nautical miles
CEP	float	%9.0g		Circular Error Probable (miles)

```
Sorted by:  country  year
```

Variables in *missile.dta* include an accuracy measure called the "Circular Error Probable" (*CEP*). *CEP* represents the radius of a bulls eye within which 50% of the missile's warheads should land. Year by year, scientists on both sides worked to improve accuracy (Figure 8.1).

```
. graph twoway mband CEP year, bands(8)
      || scatter CEP year
      || , ytitle("Circular Error Probable, miles") legend(off)
```

Figure 8.1

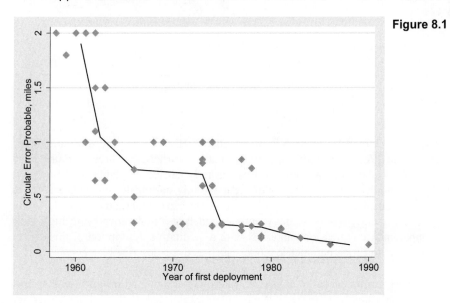

Figure 8.1 shows *CEP* declining (accuracy increasing) over time. The option **bands(8)** instructs **graph twoway mband** to divide the scatterplot into 8 equal-width vertical bands and draw line segments connecting the points (median *x*, median *y*) within each band. This curve traces how the median of *CEP* changes with *year*.

Nonparametric regression does not require the analyst to specify a relationship's functional form in advance. Instead, it allows us to explore the data with an "open mind." This process often uncovers interesting results, such as when we view trends in U.S. and Soviet missile accuracy separately (Figure 8.2). The **by(***country***)** option in the following command produces separate plots for each country, each with an overlaid band-regression curve and scatterplot. Within the **by( )** option are suboptions controlling the legend and note.

```
. graph twoway mband CEP year, bands(8)
       ||    scatter CEP year
       ||    , ytitle("Circular Error Probable, miles")
          by(country, legend(off) note(""))
```

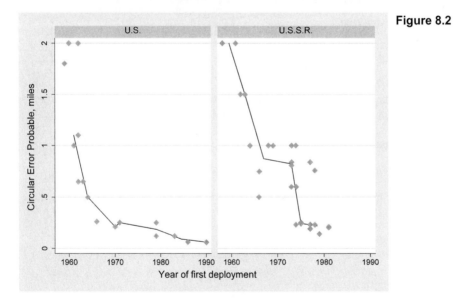

Figure 8.2

The shapes of the two curves in Figure 8.2 differ substantially. U.S. missiles became much more accurate in the 1960s, permitting a shift to smaller warheads. Three or more small warheads would fit on the same size missile that formerly carried one large warhead. The accuracy of Soviet missiles improved more slowly, apparently stalling during the late 1960s to early 1970s, and remained a decade or so behind their American counterparts. To make up for this accuracy disadvantage, Soviet strategy emphasized larger rockets carrying high-yield warheads. Nonparametric regression can assist with a qualitative description of this sort or serve as a preliminary to fitting parametric models such as those described later.

We can add band regression curves to any scatterplot by overlaying an **mband** (or **mspline**) plot. Band regression's simplicity makes it a convenient exploratory tool, but it possesses one notable disadvantage — the bands have the same width across the range of *x* values, although some of these bands contain few or no observations. With normally distributed *x* variables, for example, data density decreases toward the extremes. Consequently, the left and right endpoints of the band regression curve (which tend to dominate its appearance) often reflect just a few data points. The next section describes a more sophisticated, computation-intensive approach.

Lowess Smoothing

The **lowess** and **graph twoway lowess** commands accomplish a form of nonparametric regression called lowess smoothing (for **lo**cally **wei**ghted **s**catterplot **s**moothing). In general the **lowess** command is more specialized and more powerful, with options that control details of the fitting process. **graph twoway lowess** has advantages of simplicity, and follows the familiar syntax of the **graph twoway** family. The following example uses **graph twoway lowess** to plot *CEP* against *year* for U.S. missiles only (*country* == 0).

```
. graph twoway lowess CEP year if country == 0, bwidth(.4)
        ||  scatter CEP year
        ||  , legend(off) ytitle("Circular Error Probable, miles")
```

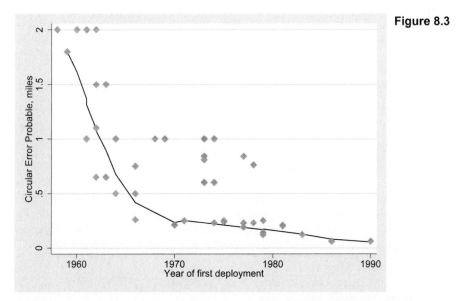

Figure 8.3

A graph very similar to Figure 8.2 would result if we had typed instead

```
. lowess CEP year if country == 0, bwidth(.4)
```

Like Figure 8.2, Figure 8.3 (above) shows U.S. missile accuracy improving rapidly during the 1960s and progressing at a more gradual rate in the 1970s and 1980s. Lowess-smoothed values of *CEP* are generated here with the name *lsCEP*. The **bwidth(.4)** option specifies the lowess bandwidth: the fraction of the sample used in smoothing each point. The default is **bwidth(.8)**. The closer bandwidth is to 1, the greater the degree of smoothing.

Lowess predicted (smoothed) y values for n observations result from n weighted regressions. Let k represent the half-bandwidth, truncated to an integer. For each y_i, a smoothed value y_i^s is obtained by weighted regression involving only those observations within the interval from $i = \max(1, i - k)$ through $i = \min(i + k, n)$. The jth observation within this interval receives weight w_j according to a tricube function:

$$w_j = (1 - | u_j |^3)^3$$

where

$$u_j = (x_i - x_j) / \Delta$$

Δ stands for the distance between x_i and its furthest neighbor within the interval. Weights equal 1 for $x_i = x_j$, but fall off to zero at the interval's boundaries. See Chambers et al. (1983) or Cleveland (1993) for more discussion and examples of lowess methods.

lowess options include the following.

mean   For running-mean smoothing. The default is running-line least squares smoothing.

noweight Unweighted smoothing. The default is Cleveland's tricube weighting function.

bwidth( ) Specifies the bandwidth. Centered subsets of approximately bwidth $\times n$ observations are used for smoothing, except towards the endpoints where smaller, uncentered bands are used. The default is **bwidth(.8)**.

logit   Transforms smoothed values to logits.

adjust   Adjusts the mean of smoothed values to equal the mean of the original y variable; like **logit**, **adjust** is useful with dichotomous y.

gen(*newvar*)   Creates *newvar* containing smoothed values of y.

nograph   Suppresses displaying the graph.

addplot( ) Add other plots to the generated graph; see **help addplot option** .

lineopts( ) Affects the rendition of the smoothed line; see **help cline options**.

Because it requires n weighted regressions, lowess smoothing proceeds slowly with large samples.

In addition to smoothing scatterplots, **lowess** can be used for exploratory time series smoothing. The file *ice.dta* contains results from the Greenland Ice Sheet 2 (GISP2) project described in Mayewski, Holdsworth, and colleagues (1993) and Mayewski, Meeker, and colleagues (1993). Researchers extracted and chemically analyzed an ice core representing more than 100,000 years of climate history. *ice.dta* includes a small fraction of this information: measured non-sea salt sulfate concentration and an index of "Polar Circulation Intensity" since AD 1500.

```
Contains data from c:\data\ice.dta
  obs:           271                          Greenland ice (Mayewski 1995)
  vars:            3                          18 May 2008 18:39
  size:        7,046 (99.9% of memory free)
─────────────────────────────────────────────────────────────────────────────
              storage   display    value
variable name   type    format     label    variable label
─────────────────────────────────────────────────────────────────────────────
year            int     %ty                  Year
sulfate         double  %10.0g               SO4 ion concentration, ppb
PCI             double  %6.0g                Polar Circulation Intensity
─────────────────────────────────────────────────────────────────────────────
Sorted by:  year
```

To retain more detail from this 271-point time series, we smooth with a relatively narrow bandwidth, only 5% of the sample. Figure 8.4 graphs the results. The smoothed curve has been drawn with "thick" width, to visually distinguish it from the raw data. (Type **help linewidthstyle** for other choices of line width.)

```
. graph twoway lowess sulfate year, bwidth(.05) clwidth(thick)
        ||   line sulfate year, clpattern(solid)
        ||   , ytitle("SO4 ion concentration, ppb")
        legend(label(1 "lowess smoothed") label(2 "raw data"))
```

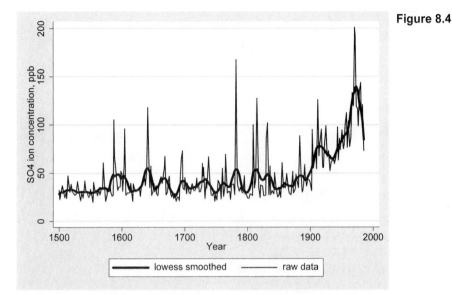

Figure 8.4

Non-sea salt sulfate (SO_4) reached the Greenland ice after being injected into the atmosphere, chiefly by volcanoes or the burning of fossil fuels such as coal and oil. Both the smoothed and raw curves in Figure 8.4 convey information. The smoothed curve shows oscillations around a slightly rising mean from 1500 through the early 1800s. After 1900, fossil fuels drive the smoothed curve upward, with temporary setbacks after 1929 (the Great Depression) and the early 1970s (combined effects of the U.S. Clean Air Act, 1970; the Arab oil embargo, 1973; and subsequent oil price hikes). Most of the sharp peaks of the raw data have been identified with known volcanic eruptions such as Iceland's Hekla (1970) or Alaska's Katmai (1912).

After smoothing time series data, it is often useful to study the smooth and rough (residual) series separately. The following commands create two new variables: lowess-smoothed values of sulfate (*smooth*) and the residuals or rough values (*rough*) calculated by subtracting the smoothed values from the raw data.

```
. lowess sulfate year, bwidth(.05) gen(smooth)
. label variable smooth "SO4 ion concentration (smoothed)"
. gen rough = sulfate - smooth
. label variable rough "SO4 ion concentration (rough)"
```

Figure 8.5 compares the *smooth* and *rough* time series in a pair of graphs annotated using the **text( )** option, and then combined.

```
. graph twoway line smooth year, ylabel(0(50)150)  xtitle("")
      ytitle("Smoothed") text(20 1540 "Renaissance")
      text(20 1900 "Industrialization")
      text(90 1860 "Great Depression 1929")
      text(150 1935 "Oil Embargo 1973") saving(fig08_05a, replace)

. graph twoway line rough year, ylabel(0(50)150) xtitle("")
      ytitle("Rough") text(75 1630 "Awu 1640",
         orientation(vertical))
      text(120 1770 "Laki 1783", orientation(vertical))
      text(90 1805 "Tambora 1815", orientation(vertical))
      text(65 1902 "Katmai 1912", orientation(vertical))
      text(80 1960 "Hekla 1970", orientation(vertical))
      yline(0) saving(fig08_05b, replace)

. graph combine fig08_05a.gph fig08_05b.gph, rows(2)
```

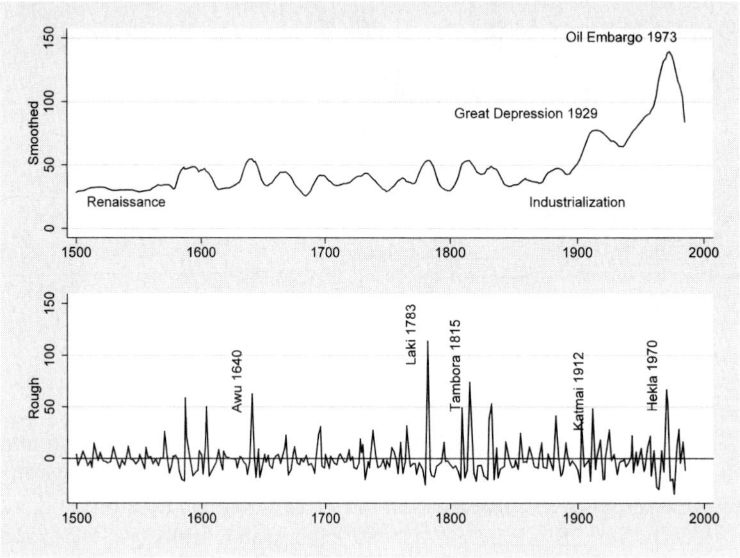

Figure 8.5

Regression with Transformed Variables — 1

By subjecting one or more variables to nonlinear transformation, and then including the transformed variable(s) in a linear regression, we implicitly fit a curvilinear model to the underlying data. Chapters 6 and 7 gave one example of this approach, polynomial regression, which incorporates second (and perhaps higher) powers of at least one *x* variable among the predictors. Logarithms also are used routinely in many fields. Other common transformations include those of the ladder of powers and Box–Cox transformations, introduced in Chapter 4.

Dataset *tornado.dta* provides a simple illustration involving U.S. tornados from 1916 to 1986 (from the Council on Environmental Quality, 1988).

```
Contains data from c:\data\tornado.dta
   obs:           71                      U.S. tornados 1916-1986 (Council
                                            on Env. Quality 1988)
   vars:           4                      18 May 2008 18:40
   size:        1,278 (99.9% of memory free)

               storage  display    value
variable name    type   format     label    variable label

year             int    %8.0g               Year
tornado          int    %8.0g               Number of tornados
lives            int    %8.0g               Number of lives lost
avlost           float  %9.0g               Average lives lost/tornado

Sorted by:  year
```

The number of fatalities decreased over this period, while the number of recognized tornados increased, because of improvements in warnings and our ability to detect more tornados, even those that do little damage. Consequently, the average lives lost per tornado (*avlost*) declined with time, but a linear regression (Figure 8.6) does a poor job of describing this trend. The scatter descends more steeply than the regression line at first, then levels off in the mid-1950s. The regression line actually predicts negative numbers of deaths in later years. Furthermore, average tornado deaths exhibit more variation in early years than later — evidence of heteroskedasticity.

```
. graph twoway scatter avlost year
        || lfit avlost year, clpattern(solid)
        || , ytitle("Average number of lives lost")
     xlabel(1920(10)1990)
     xtitle("") legend(off) ylabel(0(1)7) yline(0)
```

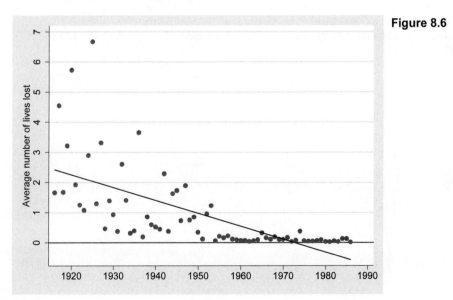

Figure 8.6

The relationship becomes linear, and heteroskedasticity vanishes if we work instead with logarithms of the average number of lives lost (Figure 8.7).

```
. generate loglost = ln(avlost)

. label variable loglost "ln(avlost)"

. regress loglost year
```

Source	SS	df	MS
Model	115.895325	1	115.895325
Residual	43.8807356	69	.63595269
Total	159.77606	70	2.28251515

```
Number of obs =      71
F(  1,     69) =  182.24
Prob > F       =  0.0000
R-squared      =  0.7254
Adj R-squared  =  0.7214
Root MSE       =  .79747
```

loglost	Coef.	Std. Err.	t	P>|t|	[95% Conf. Interval]	
year	-.0623418	.004618	-13.50	0.000	-.0715545	-.053129
_cons	120.5645	9.010312	13.38	0.000	102.5894	138.5395

```
. predict yhat2
(option xb assumed; fitted values)

. label variable yhat2 "ln(avlost) = 120.56 - .06year"

. label variable loglost "ln(avlost)"

. graph twoway scatter loglost year
        ||   mspline yhat2 year, clpattern(solid) bands(50)
        ||   , ytitle("Natural log(average lives lost)")
        xlabel(1920(10)1990) xtitle("") legend(off) ylabel(-4(1)2)
        yline(0)
```

Figure 8.7

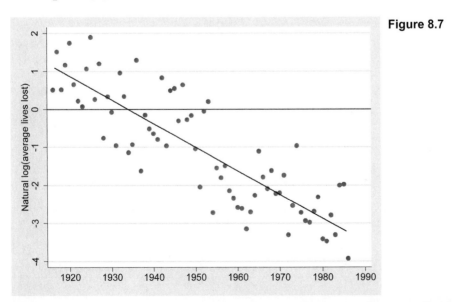

The regression model is approximately

  predicted ln(*avlost*) = 120.56 – .06*year*

Because we regressed logarithms of lives lost on *year*, the model's predicted values are also measured in logarithmic units. Return these predicted values to their natural units (lives lost) by inverse transformation, in this case exponentiating (*e* to power) *yhat2*:

```
. replace yhat2 = exp(yhat2)
(71 real changes made)
```

Graphing these inverse-transformed predicted values reveals the curvilinear regression model, which we obtained by linear regression with a transformed *y* variable (Figure 8.8). Contrast Figures 8.7 and 8.8 with Figure 8.6 to see how transformation made the analysis both simpler and more realistic.

```
. graph twoway scatter avlost year
       || mspline yhat2 year, clpattern(solid) bands(50)
       || , ytitle("Average number of lives lost") xlabel(1920(10)1990)
       xtitle("") legend(off) ylabel(0(1)7) yline(0)
```

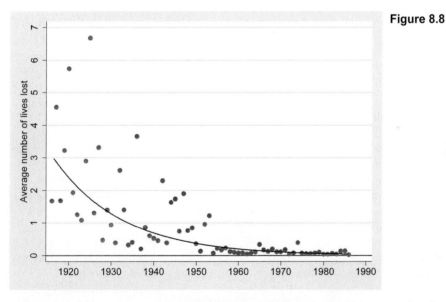

Figure 8.8

The **boxcox** command employs maximum-likelihood methods to fit curvilinear models involving Box–Cox transformations (introduced in Chapter 4). Fitting a model with Box–Cox transformation of the dependent variable (**model(lhs)** specifies left-hand side) to the tornado data, we obtain results quite similar to the model of Figures 8.7 and 8.8. The **nolog** option in the following command does not affect the model, but suppresses display of log likelihood after each iteration of the fitting process.

```
. boxcox avlost year, model(lhs) nolog
Fitting comparison model

Fitting full model
```

			Number of obs	=	71
			LR chi2(1)	=	92.28
Log likelihood = -7.7185533			Prob > chi2	=	0.000

avlost	Coef.	Std. Err.	z	P>|z|	[95% Conf. Interval]	
/theta	-.0560959	.0646727	-0.87	0.386	-.182852	.0706602

Estimates of scale-variant parameters

	Coef.
Notrans	
year	-.0661891
_cons	127.9713
/sigma	.8301177

Test H0:	Restricted log likelihood	LR statistic chi2	P-value Prob > chi2
theta = -1	-84.928791	154.42	0.000
theta = 0	-8.0941678	0.75	0.386
theta = 1	-101.50385	187.57	0.000

The **boxcox** output shows theta = −.056 as the optimal Box–Cox parameter for transforming *avlost*, in order to linearize its relationship with *year*. Therefore, the left-hand-side transformation is

$$avlost^{(-.056)} = (avlost^{-.056} - 1)/-.056$$

Box–Cox transformation by a parameter close to zero, such as −.056, produces results similar to the natural-logarithm transformation we applied earlier to this variable "by hand." It is therefore not surprising that the **boxcox** regression model

$$\text{predicted } avlost^{(-.056)} = 127.97 - .07year$$

resembles the earlier model (predicted ln(*avlost*) = 120.56 − .06*year*) drawn in Figures 8.7 and 8.8. The **boxcox** procedure assumes normal, independent, and identically distributed errors. It does not select transformations with the aim of normalizing residuals, however.

boxcox can fit several types of models, including multiple regressions in which some or all of the right-hand-side variables are transformed by a parameter different from the *y*-variable transformation. It cannot apply different transformations to each separate right-hand-side predictor. To do that, we return to a "by hand" curvilinear-regression approach, as illustrated in the next section.

Regression with Transformed Variables — 2

For a multiple-regression example, we will use data on living conditions in 109 countries found in dataset *nations.dta* (from World Bank 1987; World Resources Institute 1993).

```
Contains data from c:\data\nations.dta
  obs:           109                          Data on 109 nations, ca. 1985
  vars:           15                          2 Jan 2008 13:31
  size:         4,578 (99.9% of memory free)

              storage   display    value
variable name   type    format     label    variable label
country         str8    %9s                  Country
pop             float   %9.0g                1985 population in millions
birth           byte    %8.0g                Crude birth rate/1000 people
death           byte    %8.0g                Crude death rate/1000 people
chldmort        byte    %8.0g                Child (1-4 yr) mortality 1985
infmort         int     %8.0g                Infant (<1 yr) mortality 1985
life            byte    %8.0g                Life expectancy at birth 1985
food            int     %8.0g                Per capita daily calories 1985
energy          int     %8.0g                Per cap energy consumed, kg oil
gnpcap          int     %8.0g                Per capita GNP 1985
gnpgro          float   %9.0g                Annual GNP growth % 65-85
urban           byte    %8.0g                % population urban 1985
school1         int     %8.0g                Primary enrollment % age-group
school2         int     %8.0g                Secondary enroll % age-group
school3         byte    %8.0g                Higher ed. enroll % age-group
```

Relationships among birth rate, per capita gross national product (GNP), and child mortality are not linear, as can be seen clearly in the scatterplot matrix of Figure 8.9. The skewed *gnpcap* and *chldmort* distributions also present potential leverage and influence problems.

```
. graph matrix gnpcap chldmort birth, half
```

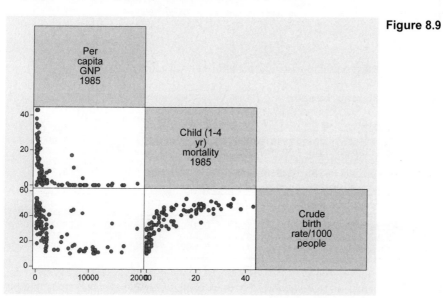

Figure 8.9

Experimenting with ladder-of-powers transformations reveals that the log of *gnpcap* and the square root of *chldmort* have distributions more symmetrical, with fewer outliers or potential leverage points than the raw variables. More importantly, these transformations largely eliminate the nonlinearities: compare the raw-data scatterplots in Figure 8.9 with their transformed-variables counterparts in Figure 8.10, below.

```
. generate loggnp = log10(gnpcap)
. label variable loggnp "Log-10 of per cap GNP"
. generate srmort = sqrt(chldmort)
. label variable srmort "Square root child mortality"
. graph matrix loggnp srmort birth, half
```

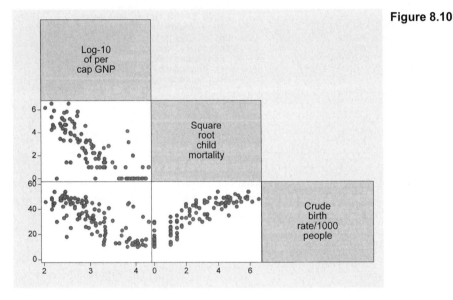

Figure 8.10

We can now apply linear regression using the transformed variables:

```
. regress birth loggnp srmort
```

Source	SS	df	MS
Model	15837.9603	2	7918.98016
Residual	4238.18646	106	39.9828911
Total	20076.1468	108	185.890248

Number of obs = 109
F(2, 106) = 198.06
Prob > F = 0.0000
R-squared = 0.7889
Adj R-squared = 0.7849
Root MSE = 6.3232

birth	Coef.	Std. Err.	t	P>|t|	[95% Conf. Interval]
loggnp	-2.353738	1.686255	-1.40	0.166	-5.696903 .9894259
srmort	5.577359	.533567	10.45	0.000	4.51951 6.635207
_cons	26.19488	6.362687	4.12	0.000	13.58024 38.80953

Unlike the raw-data regression (not shown), this transformed-variables version finds that per capita gross national product does not significantly affect birth rate once we control for child mortality. The transformed-variables regression fits slightly better: $R^2_a = .7849$ instead of

.6715.   (We can compare R^2_a across models here only because both have the same untransformed *y* variable.)   Leverage plots would confirm that transformations have much reduced the curvilinearity of the raw-data regression.

Conditional Effect Plots

Conditional effect plots trace the predicted values of *y* as a function of one *x* variable, with other *x* variables held constant at arbitrary values such as their means, medians, quartiles, or extremes. Such plots help with interpreting results from transformed-variables regression.

Continuing with the previous example, we can calculate predicted birth rates as a function of *loggnp*, with *srmort* held at its mean (2.49):

```
. generate yhat1 = _b[_cons] + _b[loggnp]*loggnp + _b[srmort]*2.49
. label variable yhat1 "birth = f(gnpcap | srmort = 2.49)
```

The _b[*varname*] terms refer to the regression coefficient on *varname* from this session's most recent regression.   _b[_cons] is the *y*-intercept.   Alternatively, we could use **adjust**:

```
. adjust srmort = 2.49, generate(yhat1)
```

For a conditional effect plot, graph *yhat1* (after inverse transformation if needed, although it is not needed here) against the untransformed *x* variable (Figure 8.11).   Because conditional effect plots do not show the scatter of data, it can be useful to add reference lines such as the *x* variable's 10th and 90th percentiles, as shown in Figure 8.11.

```
. graph twoway line yhat1 gnpcap, sort xlabel(,grid) xline(230 10890)
       saving(fig08_11)
```

Figure 8.11

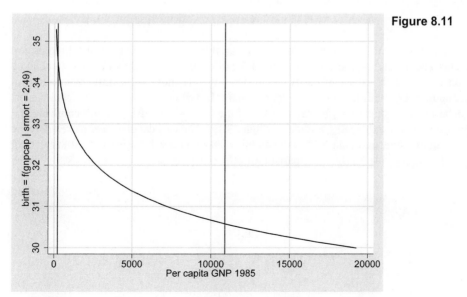

Similarly, Figure 8.12 depicts predicted birth rates as a function of *srmort*, with *loggnp* held at its mean (3.09):

```
. generate yhat2 = _b[_cons] + _b[loggnp]*3.09 + _b[srmort]*srmort
```

Or equivalently, we could obtain the same predicted values through an **adjust** command:

```
. adjust loggnp = 3.09, generate(yhat2)
```

```
. label variable yhat2 "birth = f(chldmort | loggnp = 3.09)"
. graph twoway line yhat2 chldmort, sort xlabel(,grid) xline(0 27)
      saving(fig08_12)
```

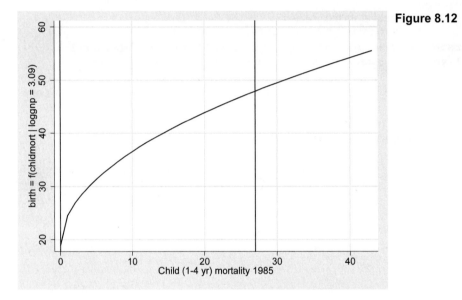

Figure 8.12

How can we compare the strength of different *x* variables' effects? Standardized regression coefficients (beta weights) are sometimes used for this purpose, but they imply a specialized definition of "strength" and can easily be misleading. A more substantively meaningful comparison might come from looking at conditional effect plots drawn with identical *y* scales. This can be accomplished easily by saving each of the graphs, then using **graph combine** and specifying common *y*-axis scales, as done in Figure 8.13. The vertical distances traveled by the predicted values curve, particularly over the middle 80% of the *x* values (between 10th and 90th percentile lines), provide a visual comparison of effect magnitude.

```
.  graph combine fig08_11.gph fig08_12.gph, ycommon cols(2)
      scale(1.25)
```

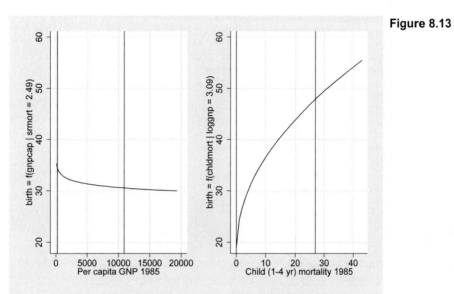

Figure 8.13

Combining several conditional effect plots into one image with common vertical scales, as done in Figure 8.13, allows quick visual comparison of the strength of different effects. Figure 8.13 makes obvious how much stronger is the effect of child mortality on birth rates — as separate plots (Figures 8.11 and 8.12) did not.

Nonlinear Regression — 1

Variable transformations allow fitting some curvilinear relationships using the familiar techniques of intrinsically linear models. Intrinsically nonlinear models, on the other hand, require a different class of fitting techniques. The **nl** command performs nonlinear regression by iterative least squares. This section introduces it using a dataset of simple examples, *nonlin.dta*:

```
Contains data from c:\data\nonlin.dta
  obs:           100                      Nonlinear model examples
                                            (artificial data)
  vars:            5                      18 May 2008 18:40
  size:         2,500 (99.9% of memory free)
```

variable name	storage type	display format	value label	variable label
x	byte	%9.0g		Independent variable
y1	float	%9.0g		y1 = 10 * 1.03^x + e
y2	float	%9.0g		y2 = 10 * (1 - .95^x) + e
y3	float	%9.0g		y3 = 5 + 25/(1+exp(-.1*(x-50))) + e
y4	float	%9.0g		y4 = 5 + 25*exp(-exp(-.1*(x-50))) + e

```
Sorted by:  x
```

The *nonlin.dta* data are manufactured, with y variables defined as various nonlinear functions of x, plus random Gaussian errors. $y1$, for example, represents the exponential growth process $y1 = 10 \times 1.03^x$. Estimating these parameters from the data, **nl** obtains $y1 = 11.20 \times 1.03^x$, which is reasonably close to the true model.

```
. nl (y1 = {b1=1}*{b2=1}^x)
(obs = 100)

Iteration 0:   residual SS =   419135.4
Iteration 1:   residual SS =   416152.4
Iteration 2:   residual SS =   409107.7
Iteration 3:   residual SS =   348535.9
Iteration 4:   residual SS =   31488.48
Iteration 5:   residual SS =   27849.49
Iteration 6:   residual SS =   26139.18
Iteration 7:   residual SS =   26138.29
Iteration 8:   residual SS =   26138.29
Iteration 9:   residual SS =   26138.29
```

Source	SS	df	MS
Model	667018.255	2	333509.128
Residual	26138.2933	98	266.717278
Total	693156.549	100	6931.56549

```
Number of obs =        100
R-squared      =     0.9623
Adj R-squared  =     0.9615
Root MSE       =   16.33148
Res. dev.      =   840.3864
```

y1	Coef.	Std. Err.	t	P>|t|	[95% Conf. Interval]	
/b1	11.20416	1.146683	9.77	0.000	8.928602	13.47971
/b2	1.028838	.0012404	829.41	0.000	1.026376	1.031299

The **predict** command obtains predicted values and residuals for a nonlinear model estimated by **nl**. Figure 8.14 graphs predicted values from the previous example, showing the close fit ($R^2 = .96$) between model and data.

```
. predict yhat1
(option yhat assumed; fitted values)

. graph twoway scatter y1 x
       || line yhat1 x, sort lpattern(solid)
       || , legend(off) ytitle("y1 = 10 * 1.03^x + e") xtitle("x")
```

Figure 8.14

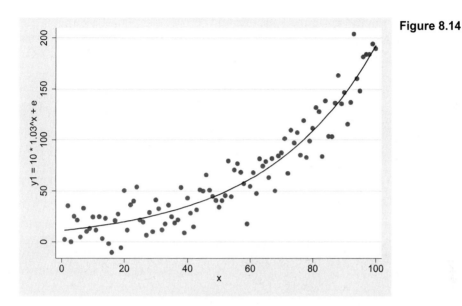

Instead of writing out the model in our **nl** command, we could have typed

```
. nl exp2: y1 x
```

The **exp2** in this command calls a brief program named *nlexp2.ado*, which defines the two-parameter exponential growth function. Stata includes several such programs, for the following functions:

    exp3 3-parameter exponential: $y = b_0 + b_1 b_2{}^x$

    exp2 2-parameter exponential: $y = b_1 b_2{}^x$

    exp2a 2-parameter negative exponential: $y = b_1 (1 - b_2{}^x)$

    log4 4-parameter logistic; b_0 starting level and $(b_0 + b_1)$ asymptotic upper limit:
$$y = b_0 + b_1 / (1 + \exp(-b_2 (x - b_3)))$$

    log3 3-parameter logistic; 0 starting level and b_1 asymptotic upper limit:
$$y = b_1 / (1 + \exp(-b_2 (x - b_3)))$$

    gom4 4-parameter Gompertz; b_0 starting level and $(b_0 + b_1)$ asymptotic upper limit:
$$y = b_0 + b_1 \exp(-\exp(-b_2 (x - b_3)))$$

    gom3 3-parameter Gompertz; 0 starting level and b_1 asymptotic upper limit:
$$y = b_1 \exp(-\exp(-b_2 (x - b_3)))$$

Users can write further nl*function* programs of their own; consult nlexp3.ado, nlgom4.ado, or others above for examples. **help nl** describes many further options for specifying and estimating models.

nonlin.dta contains examples corresponding to **exp2** (*y1*), **exp2a** (*y2*), **log4** (*y3*), and **gom4** (*y4*) functions. Figure 8.15 shows curves fit by **nl** to *y2*, *y3*, and *y4* from these data.

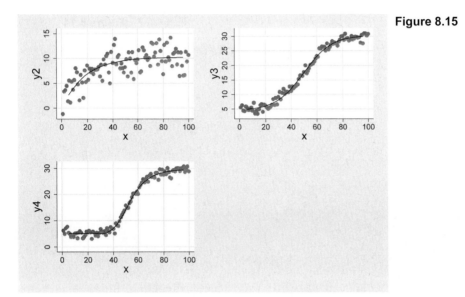

Figure 8.15

Nonlinear Regression — 2

Our second example involves observational data and illustrates some steps that can help in research. Dataset *lichen.dta* concerns measurements of lichen growth observed on the Norwegian arctic island of Svalbard (from Werner 1990). These slow-growing symbionts are often used to date rock monuments and other deposits, so their growth rates interest scientists in several fields.

```
Contains data from c:\data\lichen.dta
  obs:            11                        Lichen growth (Werner 1990)
 vars:             8                        18 May 2008 18:41
 size:           616 (99.9% of memory free)
```

variable name	storage type	display format	value label	variable label
locale	str31	%31s		Locality and feature
point	str1	%9s		Control point
date	int	%8.0g		Date
age	int	%8.0g		Age in years
rshort	float	%9.0g		Rhizocarpon short axis mm
rlong	float	%9.0g		Rhizocarpon long axis mm
pshort	int	%8.0g		P.minuscula short axis mm
plong	int	%8.0g		P.minuscula long axis mm

Lichens characteristically exhibit a period of relatively fast early growth, gradually slowing, as illustrated by the lowess-smoothed curve in Figure 8.16.

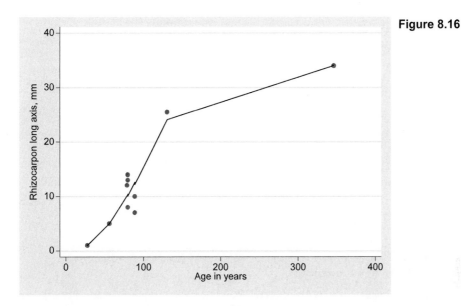

Figure 8.16

Lichenometricians seek to summarize and compare such patterns by drawing growth curves. Their growth curves might not employ an explicit mathematical model, but we can fit one here to illustrate the process of nonlinear regression. Gompertz curves are asymmetrical S-curves, which have been widely used to model biological growth, and might provide a reasonable model for lichen:

$$y = b_1 \exp(-\exp(-b_2(x - b_3)))$$

If we intend to graph a nonlinear model, the data should contain a good range of closely spaced *x* values. Actual ages of the 11 lichen samples in *lichen.dta* range from 28 to 346 years. We can create 89 additional artificial observations, with "ages" from 0 to 352 in 4-year increments, with the following commands:

```
. range newage 0 396 100
obs was 11, now 100
```

```
. replace age = newage[_n-11] if missing(age)
(89 real changes made)
```

The first command created a new variable, *newage*, with 100 values ranging from 0 to 396 in 4-year increments. In so doing, we also created 89 new artificial observations, with missing values on all variables except *newage*. The **replace** command substitutes the missing artificial-case *age* values with *newage* values, starting at 0. The first 15 observations in our data now look like this:

```
. list rlong age newage in 1/15
```

	rlong	age	newage
1.	1	28	0
2.	5	56	4
3.	12	79	8
4.	14	80	12
5.	13	80	16
6.	8	80	20
7.	7	89	24
8.	10	89	28
9.	34	346	32
10.	34	346	36
11.	25.5	131	40
12.	.	0	44
13.	.	4	48
14.	.	8	52
15.	.	12	56

```
. summarize rlong age newage
```

Variable	Obs	Mean	Std. Dev.	Min	Max
rlong	11	14.86364	11.31391	1	34
age	100	170.68	104.7042	0	352
newage	100	198	116.046	0	396

We now could **drop *newage*** . Only the original 11 observations have nonmissing *rlong* values, so only they will enter into model estimation. Stata calculates predicted values for any observation with nonmissing *x* values, however. We can therefore obtain such predictions for both the 11 real observations and the 89 artificial ones, which will allow us to graph the regression curve accurately.

Lichen growth starts with a size close to zero, so we chose the **gom3** Gompertz function rather than **gom4** (which incorporates a nonzero takeoff level, the parameter b_0). Figure 8.16 suggests an asymptotic upper limit somewhere near 34, suggesting that 34 should be a good guess or starting value of the parameter b_1. To estimate the model, type

```
. nl gom3: rlong age, initial(b1 34) nolog
(obs = 11)
```

Source	SS	df	MS		
Model	3633.16112	3	1211.05371	Number of obs =	11
Residual	77.0888815	8	9.63611018	R-squared =	0.9792
				Adj R-squared =	0.9714
				Root MSE =	3.104208
Total	3710.25	11	337.295455	Res. dev. =	52.63435

3-parameter Gompertz function, rlong = b1*exp(-exp(-b2*(age - b3)))

rlong	Coef.	Std. Err.	t	P>|t|	[95% Conf. Interval]	
/b1	34.36637	2.267185	15.16	0.000	29.13824	39.59451
/b2	.0217685	.0060806	3.58	0.007	.0077465	.0357904
/b3	88.79701	5.632537	15.77	0.000	75.80836	101.7857

A **nolog** option suppresses displaying a log of iterations with the output. All three parameter estimates differ significantly from 1. We could have fit the same model with the command

```
. nl (rlong = {b1=34}*exp(-1*exp(-1*{b2}*(age-{b3})))))
```

Starting from either approach, we next obtain predicted values using **predict**, and graph these to see the regression curve. The **yline** option is used to display the lower and estimated upper limits (0 and 34.366) of this curve in Figure 8.17.

```
. predict yhat
(option yhat assumed; fitted values)

. graph twoway scatter rlong age
       || mspline yhat age, clpattern(solid) bands(50)
       || , legend(off) yline(0 34.366)
       ytitle("Rhizocarpon long axis, mm") xlabel(0(100)400, grid)
```

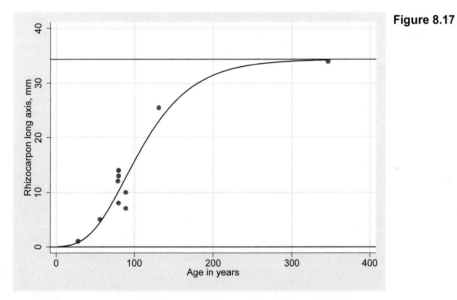

Figure 8.17

Especially when working with sparse data or a relatively complex model, nonlinear regression programs can be quite sensitive to their initial parameter estimates. Previous experience with similar data, or publications by other researchers, could help supply suitable initial values. Alternatively, we could estimate through trial and error by employing **generate** to calculate predicted values based on arbitrarily-chosen sets of parameter values and **graph** to compare the resulting predictions with the data.

9

Robust Regression

Stata's basic **regress** and **anova** commands perform ordinary least squares (OLS) regression. The popularity of OLS derives in part from its theoretical advantages given "ideal" data. If errors are normally, independently, and identically distributed (normal i.i.d.), then OLS is more efficient than any other unbiased estimator. The flip side of this statement often gets overlooked: if errors are not normal, or not i.i.d., then other unbiased estimators might outperform OLS. In fact, the efficiency of OLS degrades quickly in the face of heavy-tailed (outlier-prone) error distributions. Yet such distributions are common in many fields.

OLS tends to track outliers, fitting them at the expense of the rest of the sample. Over the long run, this leads to greater sample-to-sample variation or inefficiency when samples often contain outliers. Robust regression methods aim to achieve almost the efficiency of OLS with ideal data and substantially better-than-OLS efficiency in non-ideal (for example, nonnormal errors) situations. "Robust regression" encompasses a variety of different techniques, each with advantages and drawbacks for dealing with problematic data. This chapter introduces two varieties of robust regression, **rreg** and **qreg**, and briefly compares their results with those of OLS (**regress**).

rreg and **qreg** resist the pull of outliers, giving them better-than-OLS efficiency in the face of nonnormal, heavy-tailed error distributions. They share the OLS assumption that errors are independent and identically distributed, however. As a result, their standard errors, tests, and confidence intervals are not trustworthy in the presence of heteroskedasticity or correlated errors. To relax the assumption of independent, identically distributed errors when using **regress** or certain other modeling commands (although not **rreg** or **qreg**), Stata offers options that estimate robust standard errors.

For clarity, this chapter focuses mostly on two-variable examples, but robust multiple regression or *N*-way ANOVA are straightforward using the same commands. Chapter 16 returns to the topic of robustness, showing how we can use Monte Carlo experiments to evaluate competing statistical techniques.

Several of the techniques described in this chapter are available through menu selections:

Statistics > Linear models and related > Quantile regression

Statistics > Linear models and related > Linear regression > (SE/robust)

Example Commands

. `rreg y x1 x2 x3`

Performs robust regression of y on three predictors, using iteratively reweighted least squares with Huber and biweight functions tuned for 95% Gaussian efficiency. Given appropriately configured data, **rreg** can also obtain robust means, confidence intervals, difference of means tests, and ANOVA or ANCOVA.

. `rreg y x1 x2 x3, nolog tune(6) genwt(rweight) iterate(10)`

Performs robust regression of y on three predictors. The options shown above tell Stata not to print the iteration log, to use a tuning constant of 6 (which downweights outliers more steeply than the default 7), to generate a new variable (arbitrarily named *rweight*) holding the final-iteration robust weights for each observation, and to limit the maximum number of iterations to 10.

. `qreg y x1 x2 x3`

Performs quantile regression, also known as least absolute value (LAV) or minimum $L1$-norm regression, of y on three predictors. By default, **qreg** models the conditional .5 quantile (approximate median) of y as a linear function of the predictor variables, and thus provides "median regression."

. `qreg y x1 x2 x3, quantile(.25)`

Performs quantile regression modeling the conditional .25 quantile (first quartile) of y as a linear function of $x1$, $x2$, and $x3$.

. `bsqreg y x1 x2 x3, rep(100)`

Performs quantile regression, with standard errors estimated by bootstrap data resampling with 100 repetitions (default is **rep(20)**).

. `predict e, resid`

Calculates residual values (arbitrarily named e) after any **regress**, **rreg**, **qreg**, or **bsqreg** command. Similarly, **predict** *yhat* calculates the predicted values of y. Other **predict** options apply, with some restrictions.

. `regress y x1 x2 x3, vce(robust)`

Performs OLS regression of y on three predictors. The coefficient variance-covariance matrix, and hence standard errors, are estimated by a robust method (Huber/White or sandwich) that does not assume identically distributed errors. With the **vce(cluster** *clustervar***)** option, one source of correlation among the errors can be accommodated as well. The *User's Guide* describes the reasoning behind these methods.

Regression with Ideal Data

To clarify the issue of robustness, we will explore the small ($n = 20$) contrived dataset *robust1.dta*:

```
Contains data from c:\data\robust1.dta
  obs:            20                       Robust regression examples 1
                                             (artificial data)
  vars:           10                       19 May 2008 19:59
  size:          960 (99.9% of memory free)

                storage  display   value
variable name    type    format    label    variable label

x               float    %9.0g              Normal X
e1              float    %9.0g              Normal errors
y1              float    %9.0g              y1 = 10 + 2*x + e1
e2              float    %9.0g              Normal errors with 1 outlier
y2              float    %9.0g              y2 = 10 + 2*x + e2
x3              float    %9.0g              Normal X with 1 leverage obs.
e3              float    %9.0g              Normal errors with 1 extreme
y3              float    %9.0g              y3 = 10 + 2*x3 + e3
e4              float    %9.0g              Skewed errors
y4              float    %9.0g              y4 = 10 + 2*x + e4
```

The variables *x* and *e1* each contain 20 random values from independent standard normal distributions. *y1* contains 20 values produced by the regression model:

$$y1 = 10 + 2x + e1$$

The commands that manufactured these first three variables are

```
. clear
. set obs 20
. generate x = invnormal(uniform())
. generate e1 = invnormal(uniform())
. generate y1 = 10 + 2*x + e1
```

With real data, coding mistakes and measurement errors sometimes create wildly incorrect values. To simulate this, we might shift the second observation's error from –0.89 to 19.89:

```
. generate e2 = e1
. replace e2 = 19.89 in 2
. generate y2 = 10 + 2*x + e2
```

Similar manipulations produce the other variables in *robust1.dta*.

y1 and *x* present an ideal regression problem: the expected value of *y1* really is a linear function of *x*, and errors come from normal, independent, and identical distributions because we defined them that way. OLS does a good job of estimating the true intercept (10) and slope (2), obtaining the line shown in Figure 9.1.

```
. regress y1 x
```

Source	SS	df	MS
Model	134.059351	1	134.059351
Residual	22.29157	18	1.23842055
Total	156.350921	19	8.22899586

Number of obs =	20
F(1, 18) =	108.25
Prob > F =	0.0000
R-squared =	0.8574
Adj R-squared =	0.8495
Root MSE =	1.1128

y1	Coef.	Std. Err.	t	P>|t|	[95% Conf. Interval]
x	2.048057	.1968465	10.40	0.000	1.634498 2.461616
_cons	9.963161	.2499861	39.85	0.000	9.43796 10.48836

```
. predict yhat1o

. graph twoway scatter y1 x
        || line yhat1o x, clpattern(solid) sort
        ||  , ytitle("y1 = 10 + 2*x + e1") legend(order(2)
           label(2 "OLS line") position(11) ring(0) cols(1))
```

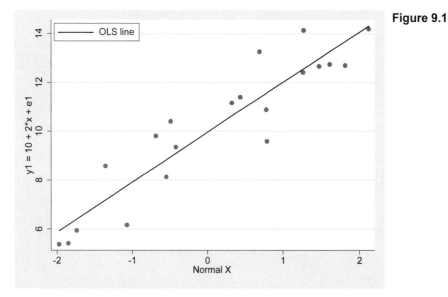

Figure 9.1

An iteratively reweighted least squares (IRLS) procedure, **rreg** , obtains robust regression estimates. The first **rreg** iteration begins with OLS. Any observations so influential as to have Cook's D values greater than 1 are automatically set aside after this first step. Next, weights are calculated for each observation using a Huber function (which downweights observations that have larger residuals) and weighted least squares is performed. After several WLS iterations, the weight function shifts to a Tukey biweight (as suggested by Li 1985), tuned for 95% Gaussian efficiency (see Hamilton 1992a for details). **rreg** estimates standard errors and tests hypotheses using a pseudovalues method (Street, Carroll and Ruppert 1988) that does not assume normality.

```
. rreg y1 x
```

```
     Huber iteration 1:   maximum difference in weights = .35774407
     Huber iteration 2:   maximum difference in weights = .02181578
  Biweight iteration 3:   maximum difference in weights = .14421371
  Biweight iteration 4:   maximum difference in weights = .01320276
  Biweight iteration 5:   maximum difference in weights = .00265408
```

```
Robust regression                              Number of obs =      20
                                               F(  1,    18) =   79.96
                                               Prob > F      =  0.0000
```

y1	Coef.	Std. Err.	t	P>|t|	[95% Conf. Interval]	
x	2.047813	.2290049	8.94	0.000	1.566692	2.528935
_cons	9.936163	.2908259	34.17	0.000	9.325161	10.54717

This "ideal data" example includes no serious outliers, so here **rreg** is unneeded. The **rreg** intercept and slope estimates resemble those obtained by **regress** (and are not far from the true values 10 and 2), but they have slightly larger estimated standard errors. Given normal i.i.d. errors, as in this example, **rreg** theoretically possesses about 95% of the efficiency of OLS.

rreg and **regress** both belong to the family of *M*-estimators (for maximum-likelihood). An alternative order-statistic strategy called *L*-estimation fits quantiles of *y*, rather than its expectation or mean. For example, we could model how the median (.5 quantile) of *y* changes with *x*. **qreg** (an *L1*-type estimator) accomplishes such quantile regression. Like **rreg**, **qreg** has good resistance to outliers:

```
. qreg y1 x
Iteration  1:  WLS sum of weighted deviations =   17.335321

Iteration  1: sum of abs. weighted deviations =   17.130001
Iteration  2: sum of abs. weighted deviations =   16.858602
```

```
Median regression                              Number of obs =      20
  Raw sum of deviations     46.84 (about 10.4)
  Min sum of deviations   16.8586               Pseudo R2     =  0.6401
```

y1	Coef.	Std. Err.	t	P>|t|	[95% Conf. Interval]	
x	2.139896	.2590447	8.26	0.000	1.595664	2.684129
_cons	9.65342	.3564108	27.09	0.000	8.904628	10.40221

Although **qreg** obtains reasonable parameter estimates, its standard errors here exceed those of **regress** (OLS) and **rreg**. Given ideal data, **qreg** is the least efficient of these three estimators. The following sections view their performance with more troublesome data.

Y Outliers

The variable *y2* is identical to *y1*, but with one outlier caused by the "wild" error of observation #2. OLS has little resistance to outliers, so this shift in observation #2 (at upper left in Figure 9.2) substantially changes the **regress** results:

```
. regress y2 x
```

Source	SS	df	MS		
Model	18.764271	1	18.764271		
Residual	348.233471	18	19.3463039		
Total	366.997742	19	19.3156706		

	Number of obs =	20
	F(1, 18) =	0.97
	Prob > F =	0.3378
	R-squared =	0.0511
	Adj R-squared =	-0.0016
	Root MSE =	4.3984

y2	Coef.	Std. Err.	t	P>|t|	[95% Conf. Interval]	
x	.7662304	.7780232	0.98	0.338	-.8683356	2.400796
_cons	11.1579	.9880542	11.29	0.000	9.082078	13.23373

```
. predict yhat2o
(option xb assumed; fitted values)
. label variable yhat2o "OLS line (regress)"
```

The outlier raises the OLS intercept (from 9.936 to 11.1579) and lessens the slope (from 2.048 to 0.766). R^2 has dropped from .8574 to .0511. Standard errors quadrupled, and the OLS slope (solid line in Figure 9.2) no longer significantly differs from zero.

The outlier has little impact on **rreg**, however, as shown by the dashed line in Figure 9.2. The robust coefficients barely change, and remain close to the true parameters 10 and 2; nor do the robust standard errors increase much.

```
. rreg y2 x, nolog genwt(rweight2)
```

Robust regression

	Number of obs =	19
	F(1, 17) =	63.01
	Prob > F =	0.0000

y2	Coef.	Std. Err.	t	P>|t|	[95% Conf. Interval]	
x	1.979015	.2493146	7.94	0.000	1.453007	2.505023
_cons	10.00897	.3071265	32.59	0.000	9.360986	10.65695

```
. predict yhat2r
(option xb assumed; fitted values)

. label variable yhat2r "robust regression (rreg)"
```

```
. graph twoway scatter y2 x
        || line yhat2o x, clpattern(solid) sort
        || line yhat2r x, clpattern(longdash) sort
        || , ytitle("y2 = 10 + 2*x + e2")
    legend(order(2 3) position(1) ring(0) cols(1) margin(sides))
```

Figure 9.2

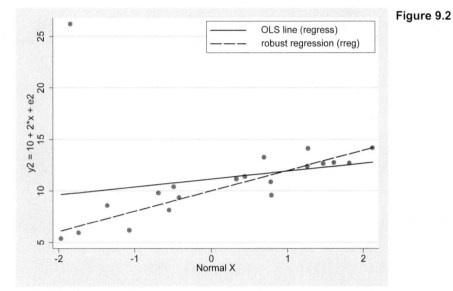

The **nolog** option above caused Stata not to print the iteration log. The **genwt(*rweight2*)** option saved robust weights as a variable named *rweight2*.

```
. predict resid2r, resid
. list y2 x resid2r rweight2
```

	y2	x	resid2r	rweight2
1.	5.37	-1.97	-.7403071	.94644465
2.	26.19	-1.85	19.84221	.
3.	5.93	-1.74	-.6354806	.96037073
4.	8.58	-1.36	1.262494	.8493384
5.	6.16	-1.07	-1.731421	.7257631
6.	9.8	-.69	1.156554	.87273631
7.	8.12	-.55	-.8005085	.93758391
8.	10.4	-.49	1.36075	.82606386
9.	9.35	-.42	.17222	.99712388
10.	11.16	.33	.4979582	.97581674
11.	11.4	.44	.5202664	.97360863
12.	13.26	.69	1.885513	.68048066
13.	10.88	.78	-.6725982	.95572833
14.	9.58	.79	-1.992389	.64644918
15.	12.41	1.26	-.0925257	.99913568

```
16. │ 14.14    1.27    1.617685    .75887073
17. │ 12.66    1.47   -.2581189    .99338589
18. │ 12.74    1.61   -.4551811    .97957817
19. │  12.7    1.81   -.8909839    .92307041
20. │ 14.19    2.12   -.0144787    .99997651
```

Residuals near zero produce weights near one; farther-out residuals get progressively lower weights. Observation #2 has been automatically set aside as too influential because of Cook's $D > 1$. **rreg** assigns its *rweight2* as "missing," so this observation has no effect on the final estimates. The same final estimates, although not the correct standard errors or tests, could be obtained using **regress** with analytical weights (results not shown):

. **regress** *y2 x* **[aweight = *rweight2*]**

Applied to the regression of *y2* on *x*, **qreg** also resists the outlier's influence and performs better than **regress**, but not as well as **rreg**. **qreg** appears less efficient than **rreg**, and in this sample its coefficient estimates are slightly farther from the true values of 10 and 2.

. **qreg** *y2 x*, **nolog**

Median regression   Number of obs =   20
  Raw sum of deviations   56.68 (about **10.88**)
  Min sum of deviations 36.20036   Pseudo R2   =   0.3613

y2	Coef.	Std. Err.	t	P>|t|	[95% Conf. Interval]	
x	1.821428	.4105945	4.44	0.000	.9588013	2.684055
_cons	10.115	.5088526	19.88	0.000	9.04594	11.18406

Monte Carlo researchers have also noticed that the standard errors calculated by **qreg** sometimes underestimate the true sample-to-sample variation, particularly with smaller samples. As an alternative, Stata provides the command **bsqreg**, which performs the same median or quantile regression as **qreg**, but employs bootstrapping (data resampling) to estimate the standard errors. The option **rep( )** controls the number of repetitions. Its default is **rep(20)**, which is enough for exploratory work. Before reaching "final" conclusions, we might take the time to draw 200 or more bootstrap samples. Both **qreg** and **bsqreg** fit identical models. In the example below, **bsqreg** also obtains similar standard errors. Chapter 16 returns to the topic of bootstrapping.

. **bsqreg** *y2 x*, **rep(50)**
(fitting base model)
(bootstrapping ....)

Median regression, bootstrap(50) SEs   Number of obs =   20
  Raw sum of deviations   56.68 (about **10.88**)
  Min sum of deviations 36.20036   Pseudo R2   =   0.3613

y2	Coef.	Std. Err.	t	P>|t|	[95% Conf. Interval]	
x	1.821428	.4056204	4.49	0.000	.9692515	2.673605
_cons	10.115	.4723573	21.41	0.000	9.122614	11.10739

X Outliers (Leverage)

rreg, **qreg**, and **bsqreg** deal comfortably with *y*-outliers, unless the observations with unusual *y* values have unusual *x* values (leverage) too. The *y3* and *x3* variables in *robust1.dta* present an extreme example of leverage. Variable *y3* is identical to *y2*. *x3* equals *x* for every observation except #2, which has an outlying *x3* value that results in high leverage.

The high leverage of observation #2, combined with its exceptional *y3* value, make it influential: **regress** and **qreg** both track this outlier, reporting that the "best-fitting" line has a negative slope (Figure 9.3).

```
. regress y3 x3
```

Source	SS	df	MS
Model	139.306724	1	139.306724
Residual	227.691018	18	12.649501
Total	366.997742	19	19.3156706

Number of obs = 20
F(1, 18) = 11.01
Prob > F = 0.0038
R-squared = 0.3796
Adj R-squared = 0.3451
Root MSE = 3.5566

y3	Coef.	Std. Err.	t	P>|t|	[95% Conf. Interval]
x3	-.6212248	.1871973	-3.32	0.004	-1.014512 -.227938
_cons	10.80931	.8063436	13.41	0.000	9.115244 12.50337

```
. predict yhat3o
. label variable yhat3o "OLS regression (regress)"

. qreg y3 x3, nolog
```

Median regression
  Raw sum of deviations   56.68 (about 10.88)
  Min sum of deviations 56.19466

Number of obs = 20

Pseudo R2 = 0.0086

y3	Coef.	Std. Err.	t	P>|t|	[95% Conf. Interval]
x3	-.6222217	.347103	-1.79	0.090	-1.351458 .1070146
_cons	11.36533	1.419214	8.01	0.000	8.383676 14.34699

```
. predict yhat3q
. label variable yhat3q "median regression (qreg)"

. rreg y3 x3, nolog
```

Robust regression

Number of obs = 19
F(1, 17) = 63.01
Prob > F = 0.0000

y3	Coef.	Std. Err.	t	P>|t|	[95% Conf. Interval]
x3	1.979015	.2493146	7.94	0.000	1.453007 2.505023
_cons	10.00897	.3071265	32.59	0.000	9.360986 10.65695

```
. predict yhat3r
. label variable yhat3r "robust regression (rreg)"
. graph twoway scatter y3 x3
        ||   line yhat3o x3, clpattern(solid) sort
        ||   line yhat3r x3, clpattern(longdash) sort
        ||   line yhat3q x3, clpattern(shortdash) sort ,
     ytitle("y3 = 10 + 2*x + e3") legend(order(4 3 2) position(5)
         ring(0) cols(1) margin(sides)) ylabel(-30(10)30)
```

Figure 9.3

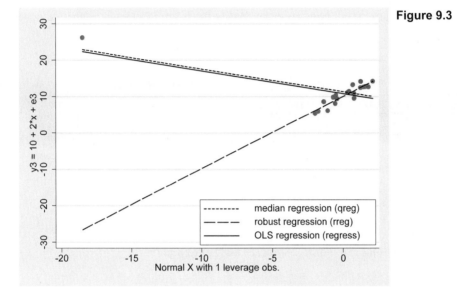

Figure 9.3 illustrates that **regress** and **qreg** are not robust against leverage (*x*-outliers). The **rreg** program, however, not only downweights large-residual observations (which by itself gives little protection against leverage), but also automatically sets aside observations with Cook's *D* (influence) statistics greater than 1. This happened when we regressed *y3* on *x3*; **rreg** ignored the one influential observation and produced a more reasonable regression line with a positive slope, based on the remaining 19 observations.

Setting aside high-influence observations, as done by **rreg**, provides a simple but not foolproof way to deal with leverage. More comprehensive methods, termed bounded-influence regression, also exist and could be implemented in a Stata program.

The examples in Figures 9.2 and 9.3 involve single outliers, but robust procedures can handle more. Too many severe outliers, or a cluster of similar outliers, might cause them to break down. But in such situations, which are often noticeable in diagnostic plots, the analyst must question whether fitting a linear model makes sense. It might be worthwhile to seek an explicit model for what is causing the outliers to be different.

Monte Carlo experiments (illustrated in Chapter 16) confirm that estimators like **rreg** and **qreg** generally remain unbiased, with better-than-OLS efficiency, when applied to heavy-tailed (outlier-prone) but symmetrical error distributions. The next section illustrates what can happen when errors have asymmetrical distributions.

Asymmetrical Error Distributions

The variable *e4* in *robust1.dta* has a skewed and outlier-filled distribution: *e4* equals *e1* (a standard normal variable) raised to the fourth power, and then adjusted to have 0 mean. These skewed errors, plus the linear relationship with *x*, define the variable *y4* = 10 + 2*x* + *e4*. Regardless of an error distribution's shape, OLS remains an unbiased estimator. Over the long run, its estimates should center on the true parameter values.

```
. regress y4 x
```

Source	SS	df	MS		Number of obs =	20
					F(1, 18) =	6.97
Model	155.870383	1	155.870383		Prob > F =	0.0166
Residual	402.341909	18	22.3523283		R-squared =	0.2792
					Adj R-squared =	0.2392
Total	558.212291	19	29.3795943		Root MSE =	4.7278

y4	Coef.	Std. Err.	t	P>|t|	[95% Conf. Interval]	
x	2.208388	.8362862	2.64	0.017	.4514157	3.96536
_cons	9.975681	1.062046	9.39	0.000	7.744406	12.20696

The same is not true for most robust estimators. Unless errors are symmetrical, the median line fit by **qreg**, or the biweight line fit by **rreg**, does not theoretically coincide with the expected-*y* line estimated by **regress**. So long as the errors' skew reflects only a small fraction of their distribution, **rreg** might exhibit little bias. But when the entire distribution is skewed, as with *e4*, **rreg** will downweight mostly one side, resulting in noticeably biased *y*-intercept estimates.

```
. rreg y4 x, nolog
```

Robust regression

					Number of obs =	20
					F(1, 18) =	1319.29
					Prob > F =	0.0000

y4	Coef.	Std. Err.	t	P>|t|	[95% Conf. Interval]	
x	1.952073	.0537435	36.32	0.000	1.839163	2.064984
_cons	7.476669	.0682518	109.55	0.000	7.333278	7.620061

Although the **rreg** *y*-intercept in Figure 9.4 is too low, the slope remains parallel to the OLS line and the true model. In fact, being less affected by outliers, the **rreg** slope (1.95) is closer to the true slope (2) and has a much smaller standard error than that of **regress**. This illustrates the tradeoff of using **rreg** or similar estimators with skewed errors: we risk getting biased estimates of the *y*-intercept, but can still expect unbiased and relatively precise estimates of other regression coefficients. In many applications, such coefficients are substantively more interesting than the *y*-intercept, making the tradeoff worthwhile. Moreover, the robust *t* and *F* tests, unlike those of OLS, do not assume normal errors.

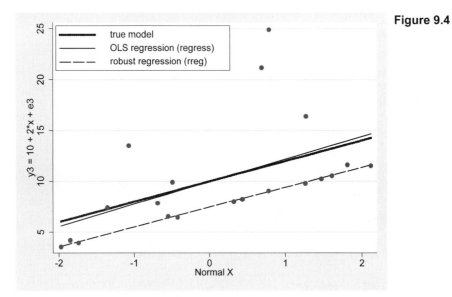

Figure 9.4

Robust Analysis of Variance

rreg can also perform robust analysis of variance or covariance once the model is recast in regression form. For illustration, consider the data on college faculty salaries in *faculty.dta*.

```
Contains data from c:\data\faculty.dta.
  obs:           226                          College faculty salaries
  vars:            6                          19 May 2008 20:00
  size:        3,842 (99.9% of memory free)

              storage  display    value
variable name  type    format     label      variable label

rank           byte    %8.0g      rank       Academic rank
gender         byte    %8.0g      sex        Gender (dummy variable)
female         byte    %8.0g                 Gender (effect coded)
assoc          byte    %8.0g                 Assoc Professor (effect coded)
full           byte    %8.0g                 Full Professor (effect coded)
pay            float   %9.0g                 Annual salary
```

Faculty salaries increase with rank. In this sample, men have higher average salaries:

```
. table gender rank, contents(mean pay)
```

Gender (dummy variable)	Academic rank		
	Assist	Assoc	Full
Male	29280	38622.22	52084.9
Female	28711.04	38019.05	47190

An ordinary (OLS) analysis of variance indicates that both *rank* and *gender* significantly affect salary. Their interaction is not significant.

`. anova pay rank gender rank*gender`

```
                    Number of obs =      226    R-squared     =  0.7305
                    Root MSE      = 5108.21    Adj R-squared =  0.7244
```

Source	Partial SS	df	MS	F	Prob > F
Model	1.5560e+10	5	3.1120e+09	119.26	0.0000
rank	7.6124e+09	2	3.8062e+09	145.87	0.0000
gender	127361829	1	127361829	4.88	0.0282
rank*gender	87997720.1	2	43998860.1	1.69	0.1876
Residual	5.7406e+09	220	26093824.5		
Total	2.1300e+10	225	94668810.3		

But salary is not normally distributed, and the senior-rank averages reflect the influence of a few highly paid outliers. Suppose we want to check these results by performing a robust analysis of variance. We need effect-coded versions of the *rank* and *gender* variables which this dataset also contains.

`. tabulate gender female`

Gender (dummy variable)	Gender (effect coded)		Total
	-1	1	
Male	149	0	149
Female	0	77	77
Total	149	77	226

`. tabulate rank assoc`

Academic rank	Assoc Professor (effect coded)			Total
	-1	0	1	
Assist	64	0	0	64
Assoc	0	0	105	105
Full	0	57	0	57
Total	64	57	105	226

```
. tab rank full
```

Academic rank	Full Professor (effect coded) -1	0	1	Total
Assist	64	0	0	64
Assoc	0	105	0	105
Full	0	0	57	57
Total	64	105	57	226

If *faculty.dta* did not already have these effect-coded variables (*female*, *assoc*, and *full*), we could create them from *gender* and *rank* using a series of **generate** and **replace** statements. We also need two interaction terms representing female associate professors and female full professors:

```
. generate femassoc = female*assoc
. generate femfull = female*full
```

Males and assistant professors are "omitted categories" in this example. Now we can duplicate the previous ANOVA using regression:

```
. regress pay assoc full female femassoc femfull
```

Source	SS	df	MS		Number of obs =	226
					F(5, 220) =	119.26
Model	1.5560e+10	5	3.1120e+09		Prob > F =	0.0000
Residual	5.7406e+09	220	26093824.5		R-squared =	0.7305
					Adj R-squared =	0.7244
Total	2.1300e+10	225	94668810.3		Root MSE =	5108.2

pay	Coef.	Std. Err.	t	P>|t|	[95% Conf. Interval]	
assoc	-663.8995	543.8499	-1.22	0.223	-1735.722	407.9229
full	10652.92	783.9227	13.59	0.000	9107.957	12197.88
female	-1011.174	457.6938	-2.21	0.028	-1913.199	-109.1483
femassoc	709.5864	543.8499	1.30	0.193	-362.236	1781.409
femfull	-1436.277	783.9227	-1.83	0.068	-2981.237	108.6819
_cons	38984.53	457.6938	85.18	0.000	38082.51	39886.56

```
. test assoc full

 ( 1)   assoc = 0
 ( 2)   full = 0

       F(  2,   220) =   145.87
            Prob > F =    0.0000
```

```
. test female

 ( 1)   female = 0

       F(  1,   220) =     4.88
            Prob > F =    0.0282
```

```
. test femassoc femfull

 ( 1)   femassoc = 0
 ( 2)   femfull = 0

      F(  2,   220) =     1.69
           Prob > F =     0.1876
```

regress followed by the appropriate **test** commands obtains exactly the same R^2 and F test results that we found earlier using **anova**. Predicted values from this regression equal the mean salaries.

```
. predict predpay1
(option xb assumed; fitted values)
. label variable predpay1 "OLS predicted salary"
. table gender rank, contents(mean predpay1)
```

Gender (dummy variable)	Academic rank		
	Assist	Assoc	Full
Male	29280	38622.22	52084.9
Female	28711.04	38019.05	47190

Predicted values (means), R^2, and F tests would also be the same regardless of which categories we chose to omit from the regression. Our "omitted categories," males and assistant professors, are not really absent. Their information is implied by the included categories: if a faculty member is not female, he must be male, and so forth.

To perform a robust analysis of variance, apply **rreg** to this model:

```
. rreg pay assoc full female femassoc femfull, nolog
```

```
Robust regression                           Number of obs =      226
                                            F(  5,   220) =   138.25
                                            Prob > F      =   0.0000
```

pay	Coef.	Std. Err.	t	P>|t|	[95% Conf. Interval]	
assoc	-315.6463	458.1588	-0.69	0.492	-1218.588	587.2956
full	9765.296	660.4048	14.79	0.000	8463.767	11066.83
female	-749.4949	385.5778	-1.94	0.053	-1509.394	10.40397
femassoc	197.7833	458.1588	0.43	0.666	-705.1587	1100.725
femfull	-913.348	660.4048	-1.38	0.168	-2214.878	388.1815
_cons	38331.87	385.5778	99.41	0.000	37571.97	39091.77

```
. test assoc full

 ( 1)   assoc = 0
 ( 2)   full = 0

      F(  2,   220) =   182.67
           Prob > F =     0.0000
```

```
. test female
```

(1)   female = 0

```
        F(  1,   220) =     3.78
              Prob > F =     0.0532
```

```
. test femassoc femfull
```

(1)   femassoc = 0
(2)   femfull = 0

```
        F(  2,   220) =     1.16
              Prob > F =     0.3144
```

rreg downweights several outliers, mainly highly-paid male full professors. To see the robust means, again use predicted values:

```
. predict predpay2
(option xb assumed; fitted values)
. label variable predpay2 "Robust predicted salary"
. table gender rank, contents(mean predpay2)
```

Gender (dummy variable)	Academic rank		
	Assist	Assoc	Full
Male	28916.15	38567.93	49760.01
Female	28848.29	37464.51	46434.32

The male–female salary gap among assistant and full professors appears smaller if we use robust means. It does not entirely vanish, however, and the gender gap among associate professors slightly widens.

With effect coding and suitable interaction terms, **regress** can duplicate ANOVA exactly. **rreg** can do parallel analyses, testing for differences among robust means instead of ordinary means (as **regress** and **anova** do). Used in similar fashion, **qreg** opens the third possibility of testing for differences among medians. For comparison, here is a quantile regression version of the faculty pay analysis:

```
. qreg pay assoc full female femassoc femfull, nolog
```

```
Median regression                             Number of obs =        226
  Raw sum of deviations   1738010 (about 37360)
  Min sum of deviations    798870              Pseudo R2     =     0.5404
```

pay	Coef.	Std. Err.	t	P>|t|	[95% Conf. Interval]	
assoc	-760	440.1693	-1.73	0.086	-1627.488	107.4881
full	10335	615.7735	16.78	0.000	9121.43	11548.57
female	-623.3333	365.1262	-1.71	0.089	-1342.926	96.25942
femassoc	-156.6667	440.1693	-0.36	0.722	-1024.155	710.8215
femfull	-691.6667	615.7735	-1.12	0.263	-1905.236	521.9032
_cons	38300	365.1262	104.90	0.000	37580.41	39019.59

```
. test assoc full

( 1)   assoc = 0
( 2)   full = 0

        F(  2,    220) =   208.94
             Prob > F =    0.0000


. test female

( 1)   female = 0

        F(  1,    220) =     2.91
             Prob > F =    0.0892


. test femassoc femfull

( 1)   femassoc = 0
( 2)   femfull = 0

        F(  2,    220) =     1.60
             Prob > F =    0.2039


. predict predpay3
(option xb assumed; fitted values)
. label variable predpay3 "Median predicted salary"
. table gender rank, contents(mean predpay3)
```

Gender (dummy variable)	Academic rank		
	Assist	Assoc	Full
Male	28500	38320	49950
Female	28950	36760	47320

Predicted values from this quantile regression closely resemble the median salaries in each subgroup, as we can verify directly:

```
. table gender rank, contents(median pay)
```

Gender (dummy variable)	Academic rank		
	Assist	Assoc	Full
Male	28500	38320	49950
Female	28950	36590	46530

qreg thus allows us to fit models analogous to *N*-way ANOVA or ANCOVA, but involving .5 quantiles or approximate medians instead of the usual means. In theory, .5 quantiles and medians are the same. In practice, quantiles are approximated from actual sample data values, whereas the median is calculated by averaging the two central values, if a subgroup contains an even number of observations. The sample median and .5 quantile approximations then can be different, but in a way that does not much affect model interpretation.

Further rreg and qreg Applications

Diagnostic statistics and plots (Chapter 7) and nonlinear transformations (Chapter 8) extend the usefulness of robust procedures as they do in ordinary regression. With transformed variables, **rreg** or **qreg** fit curvilinear regression models. **rreg** can also robustly perform simpler types of analysis. To obtain a 90% confidence interval for the mean of a single variable, y, we could type either the usual confidence-interval command **ci**:

```
. ci y, level(90)
```

or, we could get exactly the same mean and interval through a regression with no x variables:

```
. regress y, level(90)
```

Similarly, we can obtain robust mean with 90% confidence interval by typing

```
. rreg y, level(90)
```

qreg could be used in the same way, but keep in mind the previous section's note about how a .5 quantile found by **qreg** might differ from a sample median. In any of these commands, the **level( )** option specifies the desired degree of confidence. If we omit this option, Stata automatically displays a 95% confidence interval.

To compare two means, analysts typically employ a two-sample t test (**ttest**) or one-way analysis of variance (**oneway** or **anova**). As seen earlier, we can perform equivalent tests (yielding identical t and F statistics) with regression, for example, by regressing the measurement variable on a dummy variable (here called *group*) representing the two categories:

```
. regress y group
```

A robust version of this test results from typing the following command:

```
. rreg y group
```

qreg performs median regression by default, but it is actually a more general tool. It can fit linear models for any quantile of y, not just the median (.5 quantile). For example, commands such as the following analyze how the first quartile (.25 quantile) of y changes with x.

```
. qreg y x, quant(.25)
```

Assuming constant error variance, the slopes of the .25 and .75 quantile lines should be roughly the same. **qreg** thus could perform a check for heteroskedasticity or subtle kinds of nonlinearity.

Robust Estimates of Variance — 1

Both **rreg** and **qreg** tend to perform better than OLS (**regress** or **anova**) in the presence of outlier-prone, nonnormal errors. All of these procedures share the common assumption that errors follow independent and identical distributions, however. If the distributions of errors vary across x values or observations, then the standard errors calculated by **anova**, **regress**, **rreg**, or **qreg** probably will understate the true sample-to-sample variation, and yield unrealistically narrow confidence intervals.

regress and some other model fitting commands (although not **rreg** or **qreg**) have an option that estimates standard errors without relying on the strong and sometimes implausible assumptions of independent, identically distributed errors. This option uses an approach derived independently by Huber, White, and others that is sometimes referred to as a sandwich estimator of variance. (Type **help vce option**, or see vce_option in the Stata *Reference Manual*, for technical details.) The artificial dataset (*robust2.dta*) provides our first illustration.

```
Contains data from c:\data\robust2.dta
  obs:          500                    Robust regression examples 2
                                         (artificial data)
  vars:          12                    19 May 2008 19:59
  size:      26,500 (99.9% of memory free)
```

variable name	storage type	display format	value label	variable label
x	float	%9.0g		Standard normal x
e5	float	%9.0g		Standard normal errors
y5	float	%9.0g		y5 = 10 + 2*x + e5 (normal i.i.d. errors)
e6	float	%9.0g		Contaminated normal errors: 95% N(0,1), 5%(N(0,10)
y6	float	%9.0g		y6 = 10 + 2*x + e6 (Contaminated normal errors)
e7	float	%9.0g		Centered chi-square(1) errors
y7	float	%9.0g		y7 = 10 + 2*x + e7 (skewed errors)
e8	float	%9.0g		Normal errors, variance increases with x
y8	float	%9.0g		y8 = 10 + 2*x + e8 (heteroskedasticity)
group	byte	%9.0g		
e9	float	%9.0g		Normal errors, variance increases with x, mean & variance increase with cluster
y9	float	%9.0g		y9 = 10 + 2*x + e9 (heteroskedasticity & correlated errors)

When we regress *y8* on *x*, we obtain a significant positive slope. A scatterplot shows strong heteroskedasticity, however (Figure 9.5). Variation around the regression line increases with *x*. Because errors do not appear to be identically distributed at all values of *x*, the standard errors, confidence intervals, and tests printed by **regress** are untrustworthy. **rreg** or **qreg** would face the same problem.

```
. regress y8 x
```

Source	SS	df	MS
Model	1607.35658	1	1607.35658
Residual	5975.19162	498	11.9983767
Total	7582.5482	499	15.1954874

```
Number of obs =     500
F(  1,    498) =  133.96
Prob > F       =  0.0000
R-squared      =  0.2120
Adj R-squared  =  0.2104
Root MSE       =  3.4639
```

y8	Coef.	Std. Err.	t	P>|t|	[95% Conf. Interval]	
x	1.819032	.1571612	11.57	0.000	1.510251	2.127813
_cons	10.06642	.154919	64.98	0.000	9.762047	10.3708

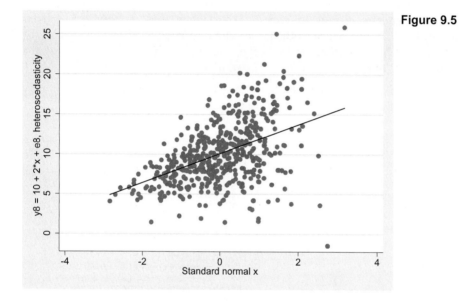

Figure 9.5

More credible standard errors and confidence intervals for this OLS regression can be obtained by using the **vce(robust)** option:

```
. regress y8 x, vce(robust)
```

Linear regression

```
Number of obs =     500
F(  1,    498) =   83.80
Prob > F       =  0.0000
R-squared      =  0.2120
Root MSE       =  3.4639
```

y8	Coef.	Robust Std. Err.	t	P>|t|	[95% Conf. Interval]	
x	1.819032	.1987122	9.15	0.000	1.428614	2.209449
_cons	10.06642	.1561846	64.45	0.000	9.759561	10.37328

Although the fitted model remains unchanged, the robust standard error for the slope is 27% larger (.199 vs. .157) than its nonrobust counterpart. With the **vce(robust)** option, the regression output does not show the usual ANOVA sums of squares because these no longer have their customary interpretation.

The rationale underlying these robust standard-error estimates is explained in the *User's Guide*. Briefly, we give up on the classical goal of estimating true population parameters (β's) for a model such as

$$y_i = \beta_0 + \beta_1 x_i + \epsilon_i$$

Instead, we pursue the less ambitious goal of simply estimating the sample-to-sample variation that our b coefficients might have, if we drew many random samples and applied OLS repeatedly to calculate b values for a model such as

$$y_i = b_0 + b_1 x_i + e_i$$

We do not assume that these b estimates will converge on some "true" population parameter. Confidence intervals formed using the robust standard errors therefore lack the classical interpretation of having a certain likelihood (across repeated sampling) of containing the true value of β. Rather, the robust confidence intervals have a certain likelihood (across repeated sampling) of containing b, defined as the value upon which sample b estimates converge. Thus, we pay for relaxing the identically-distributed-errors assumption by settling for a less impressive conclusion.

Robust Estimates of Variance — 2

Another robust-variance option, **vce(cluster *clustervar*)**, allows us to relax the independent-errors assumption in a limited way, when errors are correlated within subgroups or clusters of the data. The data in *attract.dta* describe an undergraduate social experiment that can be used for illustration. In this experiment, 51 college students were asked to individually rate the attractiveness, on a scale from 1 to 10, of photographs of unknown men and women. The rating exercise was repeated by each participant, given the same photos shuffled in random order, on four occasions during evening social events. Variable *ratemale* is the mean rating each participant gave to all the male photos in one sitting, and *ratefem* is the mean rating given to female photos. *gender* records the participant's (rater's) own gender, and *bac* his or her blood alcohol content at the time, measured by Breathalyzer.

```
Contains data from c:\data\attract.dta
  obs:           204                    Perceived attractiveness and
                                          drinking (D. C. Hamilton 2003)
  vars:            8                    19 May 2008 19:59
  size:        6,324 (99.9% of memory free)
```

variable name	storage type	display format	value label	variable label
id	byte	%9.0g		Participant number
gender	byte	%9.0g	sex	Participant gender (female)
bac	float	%9.0g		Blood alchohol content
genbac	float	%9.0g		gender*bac interaction
relstat	byte	%9.0g	rel	Relationship status (single)
drinkfrq	float	%9.0g		Days drinking in previous week
ratefem	float	%9.0g		Rated attractiveness of females
ratemale	float	%9.0g		Rated attractiveness of males

```
Sorted by:  id
```

Although the data contain 204 observations, these represent only 51 individual participants. It seems reasonable to assume that disturbances (unmeasured influences on the ratings) were correlated across the repetitions by each individual. Viewing each participant's four rating sessions as a cluster should yield more realistic standard error estimates. Adding the option **vce(cluster** *id***)** to a regression command, as seen below, obtains robust standard errors across clusters defined by *id* (individual participant).

```
. regress ratefem bac gender genbac, vce(cluster id)
```

```
Linear regression                          Number of obs =      204
                                           F(  3,    50) =     7.75
                                           Prob > F      =   0.0002
                                           R-squared     =   0.1264
                                           Root MSE      =   1.1219

                        (Std. Err. adjusted for 51 clusters in id)
```

		Robust					
ratefem	Coef.	Std. Err.	t	P>|t|	[95% Conf.	Interval]
bac	2.896741	.8543378	3.39	0.001	1.180753	4.612729
gender	-.7299888	.3383096	-2.16	0.036	-1.409504	-.0504741
genbac	.2080538	1.708146	0.12	0.904	-3.222859	3.638967
_cons	6.486767	.229689	28.24	0.000	6.025423	6.94811

Blood alcohol content (*bac*) has a significant positive effect: as *bac* goes up, predicted attractiveness rating of female photos increases as well. Gender (female) has a negative effect: female participants tended to rate female photos as somewhat less attractive (about .73 lower) than male participants did. The interaction of *gender* and *bac* is weak (.21). The intercept- and slope-dummy variable regression model, approximately

$$\text{predicted } ratefem = 6.49 + 2.90bac - .73gender + .21genbac$$

can be reduced for male participants (*gender* = 0) to

$$\text{predicted } ratefem = 6.49 + 2.90bac - (.73 \times 0) + (.21 \times 0 \times bac)$$
$$= 6.49 + 2.90bac$$

and for female participants (*gender* = 1) to

$$\text{predicted } ratefem = 6.49 + 2.90bac - (.73 \times 1) + (.21 \times 1 \times bac)$$
$$= 6.49 + 2.90bac - .73 + .21bac$$
$$= 5.76 + 3.11bac$$

The slight difference between the effects of alcohol on males (2.90) and females (3.11) equals the interaction coefficient, .21.

Attractiveness ratings for photographs of males were likewise positively affected by blood alcohol content. Gender has a stronger effect on the ratings of male photos: female participants tended to give male photos much higher ratings than male participants did. For male-photo ratings, the *gender* × *bac* interaction is substantial (–4.36), although it falls short of the .05 significance level.

```
. regress ratemal bac gender genbac, vce(cluster id)
```

Linear regression

```
                                          Number of obs =      201
                                          F( 3,    50) =    10.96
                                          Prob > F      =   0.0000
                                          R-squared     =   0.3516
                                          Root MSE      =   1.3931
```

(Std. Err. adjusted for **51** clusters in id)

ratemale	Coef.	Robust Std. Err.	t	P>|t|	[95% Conf. Interval]	
bac	4.246042	2.261792	1.88	0.066	-.2969005	8.788985
gender	2.443216	.4529047	5.39	0.000	1.53353	3.352902
genbac	-4.364301	3.573689	-1.22	0.228	-11.54227	2.813663
_cons	3.628043	.2504253	14.49	0.000	3.125049	4.131037

The regression equation for ratings of male photos by male participants is approximately

$$\text{predicted } ratemale = 3.63 + 4.25 bac + (2.44 \times 0) - (4.36 \times 0 \times bac)$$
$$= 3.63 + 4.25 bac$$

and for rating of male photos by female participants,

$$\text{predicted } ratemale = 3.63 + 4.25 bac + (2.44 \times 1) - (4.36 \times 1 \times bac)$$
$$= 6.07 - 0.11 bac$$

The difference between the substantial alcohol effect on male participants (4.25) and the near-zero alcohol effect on females (–0.11) equals the interaction coefficient, –4.36. In this sample, males' ratings of male photos increase steeply, and females' ratings of male photos remain virtually steady, as the rater's *bac* increases.

Figure 9.6 visualizes these results in a graph. We see positive *rating–bac* relationships across all subplots except for females rating males. The graphs also show other gender differences, including higher *bac* values among male participants.

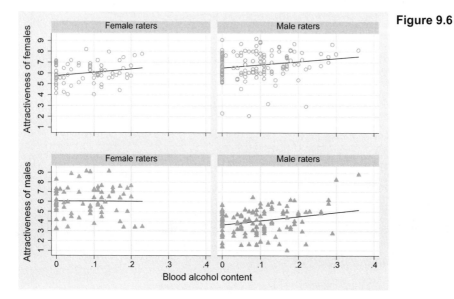

Figure 9.6

OLS regression with robust standard errors, estimated by **regress** with the **vce(robust)** option, should not be confused with the robust regression estimated by **rreg**. Despite similar-sounding names, the two procedures are unrelated, and solve different problems.

10

Logistic Regression

The regression and ANOVA methods described in Chapters 5 through 9 require measured dependent variables. Stata also offers a full range of techniques for modeling categorical, ordinal, and censored dependent variables. The following list gives some idea of the variety of methods available. For more details on specific commands, type **help** *command*. Long and Freese (2006) provide an excellent source on Stata's main methods for limited dependent variables; also see Hosmer and Lemeshow (2000).

asclogit   Alternative-specific conditional logit (McFadden's choice)

asmprobit   Alternative-specific multinomial probit regression.

asroprobit   Alternative-specific rank-ordered probit regression.

bilogit   Bivariate probit regression.

binreg   Binomial regression (generalized linear models).

blogit   Logit estimation with grouped (blocked) data.

bprobit   Probit estimation with grouped (blocked) data.

clogit   Conditional fixed-effects logistic regression.

cloglog   Complementary log-log estimation.

cnreg   Censored-normal regression, assuming that y follows a Gaussian distribution but is censored at a point that might vary from observation to observation.

constraint   Defines, lists, and drops linear constraints.

dprobit   Probit regression giving changes in probabilities instead of coefficients.

exlogistic   Exact logistic regression.

glm   Generalized linear models. Includes option to model logistic, probit, or complementary log-log links. Allows response variable to be binary or proportional for grouped data.

glogit   Logit regression for grouped data.

gprobit   Probit regression for grouped data.

heckprob   Probit estimation with selection.

hetprob   Heteroskedastic probit estimation.

intreg   Interval regression, where y is either point data, interval data, left-censored data, or right-censored data.

ivprobit   Probit with continuous exogenous regressors.

logistic   Logistic regression, giving odds ratios.

logit   Logistic regression — similar to **logistic**, but giving coefficients instead of odds ratios.

mlogit   Multinomial logistic regression, for polytomous y variables.

nlogit   Nested logit estimation.

ologit   Logistic regression for ordinal y variables.

oprobit   Probit regression for ordinal y variables.

probit	Probit regression, for dichotomous *y* variable.
rologit	Rank-ordered logit model for rankings (also known as the Plackett–Luce model, exploded logit model, or choice-based conjoint analysis).
scobit	Skewed probit estimation.
slogit	Stereotype logistic regression.
svy: logit	Logistic regression with complex survey data. Survey (**svy**) versions of many other categorical-variable modeling commands also exist (Chapter 14).
tobit	Tobit regression, assuming *y* follows a Gaussian distribution but is censored at a known, fixed point (see **help cnreg** for a more general version).
xtmelogit	Binary logit mixed or multilevel models, with fixed and random effects. Type **help xtmelogit** for more about this powerful new command, which is illustrated in Chapter 15. Many other panel-data commands exist; type **help xt** for a list.

After most model-fitting commands, **predict** can calculate predicted values or probabilities. **predict** also obtains appropriate diagnostic statistics, such as those described for logistic regression in Hosmer and Lemeshow (2000). Specific **predict** options depend on the type of model just fitted. A different post-fitting command, **predictnl**, obtains nonlinear predictions and their confidence intervals (see **help predictnl**).

Examples of several of these commands appear in the next section. The diverse methods for modeling categorical or limited dependent variables can be accessed under a number of different menus, including

Statistics > Binary outcomes

Statistics > Ordinal outcomes

Statistics > Categorical outcomes

Statistics > Generalized linear models

Statistics > Longitudinal/panel data

Statistics > Linear models and related

Statistics > Multilevel mixed-effects models

After the Example Commands section, the remainder of this chapter concentrates on an important family of methods called logit or logistic regression. We review basic logit methods for dichotomous, ordinal, and polytomous dependent variables.

Example Commands

. logistic *y x1 x2 x3*

Performs logistic regression of {0,1} variable *y* on predictors *x1*, *x2*, and *x3*. Predictor variable effects are reported as odds ratios. A closely related command, **logit**, performs essentially the same analysis, but reports effects as log-odds regression coefficients. The underlying models fit by **logistic** and **logit** are the same, so subsequent predictions or diagnostic tests will be identical.

. estat gof

Presents a Pearson chi-squared goodness-of-fit test for the fitted logistic model: observed versus expected frequencies of *y* = 1, using cells defined by the covariate (*x*-variable) patterns. When a large number of *x* patterns exist, we might want to group them according to estimated probabilities. **estat gof, group(10)** would perform the test with 10 approximately equal-size groups.

. estat classification

Presents classification statistics and classification table. **estat classification**, **lroc**, and **lsens** (see below) are particularly useful when the point of analysis is classification. These commands all refer to the previously-fit logistic model.

. lroc

Graphs the receiver operating characteristic (ROC) curve, and calculates area under the curve.

. lsens

Graphs both sensitivity and specificity versus the probability cutoff.

. predict *phat*

Generates a new variable (here arbitrarily named *phat*) equal to predicted probabilities that *y* = 1 based on the most recent **logistic** model.

. predict *dX2*, **dx2**

Generates a new variable (arbitrarily named *dX2*), the diagnostic statistic measuring change in Pearson chi-squared, from the most recent **logistic** analysis.

. mlogit *y x1 x2 x3*, **base(3) rrr nolog**

Performs multinomial logistic regression of multiple-outcome variable *y* on three *x* variables. Option **base(3)** specifies *y* = 3 as the base category for comparison; **rrr** calls for relative risk ratios instead of regression coefficients; and **nolog** suppresses display of the log likelihood on each iteration.

. predict *P2*, **outcome(2)**

Generates a new variable (arbitrarily named *P2*) representing the predicted probability that *y* = 2, based on the most recent **mlogit** analysis.

. **glm** *success x1 x2 x3*, **family(binomial** *trials***) eform**

Performs a logistic regression via generalized linear modeling using tabulated rather than individual-observation data. The variable *success* gives the number of times that the outcome of interest occurred, and *trials* gives the number of times it could have occurred for each combination of the predictors *x1*, *x2*, and *x3*. That is, *success / trials* would equal the proportion of times that an outcome such as "patient recovers" occurred. The **eform** option asks for results in the form of odds ratios ("exponentiated form") rather than logit coefficients.

. **cnreg** *y x1 x2 x3*, **censored(***cen***)**

Performs censored-normal regression of measurement variable *y* on three predictors *x1*, *x2*, and *x3*. If an observation's true *y* value is unknown due to left or right censoring, it is replaced for this regression by the nearest *y* value at which censoring occurs. The censoring variable *cen* is a {−1,0,1} indicator of whether each observation's value of *y* has been left censored, not censored, or right censored.

Space Shuttle Data

Our main example for this chapter, *shuttle.dta*, involves data covering the first 25 flights of the U.S. space shuttle. These data contain evidence that, if properly analyzed, might have persuaded NASA officials not to launch *Challenger* on its fatal flight in 1985 (the 25th shuttle flight, designated STS 51-L). The data are drawn from the *Report of the Presidential Commission on the Space Shuttle Challenger Accident* (1986) and from Tufte (1997). Tufte's book contains an excellent discussion about data and analytical issues. His comments regarding specific shuttle flights are included as a string variable in these data.

```
Contains data from c:\data\shuttle.dta
  obs:            25                          First 25 space shuttle flights
  vars:            8                          21 May 2008 07:36
  size:        1,775 (99.9% of memory free)
```

variable name	storage type	display format	value label	variable label
flight	byte	%8.0g	flbl	Flight
month	byte	%8.0g		Month of launch
day	byte	%8.0g		Day of launch
year	int	%8.0g		Year of launch
distress	byte	%8.0g	dlbl	Thermal distress incidents
temp	byte	%8.0g		Joint temperature, degrees F
damage	byte	%9.0g		Damage severity index (Tufte 1997)
comments	str55	%55s		Comments (Tufte 1997)

. **list** *flight-temp*, **sepby(***year***)**

	flight	month	day	year	distress	temp
1.	STS-1	4	12	1981	none	66
2.	STS-2	11	12	1981	1 or 2	70
3.	STS-3	3	22	1982	none	69
4.	STS-4	6	27	1982	.	80
5.	STS-5	11	11	1982	none	68

6.	STS-6	4	4	1983	1 or 2	67
7.	STS-7	6	18	1983	none	72
8.	STS-8	8	30	1983	none	73
9.	STS-9	11	28	1983	none	70
10.	STS_41-B	2	3	1984	1 or 2	57
11.	STS_41-C	4	6	1984	3 plus	63
12.	STS_41-D	8	30	1984	3 plus	70
13.	STS_41-G	10	5	1984	none	78
14.	STS_51-A	11	8	1984	none	67
15.	STS_51-C	1	24	1985	3 plus	53
16.	STS_51-D	4	12	1985	3 plus	67
17.	STS_51-B	4	29	1985	3 plus	75
18.	STS_51-G	6	17	1985	3 plus	70
19.	STS_51-F	7	29	1985	1 or 2	81
20.	STS_51-I	8	27	1985	1 or 2	76
21.	STS_51-J	10	3	1985	none	79
22.	STS_61-A	10	30	1985	3 plus	75
23.	STS_61-B	11	26	1985	1 or 2	76
24.	STS_61-C	1	12	1986	3 plus	58
25.	STS_51-L	1	28	1986	.	31

This chapter examines three of the *shuttle.dta* variables:

distress   The number of "thermal distress incidents," in which hot gas blow-through or charring damaged joint seals of a flight's booster rockets. Burn-through of a booster joint seal precipitated the *Challenger* disaster. Many previous flights had experienced less severe damage, so the joint seals were known to be a source of possible danger.

temp   The calculated joint temperature at launch time, in degrees Fahrenheit. Temperature depends largely on weather. Rubber O-rings sealing the booster rocket joints become less flexible when cold.

date   Date, measured in days elapsed since January 1, 1960 (an arbitrary starting point). *date* is generated from the month, day, and year of launch using the **mdy** function (month-day-year to elapsed time; see **help dates**):

```
. generate date = mdy(month, day, year)
. format %td date
. label variable date "Date (days since 1/1/60)"
```

Elapsed date matters because several changes over the course of the shuttle program might have made it riskier. Booster rocket walls were thinned to save weight and increase payloads, and joint seals were subjected to higher-pressure testing. Furthermore, the reusable shuttle hardware was aging. So we might ask, did the probability of booster joint damage (one or more distress incidents) increase with launch date?

distress is a labeled numeric variable:

```
. tabulate distress
```

Thermal distress incidents	Freq.	Percent	Cum.
none	9	39.13	39.13
1 or 2	6	26.09	65.22
3 plus	8	34.78	100.00
Total	23	100.00	

Ordinarily, **tabulate** displays the labels, but the **nolabel** option reveals that the underlying numeric codes are 0 = "none", 1 = "1 or 2", and 2 = "3 plus".

```
. tabulate distress, nolabel
```

Thermal distress incidents	Freq.	Percent	Cum.
0	9	39.13	39.13
1	6	26.09	65.22
2	8	34.78	100.00
Total	23	100.00	

We can use these codes to create a new dummy variable, *any*, coded 0 for no distress and 1 for one or more distress incidents:

```
. generate any = distress
(2 missing values generated)
. replace any = 1 if distress == 2
(8 real changes made)
. label variable any "Any thermal distress"
```

To check what these **generate** and **replace** commands accomplished, and be sure that missing values were handled correctly,

```
. tabulate distress any, miss
```

Thermal distress incidents	Any thermal distress 0	1	.	Total
none	9	0	0	9
1 or 2	0	6	0	6
3 plus	0	8	0	8
.	0	0	2	2
Total	9	14	2	25

Logistic regression models how a {0,1} dichotomy such as *any* depends on one or more x variables. The syntax of **logit** resembles that of **regress** and most other model-fitting commands, with the dependent variable listed first.

```
. logit any date, coef

Iteration 0:   log likelihood = -15.394543
Iteration 1:   log likelihood =  -13.01923
Iteration 2:   log likelihood = -12.991146
Iteration 3:   log likelihood = -12.991096
```

```
Logistic regression                        Number of obs   =        23
                                           LR chi2(1)      =      4.81
                                           Prob > chi2     =    0.0283
Log likelihood = -12.991096                Pseudo R2       =    0.1561
```

any	Coef.	Std. Err.	z	P>|z|	[95% Conf. Interval]	
date	.0020907	.0010703	1.95	0.051	-6.93e-06	.0041884
_cons	-18.13116	9.517217	-1.91	0.057	-36.78456	.5222396

The **logit** iterative estimation procedure maximizes the log of the likelihood function, shown above the logistic table. At iteration 0, the log likelihood describes the fit of a model including only the constant. The last log likelihood describes the fit of the final model,

$$L = -18.13116 + .0020907 date \qquad [10.1]$$

where L represents the predicted logit, or log odds, of any distress incidents:

$$L = \ln[P(any = 1) / P(any = 0)] \qquad [10.2]$$

An overall χ^2 test at the upper right evaluates the null hypothesis that all coefficients in the model, except the constant, equal zero,

$$\chi^2 = -2(\ln \mathcal{L}_i - \ln \mathcal{L}_f) \qquad [10.3]$$

where $\ln \mathcal{L}_i$ is the initial or iteration 0 (model with constant only) log likelihood, and $\ln \mathcal{L}_f$ is the final iteration's log likelihood. Here,

$$\chi^2 = -2[-15.394543 - (-12.991096)]$$
$$= 4.81$$

The probability of a greater χ^2, with 1 degree of freedom (the difference in complexity between initial and final models), is low enough (.0283) to reject the null hypothesis in this example. Consequently, *date* does have a significant effect.

Less accurate, though convenient, tests are provided by the asymptotic z (standard normal) statistics displayed with **logit** results. With one predictor variable, that predictor's z statistic and the overall χ^2 statistic test equivalent hypotheses, analogous to the usual t and F statistics in simple OLS regression. Unlike their OLS counterparts, the logit z approximation and χ^2 tests sometimes disagree (they do here). The χ^2 test has more general validity.

Like some other Stata maximum-likelihood procedures, **logit** displays a pseudo R^2:

$$\text{pseudo } R^2 = 1 - \ln \mathcal{L}_f / \ln \mathcal{L}_i \qquad [10.4]$$

For this example,

$$\text{pseudo } R^2 = 1 - (-12.991096) / (-15.394543)$$
$$= .1561$$

Although pseudo R^2 statistics provide a quick way to describe or compare the fit of different models for the same dependent variable, they lack the straightforward explained-variance interpretation of true R^2 in OLS regression.

After **logit**, the **predict** command (with no options) obtains predicted probabilities,

$$Phat = 1 / (1 + e^{-L})$$ [10.5]

Graphed against *date*, these predicted probabilities follow an S-shaped logistic curve. Because we specified **format %td *date*** earlier after we defined variable *date*, values are appropriately labeled on the horizontal or time axis in Figure 10.1.

```
. predict Phat
(option pr assumed; Pr(any))
. label variable Phat "Predicted P(distress >= 1)"
. graph twoway connected Phat date, xtitle("Launch date") sort
```

Figure 10.1

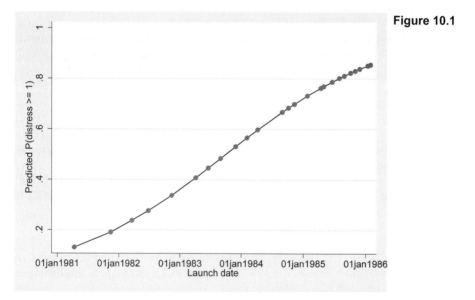

The coefficient in this **logit** example (.0020907) describes *date*'s effect on the logit or log odds that any thermal distress incidents occur. Each additional day increases the predicted log odds of thermal distress by .0020907. Equivalently, we could say that each additional day multiplies predicted odds of thermal distress by $e^{.0020907} = 1.0020929$; each 100 days therefore multiplies the odds by $(e^{.0020907})^{100} = 1.23$. ($e \approx 2.71828$, the base number for natural logarithms.) Stata can make these calculations utilizing the _b[*varname*] coefficients stored after any estimation:

```
. display exp(_b[date])
1.0020929
```

```
. display exp(_b[date])^100
1.2325359
```

We could also simply include an **or** (odds ratio) option on the **logit** command line. A third alternative way to obtain odds ratios employs the **logistic** command, described in the next section. **logistic** fits exactly the same model as **logit**, but its default output table displays odds ratios rather than coefficients.

Using Logistic Regression

Here is the same regression seen earlier, but using **logistic** instead of **logit**:

```
. logistic any date
```

Logistic regression

```
Number of obs   =        23
LR chi2(1)      =      4.81
Prob > chi2     =    0.0283
Pseudo R2       =    0.1561
```

Log likelihood = -12.991096

any	Odds Ratio	Std. Err.	z	P>|z|	[95% Conf. Interval]
date	1.002093	.0010725	1.95	0.051	.9999931 1.004197

Note the identical log likelihoods and χ^2 statistics. Instead of coefficients (b), **logistic** displays odds ratios (e^b). The numbers in the "Odds Ratio" column of the **logistic** output are amounts by which the odds favoring $y = 1$ are multiplied, with each 1-unit increase in that x variable (if other x variables' values stay the same).

After fitting a model, we can obtain a classification table and related statistics by typing

```
. estat class
```

Logistic model for any

Classified	True D	~D	Total
+	12	4	16
-	2	5	7
Total	14	9	23

Classified + if predicted Pr(D) >= .5
True D defined as any != 0

Sensitivity	Pr(+| D)	85.71%
Specificity	Pr(-|~D)	55.56%
Positive predictive value	Pr(D| +)	75.00%
Negative predictive value	Pr(~D| -)	71.43%
False + rate for true ~D	Pr(+|~D)	44.44%
False - rate for true D	Pr(-| D)	14.29%
False + rate for classified +	Pr(~D| +)	25.00%
False - rate for classified -	Pr(D| -)	28.57%
Correctly classified		73.91%

By default, **estat class** employs a probability of .5 as its cutoff (although we can change this by adding a **cutoff( )** option). Symbols in the classification table have the following meanings:

D   The event of interest did occur (that is, $y = 1$) for that observation. In this example, D indicates that thermal distress occurred.

~D   The event of interest did not occur (that is, $y = 0$) for that observation. In this example, ~D corresponds to flights having no thermal distress.

+   The model's predicted probability is greater than or equal to the cutoff point. Since we used the default cutoff, + here indicates that the model predicts a .5 or higher probability of thermal distress.

−   The predicted probability is less than the cutoff. Here, − means a predicted probability of thermal distress below .5.

Thus for 12 flights, classifications are accurate in the sense that the model estimated at least a .5 probability of thermal distress, and distress did in fact occur. For 5 other flights, the model predicted less than a .5 probability, and distress did not occur. The overall "correctly classified" rate is therefore $12 + 5 = 17$ out of 23, or 73.91%. The table also gives conditional probabilities such as "sensitivity" or the percentage of observations with $P \geq .5$ given that thermal distress occurred (12 out of 14 or 85.71%).

After **logistic** or **logit** , the postestimation command **predict** calculates various prediction and diagnostic statistics. Explanations for logit model diagnostic statistics can be found in Hosmer and Lemeshow (2000).

predict *newvar*	Predicted probability that $y = 1$
predict *newvar*, **xb**	Linear prediction (predicted log odds that $y = 1$)
predict *newvar*, **stdp**	Standard error of the linear prediction
predict *newvar*, **dbeta**	ΔB influence statistic, analogous to Cook's D
predict *newvar*, **deviance**	Deviance residual for jth x pattern, d_j
predict *newvar*, **dx2**	Change in Pearson χ^2, written as $\Delta\chi^2$ or $\Delta\chi^2_P$
predict *newvar*, **ddeviance**	Change in deviance χ^2, written as ΔD or $\Delta\chi^2_D$
predict *newvar*, **hat**	Leverage of the jth x pattern, h_j
predict *newvar*, **number**	Assigns numbers to x patterns, $j = 1,2,3 ... J$
predict *newvar*, **resid**	Pearson residual for jth x pattern, r_j
predict *newvar*, **rstandard**	Standardized Pearson residual
predict *newvar*, **score**	1st derivative of log likelihood with respect to **Xb**.

Statistics obtained by the **dbeta**, **dx2**, **ddeviance**, and **hat** options do not measure the influence of individual observations, as their counterparts in ordinary regression do. Rather, these statistics measure the influence of "covariate patterns;" that is, the consequences of dropping all observations with that particular combination of x values. See Hosmer and Lemeshow (2000) for details. A later section of this chapter shows these statistics in use.

Does booster joint temperature also affect the probability of any distress incidents? We could investigate by including *temp* as a second predictor variable.

```
. logistic any date temp
```

Logistic regression

Log likelihood = -11.350748

				Number of obs	=	23
				LR chi2(2)	=	8.09
				Prob > chi2	=	0.0175
				Pseudo R2	=	0.2627

any	Odds Ratio	Std. Err.	z	P>|z|	[95% Conf. Interval]	
date	1.00297	.0013675	2.17	0.030	1.000293	1.005653
temp	.8408309	.0987887	-1.48	0.140	.6678848	1.058561

Including temperature as a predictor slightly improves the correct classification rate, to 78.26%.

```
. estat class
```

Logistic model for any

	True		
Classified	D	~D	Total
+	12	3	15
-	2	6	8
Total	14	9	23

Classified + if predicted Pr(D) >= .5
True D defined as any != 0

Sensitivity	Pr(+| D)	85.71%
Specificity	Pr(-|~D)	66.67%
Positive predictive value	Pr(D| +)	80.00%
Negative predictive value	Pr(~D| -)	75.00%
False + rate for true ~D	Pr(+|~D)	33.33%
False - rate for true D	Pr(-| D)	14.29%
False + rate for classified +	Pr(~D| +)	20.00%
False - rate for classified -	Pr(D| -)	25.00%
Correctly classified		78.26%

According to the fitted model, each 1-degree increase in joint temperature multiplies the odds of booster joint damage by .84 (in other words, each 1-degree warming reduces the odds of damage by about 16%). Although this effect seems strong enough to cause concern, the asymptotic z test says that it is not statistically significant ($z = -1.476$, $P = .140$). A more definitive test, however, employs the likelihood-ratio χ^2. The **lrtest** command compares nested models estimated by maximum likelihood. First, estimate a "full" model containing all variables of interest, as done above with the **logistic** *any date temp* command. Next, type an **estimates store** command, giving a name (such as *full*) to identify this first model:

```
. estimates store full
```

Now estimate a reduced model, including only a subset of the x variables from the full model. (Such reduced models are said to be "nested.") Finally, a command such as **lrtest** *full* requests a test of the nested model against the previously stored *full* model. For example (using the **quietly** prefix, because we already saw this output once),

```
. quietly logistic any date
. lrtest full
```

```
Likelihood-ratio test                        LR chi2(1)  =      3.28
(Assumption: . nested in full)                Prob > chi2 =    0.0701
```

This **lrtest** command tests the recent (presumably nested) model against the model previously saved by **estimates store**. It employs a general test statistic for nested maximum-likelihood models,

$$\chi^2 = -2(\ln \mathcal{L}_1 - \ln \mathcal{L}_0) \qquad [10.6]$$

where $\ln \mathcal{L}_0$ is the log likelihood for the first model (with all x variables), and $\ln \mathcal{L}_1$ is the log likelihood for the second model (with a subset of those x variables). Compare the resulting test statistic to a χ^2 distribution with degrees of freedom equal to the difference in complexity (number of x variables dropped) between models 0 and 1. Type **help lrtest** for more about this command, which works with any of Stata's maximum-likelihood estimation procedures (**logit**, **mlogit**, **stcox**, and many others). The overall χ^2 statistic routinely given by **logit** or **logistic** output (equation [10.3]) is a special case of [10.6].

The previous **lrtest** example performed this calculation:
$$\chi^2 = -2[-12.991096 - (-11.350748)]$$
$$= 3.28$$
with 1 degree of freedom, yielding $P = .0701$; the effect of *temp* is significant at $\alpha = .10$. Given the small sample and the disastrous consequences of a Type II error regarding space shuttle safety, $\alpha = .10$ seems a more prudent cutoff than the usual $\alpha = .05$.

Conditional Effect Plots

Conditional effect plots help in understanding what a logistic model implies about probabilities. The idea behind such plots is to draw a curve showing how the model's prediction of y changes as a function of one x variable, while holding all other x variables constant at chosen values such as their means, quartiles, or extremes. For example, we could find the predicted probability of any thermal distress incidents as a function of *temp*, holding *date* constant at its 25th percentile. One way of doing this is to run **summarize, detail** and then use result r(p25), the 25th percentile, in an **adjust** command to generate a new variable (*Phat1*) for the predicted probabilities:

```
. quietly logit any date temp
. summarize date, detail
. adjust date = `r(p25)', gen(Phat1) pr
```

The 25th percentile of *date*, found by **summarize *date*, detail**, is 8569 — that is, June 18, 1983. Similar steps find the predicted probability of any distress with *date* fixed at its 75th percentile, (r(p75), which equals 9341 or July 29, 1985).

```
. summarize date, detail
. adjust date = `r(p75)', gen(Phat2) pr
```

We can now graph the relationship between *temp* and the probability of any distress for the two levels of *date*, as shown in Figure 10.2. Using median splines with many vertical bands, **graph twoway mspline, bands(50)**, produces smooth curves in this figure, approximating the smooth logistic functions.

```
. graph twoway mspline Phat1 temp, bands(50)
        ||   mspline Phat2 temp, bands(50) lpattern(dash)
        ||   , ytitle("Probability of thermal distress")
      ylabel(0(.2)1, grid) xlabel(, grid)
        legend(label(1 "June 1983") label(2 "July 1985")
        rows(2) position(7) ring(0))
```

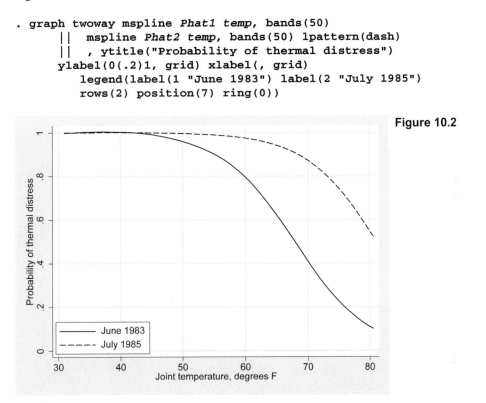

Figure 10.2

Among earlier flights (*date* = June 1983, left curve), the probability of thermal distress goes from very low, at around 80° F, to near 1, below 50° F. Among later flights (*date* = July 1985, right curve), however, the probability of any distress exceeds .5 even in warm weather, and climbs toward 1 on flights below 70° F. Note that *Challenger*'s launch temperature, 31° F, would place it at top left in Figure 10.2.

Diagnostic Statistics and Plots

As mentioned earlier, the logistic regression influence and diagnostic statistics obtained by **predict** refer not to individual observations, as do the OLS regression diagnostics of Chapter 7. Rather, logistic diagnostics refer to *x* patterns. In the space shuttle data, however, each *x* pattern is unique — no two flights share the same combination of *date* and *temp* (naturally, because no two were launched the same day). Before using **predict**, we quietly refit the recent model, to be sure that model is what we think:

```
. quietly logistic any date temp
. predict Phat3
(option pr assumed; Pr(any))
. label variable Phat3 "Predicted probability"
. predict dX2, dx2
(2 missing values generated)
. label variable dX2 "Change in Pearson chi-squared"
. predict dB, dbeta
(2 missing values generated)
. label variable dB "Influence"
. predict dD, ddeviance
(2 missing values generated)
. label variable dD "Change in deviance"
```

Hosmer and Lemeshow (2000) suggest plots that help in reading these diagnostics. To graph change in Pearson χ^2 versus probability of distress (Figure 10.3), type:

```
. graph twoway scatter dX2 Phat3
```

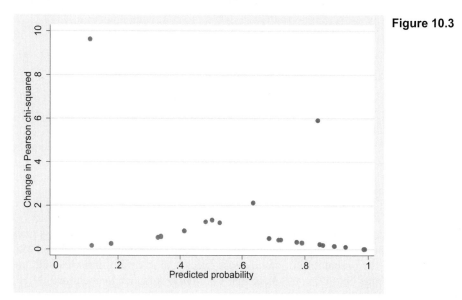

Figure 10.3

Two poorly fit *x* patterns, at upper right and left in Figure 10.3, stand out. We can visually identify the flights with high *dX2* values by adding marker labels (*flight* numbers, in this case) to the scatterplot. In Figure 10.4, only those flights with *dX2* > 2 have been labeled by adding a second overlaid scatterplot. (If we had instead labeled all of the data points, the bottom of the graph would become an unreadable mess.)

```
. graph twoway scatter dX2 Phat3
     || scatter dX2 Phat3 if dX2 > 2, mlabel(flight)
        mlabsize(medsmall)
     || , legend(off)
```

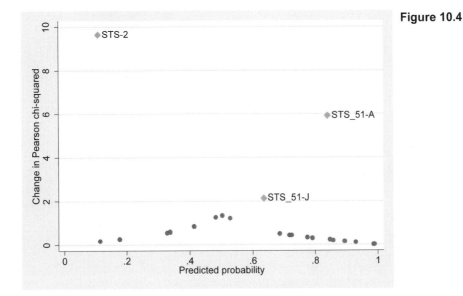

Figure 10.4

```
. list flight any date temp dX2 Phat3 if dX2 > 2
```

	flight	any	date	temp	dX2	Phat3
2.	STS-2	1	12nov1981	70	9.630337	.1091805
4.	STS-4	.	27jun1982	80		.0407113
14.	STS_51-A	0	08nov1984	67	5.899742	.8400974
21.	STS_51-J	0	03oct1985	79	2.124642	.6350927
25.	STS_51-L	.	28jan1986	31		.9999012

Flight STS 51-A experienced no thermal distress, despite a late launch date and cool temperature (see Figure 10.2). The model predicts a .84 probability of distress for this flight. All points along the up-to-right curve in Figure 10.4 experienced no thermal distress (*any* = 0). Atop the up-to-left (*any* = 1) curve, flight STS-2 experienced thermal distress despite being one of the earliest flights, and launched in slightly milder weather. The model predicts only a .109 probability of distress. Because Stata views missing values as large numbers, it lists the two missing-values flights, including *Challenger*, among those with *dX2* > 2.

Similar findings result from plotting *dD* versus predicted probability, as seen in Figure 10.5 (following page). Again, flights STS-2 (top left) and STS 51-A (top right) stand out as poorly fit. Figure 10.5 illustrates a variation on the labeled-marker scatterplot. Instead of putting the flight-number labels near the markers, as done in Figure 10.4 above, we make the markers themselves invisible and place labels where the markers would have been for every data point in Figure 10.5.
```
```

```
. graph twoway scatter dD Phat3, msymbol(i) mlabposition(0)
     mlabel(flight) mlabsize(small)
```

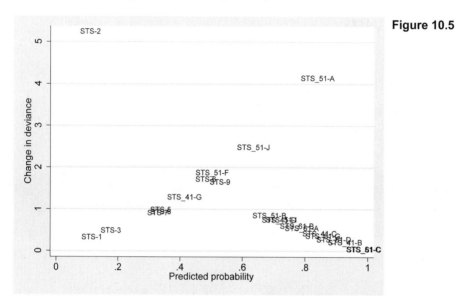

Figure 10.5

dB measures an *x* pattern's influence in logistic regression. For a logistic-regression analog to the OLS plot in Figure 7.7 (Chapter 7), we can make the plotting symbols proportional to influence as done in Figure 10.6. The two worst-fit observations are also the most influential.

```
. graph twoway scatter dD Phat3 [aweight = dB], msymbol(oh)
```

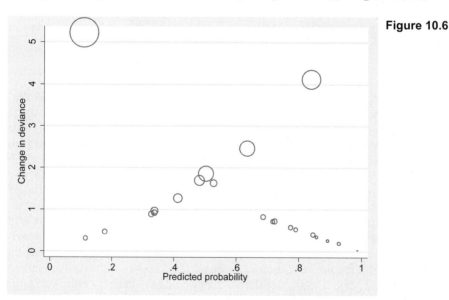

Figure 10.6

Poorly fit and influential observations deserve special attention because they both contradict the main pattern of the data and pull model estimates in their contrary direction. Of course, simply removing such outliers allows a "better fit" with the remaining data, but this is circular reasoning. A more thoughtful reaction would be to investigate what makes the outliers unusual. Why did shuttle flight STS-2, but not STS 51-A, experience booster joint damage? Seeking an answer might lead investigators to previously overlooked variables or to a better model.

Logistic Regression with Ordered-Category *y*

logit and **logistic** fit models for variables with two outcomes, coded 0 and 1. We need other methods for models in which *y* takes on more than two values. Two important possibilities are ordered and multinomial logistic regression.

ologit Ordered logistic regression, where *y* is an ordinal (ordered-category) variable. The numeric values representing the categories do not matter, except that higher numbers mean "more." For example, the *y* categories might be {1 = "poor," 2 = "fair," 3 = "excellent"}.

mlogit Multinomial logistic regression, where *y* has multiple but unordered categories such as {1 = "Democrat," 2 = "Republican," 3 = "undeclared"}.

If *y* is {0,1}, **logit** (or **logistic**), **ologit**, and **mlogit** all produce essentially the same estimates.

We earlier simplified the three-outcome ordinal variable *distress* into a dichotomy, *any*. **logit** and **logistic** require {0,1} dependent variables. **ologit**, on the other hand, is designed for ordinal variables that have more than two values. Recall that *distress* has outcomes 0 = "none," 1 = "1 or 2," and 2 = "3 plus" incidents of booster-joint distress.

Ordered logistic regression indicates that *date* and *temp* both affect *distress*, with the same signs (positive for *date*, negative for *temp*) seen in our earlier binary logit analyses:

```
. ologit distress date temp, nolog
```

```
Ordered logistic regression                    Number of obs   =        23
                                                LR chi2(2)      =     12.32
                                                Prob > chi2     =    0.0021
Log likelihood = -18.79706                      Pseudo R2       =    0.2468
```

distress	Coef.	Std. Err.	z	P>|z|	[95% Conf. Interval]	
date	.003286	.0012662	2.60	0.009	.0008043	.0057677
temp	-.1733752	.0834473	-2.08	0.038	-.336929	-.0098215
/cut1	16.42813	9.554813			-2.29896	35.15522
/cut2	18.12227	9.722293			-.9330729	37.17761

Likelihood-ratio tests are more accurate than the asymptotic *z* tests shown. First, have **estimates store** preserve in memory the results from the full model (with two predictors) just estimated. We can give this model any descriptive name, such as *date_temp*.

```
. estimates store date_temp
```

Next, fit a simpler model without *temp*, store its results with the name *notemp*, and ask for a likelihood-ratio test of whether the fit of reduced model *notemp* differs significantly from that of the full model *date_temp*:

```
. quietly ologit distress date
. estimates store notemp
. lrtest notemp date_temp
```

Likelihood-ratio test   LR chi2(1)   =   6.12
(Assumption: <u>notemp</u> nested in <u>date_temp</u>)   Prob > chi2 =   0.0133

The **lrtest** output notes its assumption that model *notemp* is nested in model *date_temp*, meaning that the parameters estimated in *notemp* are a subset of those in *date_temp*, and that both models are estimated from the same pool of observations (which can be tricky when the data contain missing values). This likelihood-ratio test indicates that *notemp*'s fit is significantly poorer. Because the presence of *temp* as a predictor in model *date_temp* is the only difference, the likelihood-ratio test thus informs us that *temp*'s contribution is significant. Similar steps confirm that *date* also has a significant effect.

```
. quietly ologit distress temp
. estimates store nodate
. lrtest nodate date_temp;
```

Likelihood-ratio test   LR chi2(1)   =   10.33
(Assumption: <u>nodate</u> nested in <u>date_temp</u>)   Prob > chi2 =   0.0013

The **estimates store** and **lrtest** commands provide flexible tools for comparing nested maximum-likelihood models. Type **help lrtest** and **help estimates** for details and options.

The ordered-logit model estimates a score, S, for each observation as a linear function of *date* and *temp*:

$$S \quad = .003286date - .1733752temp$$

Predicted probabilities depend on the value of S, plus a logistically distributed disturbance u, relative to the estimated cut points (shown in **ologit** output as cut1, cut2, etc.).

$P(distress="none") \quad = \quad P(S+u \leq cut1)$
$\qquad\qquad\qquad\qquad = \quad (1 + \exp(-cut1 + S))^{-1}$
$P(distress="1 \text{ or } 2") \quad = \quad P(cut1 < S+u \leq cut2)$
$\qquad\qquad\qquad\qquad = \quad (1 + \exp(-cut2 + S))^{-1} - (1 + \exp(-cut1 + S))^{-1}$
$P(distress="3 \text{ plus}") \quad = \quad P(cut2 < S+u)$
$\qquad\qquad\qquad\qquad = \quad 1 - (1 + \exp(-cut2 + S))^{-1}$

After **ologit**, **predict** calculates predicted probabilities for each category of the dependent variable. We supply **predict** with names for these probabilities. For example: *none* could denote the probability of no distress incidents (first category of *distress*); *onetwo* the probability of 1 or 2 incidents (second category of *distress*); and *threeplus* the probability of 3 or more incidents (third and last category of *distress*):

```
. quietly ologit distress date temp
. predict none onetwo threeplus
(option pr assumed; predicted probabilities)
```

This creates three new variables:

. **describe** *none onetwo threeplus*

variable name	storage type	display format	value label	variable label
none	float	%9.0g		Pr(distress==0)
onetwo	float	%9.0g		Pr(distress==1)
threeplus	float	%9.0g		Pr(distress==2)

Predicted probabilities for *Challenger*'s last flight, the 25th in these data, are unsettling:

. **list** *flight none onetwo threeplus* **if** *flight* == 25

	flight	none	onetwo	threep~s
25.	STS_51-L	.0000754	.0003346	.99959

Our model, based on the analysis of 23 pre-*Challenger* shuttle flights, predicts little chance (P = .000075) of *Challenger* experiencing no booster joint damage, a scarcely greater likelihood of one or two incidents (P = .0003), but virtual certainty (P = .9996) of three or more damage incidents.

A second example of ordered logistic regression appears in Chapter 14, on survey data analysis. See Long (1997) or Hosmer and Lemeshow (2000) for detailed presentations of this and related techniques. The *Base Reference Manual* explains Stata's implementation. Long and Freese (2006) provide additional Stata-focused discussion, and make available their ado-files for some useful interpretation and postestimation commands, such as Brant tests. To install these unofficial, free ado-files from the Web, type **findit brant** and follow the link under "Web resources."

Multinomial Logistic Regression

When the dependent variable's categories have no natural ordering, multinomial logit regression (also called polytomous logit regression) provides appropriate tools. If *y* has only two categories, **mlogit** fits the same model as **logistic**. Otherwise, though, an **mlogit** model is more complex. This section presents an extended example interpreting **mlogit** results, using data from a survey of high school students in Alaska's Northwest Arctic borough (dataset *NWactic.dta*, excerpted from a larger study described by Hamilton and Seyfrit, 1993).

```
Contains data from C:\data\NWarctic.dta
    obs:            259                      NW Arctic high school students
                                              (Hamilton & Seyfrit 1993)
    vars:             3                      21 May 2008 07:57
    size:         3,626 (99.9% of memory free)

                storage  display   value
variable name    type    format    label    variable label

life             byte    %8.0g     migrate  Expect to live most of life?
ties             float   %9.0g              Social ties to community scale
kotz             byte    %8.0g     kotz     Live in Kotzebue or smaller
                                              village?
```

Variable *life* indicates where students say they expect to live most of the rest of their lives: in the same region (Northwest Arctic), elsewhere in Alaska, or outside of Alaska:

. **tabulate** *life*, **plot**

```
Expect to
live most
 of life?    Freq.

     same       92  ***********************************************
 other AK      120  *******************************************************
 leave AK       47  **********************

    Total      259
```

Kotzebue (population near 3,000) is the Northwest Arctic's regional hub and largest city. More than a third of these students live in Kotzebue. The rest live in smaller villages of 200 to 700 people. The relatively cosmopolitan Kotzebue students less often expect to stay where they are (only 18%, compared with 45% of the village students), and lean more towards leaving the state (39%, compared with 7% of village students).

. **tabulate** *life kotz*, **column chi2**

```
Key

      frequency
 column percentage
```

Expect to live most of life?	Live in Kotzebue or smaller village? village	Kotzebue	Total
same	75 45.18	17 18.28	92 35.52
other AK	80 48.19	40 43.01	120 46.33
leave AK	11 6.63	36 38.71	47 18.15
Total	166 100.00	93 100.00	259 100.00

Pearson chi2(2) = 46.2992   Pr = 0.000

mlogit can replicate this simple analysis, although its likelihood-ratio chi-squared need not exactly equal the Pearson chi-squared found by **tabulate**.

`. mlogit life kotz, nolog base(1) rrr`

```
Multinomial logistic regression                 Number of obs   =        259
                                                LR chi2(2)      =      46.23
                                                Prob > chi2     =     0.0000
Log likelihood = -244.64465                     Pseudo R2       =     0.0863
```

life	RRR	Std. Err.	z	P>|z|	[95% Conf. Interval]	
other AK						
kotz	2.205882	.7304664	2.39	0.017	1.152687	4.221369
leave AK						
kotz	14.4385	6.307555	6.11	0.000	6.132946	33.99188

`(life==same is the base outcome)`

base(1) specifies that category 1 of y (*life* = "same") is the base outcome for comparison. The **rrr** option instructs **mlogit** to show relative risk ratios, which resemble the odds ratios given by **logistic**.

Referring back to the **tabulate** output, we can calculate that among Kotzebue students the odds favoring "leave Alaska" over "stay in the same area" are
$$P(\text{leave AK}) / P(\text{same}) = (36/93) / (17/93)$$
$$= 2.1176471$$
Among other students the odds favoring "leave Alaska" over "same area" are
$$P(\text{leave AK}) / P(\text{same}) = (11/166) / (75/166)$$
$$= .1466667$$
Thus, the odds favoring "leave Alaska" over "same area" are 14.4385 times higher for Kotzebue students than for others:
$$2.1176471 / .1466667 = 14.4385$$
This multiplier, a ratio of two odds, equals the relative risk ratio (14.4385) displayed by **mlogit**.

In general, the relative risk ratio for outcome j of y, and predictor x_k, equals the amount by which predicted odds favoring $y = j$ (compared with y = base) are multiplied, per 1-unit increase in x_k, other things being equal. In other words, the relative risk ratio rrr_{jk} is a multiplier such that, if all x variables except x_k stay the same,

$$\text{rrr}_{jk} \times \frac{P(y = j \mid x_k)}{P(y = \text{base} \mid x_k)} = \frac{P(y = j \mid x_k + 1)}{P(y = \text{base} \mid x_k + 1)}$$

ties is a continuous scale indicating the strength of students' social ties to family and community. The next model includes *ties* as a second predictor.

```
. mlogit life kotz ties, nolog base(1) rrr
```

```
Multinomial logistic regression                    Number of obs   =        259
                                                    LR chi2(4)      =      91.96
                                                    Prob > chi2     =     0.0000
Log likelihood = -221.77969                         Pseudo R2       =     0.1717
```

life	RRR	Std. Err.	z	P>|z|	[95% Conf. Interval]	
other AK						
kotz	2.214184	.7724996	2.28	0.023	1.117483	4.387193
ties	.4802486	.0799184	-4.41	0.000	.3465911	.6654492
leave AK						
kotz	14.84604	7.146824	5.60	0.000	5.778907	38.13955
ties	.230262	.059085	-5.72	0.000	.1392531	.38075

(life==same is the base outcome)

Asymptotic z tests here indicate that the four relative risk ratios, describing two x variables' effects, all differ significantly from 1.0. If a y variable has J categories, then **mlogit** models the effects of each predictor (x) variable with $J-1$ relative risk ratios or coefficients, and hence also employs $J-1$ z tests, evaluating two or more separate null hypotheses for each predictor. Likelihood-ratio tests evaluate the overall effect of each predictor. First, store the results from the full model, here given the name *full*:

```
. estimates store full
```

Then fit a simpler model with one of the x variables omitted, and perform a likelihood-ratio test. For example, to test the effect of *ties*, we repeat the regression with *ties* omitted:

```
. quietly mlogit life kotz
. estimates store no_ties
. lrtest no_ties full
```

```
Likelihood-ratio test                              LR chi2(2)  =      45.73
(Assumption: no_ties nested in full)               Prob > chi2 =     0.0000
```

The effect of *ties* is clearly significant. Next, we run a similar test on the effect of *kotz*:

```
. quietly mlogit life ties
. estimates store no_kotz
. lrtest no_kotz full
```

```
Likelihood-ratio test                              LR chi2(2)  =      39.05
(Assumption: no_kotz nested in full)               Prob > chi2 =     0.0000
```

If our data contained missing values, the three **mlogit** commands just shown might have analyzed three overlapping subsets of observations. The full model would use only observations with nonmissing *life*, *kotz*, and *ties* values; the *kotz*-only model would bring back in any observations missing just their *ties* values; and the *ties*-only model would bring back observations missing just *kotz* values. When this happens, Stata returns an error messages saying "observations differ." In such cases, the likelihood-ratio test would be invalid. To avoid that problem we could screen observations with **if** qualifiers attached to modeling commands. For example, we could estimate the full model and generate a variable named *usethese* from the e(sample) pseudo-function that indicates which observations were used for the full model — that is, which observations contained no missing values on any of the variables in the full model.

```
. mlogit life kotz ties, nolog base(1) rrr
. estimates store full
. gen byte usethese = e(sample)

. quietly mlogit life kotz if usethese
. estimates store no_ties
. lrtest no_ties full

. quietly mlogit life ties if usethese
. estimates store no_kotz
. lrtest no_kotz full
```

Or, we might simplify the sample by keeping only those observations with nonmissing values:

```
. keep if !missing(life, kotz, ties)
```

Dataset *NWarctic.dta* had already been screened in this fashion and contains no observations with missing values.

Both *kotz* and *ties* significantly predict *life*. What else can we say from this output? To interpret specific effects, recall that *life* = "same" is the base outcome. The relative risk ratios tell us that:

> Odds that a student expects migration to elsewhere in Alaska rather than staying in the same area are 2.21 times greater (increase about 121%) among Kotzebue students (*kotz*=1), adjusting for social ties to community.

> Odds that a student expects to leave Alaska rather than stay in the same area are 14.85 times greater (increase about 1385%) among Kotzebue students (*kotz*=1), adjusting for social ties to community.

> Odds that a student expects migration to elsewhere in Alaska rather than staying are multiplied by .48 (decrease about 52%) with each 1-unit (since *ties* is standardized, its units equal standard deviations) increase in social ties, controlling for Kotzebue/village residence.

> Odds that a student expects to leave Alaska rather than stay are multiplied by .23 (decrease about 77%) with each 1-unit increase in social ties, controlling for Kotzebue/village residence.

predict can calculate predicted probabilities from **mlogit**. The **outcome(#)** option specifies for which *y* outcome we want probabilities. For example, to get predicted probabilities that *life* = "leave AK" (outcome 3):

```
. quietly mlogit life kotz ties
. predict PleaveAK, outcome(3)
(option pr assumed; predicted probability)
. label variable PleaveAK "P(life = 3 | kotz, ties)"
```

Tabulating predicted probabilities for each value of the dependent variable shows how the model fits:

```
. table life, contents(mean PleaveAK) row
```

Expect to live most of life?	mean(PleaveAK)
same	.0811267
other AK	.1770225
leave AK	.3892264
Total	.1814672

A minority of these students (47/259 = 18%) expect to leave Alaska. The model averages only a .39 probability of leaving Alaska even for those who actually chose this response, reflecting the fact that although our predictors have significant effects, most variation in migration plans remains unexplained.

Conditional effect plots help to visualize what a model implies regarding continuous predictors. We can draw them using estimated coefficients (not risk ratios) to calculate probabilities.

```
. mlogit life kotz ties, nolog base(1)
```

Multinomial logistic regression

Number of obs = 259
LR chi2(4) = 91.96
Prob > chi2 = 0.0000
Log likelihood = -221.77969
Pseudo R2 = 0.1717

life	Coef.	Std. Err.	z	P>|z|	[95% Conf. Interval]
other AK					
kotz	.794884	.3488868	2.28	0.023	.1110784 1.47869
ties	-.7334513	.1664104	-4.41	0.000	-1.05961 -.407293
_cons	.206402	.1728053	1.19	0.232	-.1322902 .5450942
leave AK					
kotz	2.697733	.4813959	5.60	0.000	1.754215 3.641252
ties	-1.468537	.2565991	-5.72	0.000	-1.971462 -.9656124
_cons	-2.115025	.3758163	-5.63	0.000	-2.851611 -1.378439

(life==same is the base outcome)

The following commands calculate predicted logits, and then the probabilities needed for conditional effect plots. *L2villag* represents the predicted logit of *life* = 2 (other Alaska) for village students. *L3kotz* is the predicted logit of *life* = 3 (leave Alaska) for Kotzebue students, and so forth:

```
. generate L2villag = .206402 +.794884*0 -.7334513*ties
. generate L2kotz = .206402 +.794884*1 -.7334513*ties
. generate L3villag = -2.115025 +2.697733*0 -1.468537*ties
. generate L3kotz = -2.115025 +2.697733*1 -1.468537*ties
```

After fitting any model, we can access its coefficients and standard errors for use in subsequent expressions. If we had estimated a simple regression model, the coefficient on predictor variable *kotz* would be named _b[*kotz*]. With a multple-equation model such as **mlogit**, the coefficient names are similar but begin with a number identifying their equation. For example, [2]_b[*kotz*] refers to the coefficient on *kotz* in the model's second (*life* = 2) equation. Therefore, we could have generated the same predicted logits as follows. *L2v* will be identical to *L2villag* defined earlier, *L3k* the same as *L3kotz*, and so forth:

```
. generate L2v = [2]_b[_cons] +[2]_b[kotz]*0 +[2]_b[ties]*ties
. generate L2k = [2]_b[_cons] +[2]_b[kotz]*1 +[2]_b[ties]*ties
. generate L3v = [3]_b[_cons] +[3]_b[kotz]*0 + [3]_b[ties]*ties
. generate L3k = [3]_b[_cons] +[3]_b[kotz]*1 + [3]_b[ties]*ties
```

From either set of logits, we next calculate the predicted probabilities:

```
. generate P1villag = 1/(1 +exp(L2villag) +exp(L3villag))
. label variable P1villag "same area"

. generate P2villag = exp(L2villag)/(1+exp(L2villag)+exp(L3villag))
. label variable P2villag "other Alaska"

. generate P3villag = exp(L3villag)/(1+exp(L2villag)+exp(L3villag))
. label variable P3villag "leave Alaska"

. generate P1kotz = 1/(1 +exp(L2kotz) +exp(L3kotz))
. label variable P1kotz "same area"

. generate P2kotz = exp(L2kotz)/(1 +exp(L2kotz) +exp(L3kotz))
. label variable P2kotz "other Alaska"

. generate P3kotz = exp(L3kotz)/(1 +exp(L2kotz) +exp(L3kotz))
. label variable P3kotz "leave Alaska"
```

Figures 10.7 and 10.8 show conditional effect plots for village and Kotzebue students.

```
. graph twoway mspline P1villag ties, bands(50) lpattern(dash)

    || mspline P2villag ties, bands(50)

    || mspline P3villag ties, bands(50) lpattern(dot)

    || , xlabel(-3(1)3) ylabel(0(.2)1) yline(0 1) xline(0)
    legend(order(3 2 1) position(12) ring(0) label(1 "same area")
    label(2 "elsewhere Alaska") label(3 "leave Alaska") cols(1))
    ytitle("Probability")
```

Figure 10.7

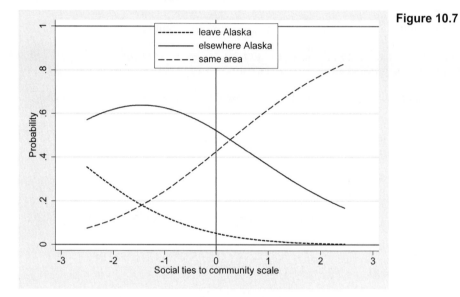

Figure 10.7 graphs probabilities of leaving or staying among village high school students. For these students, social ties increase the probability of staying rather than moving elsewhere in Alaska. Relatively few village students expect to leave Alaska.

```
. graph twoway mspline P1kotz ties, bands(50) lpattern(dash)

    ||  mspline P2kotz ties, bands(50)

    ||  mspline P3kotz ties, bands(50) lpattern(dot)

    ||  , xlabel(-3(1)3) ylabel(0(.2)1) yline(0 1) xline(0)
    legend(order(3 2 1) position(12) ring(0) label(1 "same area")
    label(2 "elsewhere Alaska") label(3 "leave Alaska") cols(1))
    ytitle("Probability")
```

Figure 10.8

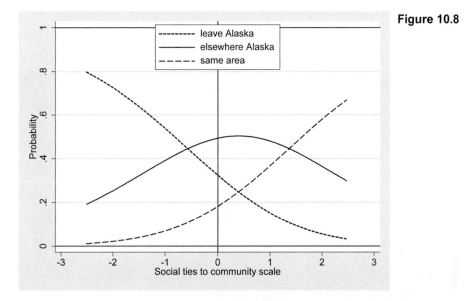

Figure 10.8 indicates that among Kotzebue high school students, *ties* particularly affects the probability of leaving Alaska, rather than simply moving elsewhere in the state. Only if they feel very strong social ties do Kotzebue students tend to favor staying put.

11

Survival and Event-Count Models

This chapter presents methods for analyzing event data. *Survival analysis* involves several related techniques that focus on times until the event of interest occurs. Although the event could be good or bad, by convention we refer to the event as a "failure." The time until failure is "survival time." Survival analysis is important in biomedical research, but it can be applied equally well to other fields from engineering to social science — for example, in modeling the time until an unemployed person gets a job, or a single person gets married. Stata offers a full range of survival analysis procedures, only a few of which are illustrated in this chapter.

We also look briefly at Poisson regression and its relatives. These methods focus not on survival times but, rather, on the rates or counts of events over a specified interval of time. Event-count methods include Poisson regression and negative binomial regression. Such models can be fit either through specialized commands or through the broader approach of generalized linear modeling (GLM).

Consult the *Survival Analysis and Epidemiological Tables Reference Manual* for more information about Stata's capabilities. Type **help st** to see an online overview. Selvin (1995) provides well-illustrated introductions to survival analysis and Poisson regression. I have borrowed (with permission) several of his examples. Other good introductions to survival analysis include the Stata-oriented volume by Cleves et al. (2008), a chapter in Rosner (1995), and comprehensive treatments by Hosmer, Lemeshow and May (2008) and Lee (1992). McCullagh and Nelder (1989) describe generalized linear models. Long (1997) has a chapter on regression models for count data (including Poisson and negative binomial), and also some material on generalized linear models. An extensive and current treatment of generalized linear models is found in Hardin and Hilbe (2007).

Stata menu groups most relevant to this chapter include:

Statistics > Survival analysis

Graphics > Survival analysis graphs

Statistics > Count outcomes

Statistics > Generalized linear models

Regarding epidemiological tables, not covered in this chapter, further information can be found by typing **help epitab** or exploring the menus for

Statistics > Epidemiology and related

Example Commands

Most of Stata's survival-analysis (**st***) commands require that the data have previously been identified as survival-time by an **stset** command (see following). **stset** need only be run once, and the data subsequently saved.

. **stset** *timevar*, **failure(***failvar***)**
   Identifies single-record survival-time data. Variable *timevar* indicates the time elapsed before either a particular event (called a "failure") occurred, or the period of observation ended ("censoring"). Variable *failvar* indicates whether a failure (*failvar* = 1) or censoring (*failvar* = 0) occurred at *timevar*. The dataset contains only one record per individual. The dataset must be **stset** before any further **st*** commands will work. If we subsequently **save** the dataset, however, the **stset** definitions are saved as well. **stset** creates new variables named *_st, _d, _t*, and *_t0* that encode information necessary for subsequent **st*** commands.

. **stset** *timevar*, **failure(***failvar***) id(***patient***) enter(time** *start***)**
   Identifies multiple-record survival-time data. In this example, the variable *timevar* indicates elapsed time before failure or censoring; *failvar* indicates whether failure (1) or censoring (0) occurred at this time. *patient* is an identification number. The same individual might contribute more than one record to the data, but always has the same identification number. *start* records the time when each individual came under observation.

. **stdescribe**
   Describes survival-time data, listing definitions set by **stset** and other characteristics of the data.

. **stsum**
   Obtains summary statistics: the total time at risk, incidence rate, number of subjects, and percentiles of survival time.

. **ctset** *time nfail ncensor nenter*, **by(***ethnic sex***)**
   Identifies count-time data. In this example, the variable *time* is a measure of time; *nfail* is the number of failures occurring at *time*. We also specified *ncensor* (number of censored observations at *time*) and *nenter* (number entering at *time*), although these can be optional. *ethnic* and *sex* are other categorical variables defining observations in these data.

. **cttost**
   Converts count-time data, previously identified by the **ctset** command, into survival-time form that can be analyzed by **st*** commands.

. `sts graph`

Graphs the Kaplan–Meier survivor function. To visually compare two or more survivor functions, such as one for each value of the categorical variable *sex*, use a **by( )** option such as **sts graph, by(*sex*)**. To adjust, through Cox regression, for the effects of a continuous independent variable such as *age*, use an **adjustfor( )** option such as **sts graph, by(*sex*) adjustfor(*age*)**. The **by( )** and **adjustfor( )** options work similarly with the other **sts** commands **sts list**, **sts generate**, and **sts test**.

. `sts list`

Lists the estimated Kaplan–Meier survivor (or failure) function.

. `sts test sex`

Tests the equality of the Kaplan–Meier survivor function across categories of *sex*.

. `sts generate survfunc = S`

Creates a new variable arbitrarily named *survfunc*, containing the estimated Kaplan–Meier survivor function.

. `stcox x1 x2 x3`

Fits a Cox proportional hazard model, regressing time to failure on continuous or dummy variable predictors *x1–x3*.

. `stcox x1 x2 x3, strata(x4) basechazard(hazard) vce(robust)`

Fits a Cox proportional hazard model, stratified by *x4*. Stores the group-specific baseline cumulative hazard function as a new variable named *hazard*. (Baseline survivor function estimates could be obtained through a **basesur(*survive*)** option.) The **vce(robust)** option requests robust standard error estimates. See Chapter 9, or for a more complete explanation of robust standard errors, consult the *User's Guide*.

. `stphplot, by(sex)`

Plots –ln(–ln(survival)) versus ln(analysis time) for each level of the categorical variable *sex*, from the previous **stcox** model. Roughly parallel curves support the Cox model assumption that the hazard ratio does not change with time. Other checks on the Cox assumptions are performed by the commands **stcoxkm** (compares Cox predicted curves with Kaplan–Meier observed survival curves) and **estat phtest** (performs test based on Schoenfeld residuals). See **help stcox** for syntax and options.

. `streg x1 x2, dist(weibull)`

Fits Weibull-distribution model regression of time-to-failure on continuous or dummy variable predictors *x1* and *x2*.

. `streg x1 x2 x3 x4, dist(exponential) vce(robust)`

Fits exponential-distribution model regression of time-to-failure on continuous or dummy predictors *x1–x4*. Obtains heteroskedasticity-robust standard error estimates. In addition to Weibull and exponential, other **dist( )** specifications for **streg** include lognormal, log-logistic, Gompertz, or generalized gamma distributions. Type **help streg** for more information.

. **stcurve, survival**

After **streg**, plots the survival function from this model at mean values of all the *x* variables.

. **stcurve, cumhaz at(*x3*=50, *x4*=0)**

After **streg**, plots the cumulative hazard function from this model at mean values of *x1* and *x2*, *x3* set at 50, and *x4* set at 0.

. **poisson *count x1 x2 x3*, irr exposure(*x4*)**

Performs Poisson regression of event-count variable *count* (assumed to follow a Poisson distribution) on continuous or dummy independent variables *x1–x3*. Independent-variable effects will be reported as incidence rate ratios (**irr**). The **exposure( )** option identifies a variable indicating the amount of exposure, if this is not the same for all observations. Note: A Poisson model assumes that the event probability remains constant, regardless of how many times an event occurs for each observation. If the probability does not remain constant, we should consider using **nbreg** (negative binomial regression) or **gnbreg** (generalized negative binomial regression) instead.

. **glm *count x1 x2 x3*, link(log) family(poisson) lnoffset(*x4*) eform**

Performs the same regression specified in the **poisson** example above, but as a generalized linear model (GLM). **glm** can fit Poisson, negative binomial, logit, and many other types of models, depending on what **link( )** (link function) and **family( )** (distribution family) options we employ.

Survival-Time Data

Survival-time data contain, at a minimum, one variable measuring how much time elapsed before a certain event occurred for each observation. The literature often terms this event of interest a "failure," regardless of its substantive meaning. When failure has not occurred to an observation by the time data collection ends, that observation is said to be "censored." The **stset** command sets up a dataset for survival-time analysis by identifying which variable measures time and (if necessary) which variable is a dummy indicating whether the observation failed or was censored. The dataset can also contain any number of other measurement or categorical variables, and individuals (for example, medical patients) can be represented by more than one observation.

To illustrate the use of **stset**, we will begin with an example from Selvin (1995:453) concerning 51 individuals diagnosed with HIV. The data initially reside in a raw-data file (*aids.raw*) that looks like this:

```
1          1          1          34
2         17          1          42
3         37          0          47
        (rows 4–50 omitted)
51        81          0          29
```

The first column values are case numbers (1, 2, 3, . . . , 51). The second column tells how many months elapsed after the diagnosis, before that person either developed symptoms of AIDS or the study ended (1, 17, 37, . . .). The third column holds a 1 if the individual developed AIDS symptoms (failure), or a 0 if no symptoms had appeared by the end of the study (censoring). The last column reports the individual's age at the time of diagnosis.

We can read the raw data into memory using **infile**, then label the variables and data and save in Stata format as file *aids1.dta*:

```
. infile case time aids age using aids.raw, clear
(51 observations read)
. label variable case "Case ID number"
. label variable time "Months since HIV diagnosis"
. label variable aids "Developed AIDS symptoms"
. label variable age "Age in years"
. label data "AIDS (Selvin 1995:453)"
. compress
case was float now byte
time was float now byte
aids was float now byte
age was float now byte
```

The next step is to identify which variable measures time and which indicates failure/ censoring. Although not necessary with these single-record data, we can also note which variable holds individual case identification numbers. In an **stset** command, the first-named variable measures time. Subsequently, we identify with **failure( )** the dummy representing whether an observation failed (1) or was censored (0). After using **stset**, we save the data again to preserve this information.

```
. stset time, failure(aids) id(case)

             id:  case
   failure event:  aids != 0 & aids < .
obs. time interval:  (time[_n-1], time]
 exit on or before:  failure

─────────────────────────────────────────────────────────
      51  total obs.
       0  exclusions
─────────────────────────────────────────────────────────
      51  obs. remaining, representing
      51  subjects
      25  failures in single failure-per-subject data
    3164  total analysis time at risk, at risk from t =        0
                           earliest observed entry t =        0
                                last observed exit t =       97
```

```
. save aids1
file c:\data\aids1.dta saved
```

stdescribe yields a brief description of how our survival-time data are structured. In this simple example we have only one record per subject, so some of this information is unneeded.

```
. stdescribe
```

```
           failure _d:  aids
     analysis time _t:  time
                   id:  case
```

| | | |— per subject —| | |
Category	total	mean	min	median	max
no. of subjects	51				
no. of records	51	1	1	1	1
(first) entry time		0	0	0	0
(final) exit time		62.03922	1	67	97
subjects with gap	0				
time on gap if gap	0	.	.	.	.
time at risk	3164	62.03922	1	67	97
failures	25	.4901961	0	0	1

The **stsum** command obtains summary statistics. We have 25 failures out of 3,164 person-months, giving an incidence rate of $25/3164 = .0079014$. The percentiles of survival time derive from a Kaplan–Meier survivor function (next section). This function estimates about a 25% chance of developing AIDS within 41 months after diagnosis, and 50% within 81 months. Over the observed range of the data (up to 97 months) the probability of AIDS does not reach 75%, so there is no 75th percentile given.

```
. stsum
```

```
           failure _d:  aids
     analysis time _t:  time
                   id:  case
```

| | | incidence | no. of | |— Survival time —| |
---	time at risk	rate	subjects	25%	50%	75%
total	3164	.0079014	51	41	81	.

If the data happen to include a grouping or categorical variable such as *sex*, we could obtain summary statistics on survival time separately for each group by a command of the following form:

```
. stsum, by(sex)
```

Later sections describe more formal methods for comparing survival times from two or more groups.

Count-Time Data

Survival-time (**st**) datasets like *aids1.dta* contain information on individual people or things, with variables indicating the time at which failure or censoring occurred for each individual. A different type of dataset called count-time (**ct**) contains aggregate data, with variables counting the number of individuals that failed or were censored at time *t*. For example, *diskdriv.dta* contains hypothetical test information on 25 disk drives. All but 5 drives failed before testing ended at 1,200 hours.

```
Contains data from c:\data\diskdriv.dta
  obs:             6                        Count-time data on disk drives
 vars:             3                        23 May 2008 12:08
 size:            72 (99.9% of memory free)

              storage  display    value
variable name   type   format     label      variable label

hours           int    %8.0g                 Hours of continuous operation
failures        byte   %8.0g                 Number of failures observed
censored        byte   %9.0g                 Number still working
```

. **list**

	hours	failures	censored
1.	200	2	0
2.	400	3	0
3.	600	4	0
4.	800	8	0
5.	1000	3	0
6.	1200	0	5

To set up a count-time dataset, we specify the time variable, the number-of-failures variable, and the number-censored variable, in that order. After **ctset**, the **cttost** command automatically converts our count-time data to survival-time format.

. **ctset *hours failures censored***

```
   dataset name:  c:\data\diskdriv.dta
          time:  hours
      no. fail:  failures
      no. lost:  censored
     no. enter:  --                      (meaning all enter at time 0)
```

```
. cttost
```

```
      failure event:  failures != 0 & failures < .
 obs. time interval:  (0, hours]
 exit on or before:  failure
             weight:  [fweight=w]
```

6	total obs.
0	exclusions

6	physical obs. remaining, equal to	
25	weighted obs., representing	
20	failures in single record/single failure data	
19400	total analysis time at risk, at risk from t =	0
	earliest observed entry t =	0
	last observed exit t =	1200

```
. list
```

	hours	failures	censored	w	_st	_d	_t	_t0
1.	200	1	0	2	1	1	200	0
2.	400	1	0	3	1	1	400	0
3.	600	1	0	4	1	1	600	0
4.	800	1	0	8	1	1	800	0
5.	1000	1	0	3	1	1	1000	0
6.	1200	0	5	5	1	0	1200	0

```
. stdescribe
```

```
       failure _d:  failures
 analysis time _t:  hours
           weight:  [fweight=w]
```

			per subject		
Category	unweighted total	unweighted mean	min	unweighted median	max
no. of subjects	6				
no. of records	6	1	1	1	1
(first) entry time		0	0	0	0
(final) exit time		700	200	700	1200
subjects with gap	0				
time on gap if gap	0				
time at risk	4200	700	200	700	1200
failures	5	.8333333	0	1	1

The **cttost** command defines a set of frequency weights, *w*, in the resulting **st**-format dataset. **st*** commands automatically recognize and use these weights in any survival-time analysis, so the data now are viewed as containing 25 observations (25 disk drives) instead of the previous 6 (six time periods).

```
. stsum
```

```
        failure _d:  failures
   analysis time _t:  hours
          weight:  [fweight=w]
```

	time at risk	incidence rate	no. of subjects	Survival time 25%	50%	75%
total	19400	.0010309	25	600	800	1000

Kaplan–Meier Survivor Functions

Let n_t represent the number of observations that have not failed, and are not censored, at the beginning of time period t. d_t represents the number of failures that occur to these observations during time period t. The Kaplan–Meier estimator of surviving beyond time t is the product of survival probabilities in t and the preceding periods:

$$S(t) = \prod_{j=t0}^{t} \{ (n_j - d_j) / n_j \} \qquad [11.1]$$

For example, in the AIDS data seen earlier, one of the 51 individuals developed symptoms only one month after diagnosis. No observations were censored this early, so the probability of "surviving" (meaning, not developing AIDS) beyond *time* = 1 is

$$S(1) = (51 - 1) / 51 = .9804$$

A second patient developed symptoms at *time* = 2, and a third at *time* = 9:

$$S(2) = .9804 \times (50 - 1) / 50 = .9608$$
$$S(9) = .9608 \times (49 - 1) / 49 = .9412$$

Graphing $S(t)$ against t produces a Kaplan–Meier survivor curve, like the one seen in Figure 11.1. Stata draws such graphs automatically with the **sts graph** command. For example,

```
. use aids, clear
(AIDS (Selvin 1995:453))
```

```
. sts graph

       failure _d:  aids
  analysis time _t:  time
             id:  case
```

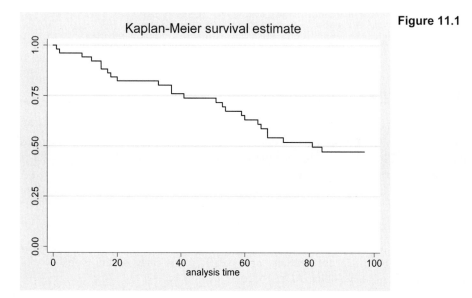

Kaplan-Meier survival estimate

Figure 11.1

For a second example of survivor functions, we turn to data in *smoking1.dta*, adapted from Rosner (1995). The observations are data on 234 former smokers, attempting to quit. Most did not succeed. Variable *days* records how many days elapsed between quitting and starting up again. The study lasted one year, and variable *smoking* indicates whether an individual resumed smoking before the end of this study (*smoking* = 1, "failure") or not (*smoking* = 0, "censored"). With new data, we should begin by using **stset** to set the data up for survival-time analysis.

```
Contains data from c:\data\smoking1.dta
  obs:           234                          Smoking (Rosner 1995:607)
  vars:            8                          18 Jun 2008 13:43
  size:        4,680  (99.9% of memory free)

              storage   display    value
variable name   type    format     label    variable label

id              int     %9.0g               Case ID number
days            int     %9.0g               Days abstinent
smoking         byte    %9.0g               Resumed smoking
age             byte    %9.0g               Age in years
sex             byte    %9.0g      sex      Sex (female)
cigs            byte    %9.0g               Cigarettes per day
co              int     %9.0g               Carbon monoxide x 10
minutes         int     %9.0g               Minutes elapsed since last cig

Sorted by:
```

```
. stset days, failure(smoking)

      failure event:  smoking != 0 & smoking < .
obs. time interval:  (0, days]
 exit on or before:  failure
```

```
    234  total obs.
      0  exclusions
```

```
    234  obs. remaining, representing
    201  failures in single record/single failure data
  18946  total analysis time at risk, at risk from t =          0
                           earliest observed entry t =          0
                              last observed exit t =          366
```

The study involved 110 men and 124 women. Incidence rates for both sexes appear to be similar:

```
. stsum, by(sex)

      failure _d:  smoking
 analysis time _t:  days
```

sex	time at risk	incidence rate	no. of subjects	Survival time 25%	50%	75%
Male	8813	.0105526	110	4	15	68
Female	10133	.0106582	124	4	15	83
total	18946	.0106091	234	4	15	73

Figure 11.2 confirms this similarity. We see little difference between the survivor functions of men and women. That is, both sexes returned to smoking at about the same rate. The survival probabilities of nonsmokers decline very steeply during the first 30 days after quitting. For either sex, there is less than a 15% chance of surviving beyond a full year.

. **sts graph, by(*sex*)**

```
        failure _d:  smoking
  analysis time _t:  days
```

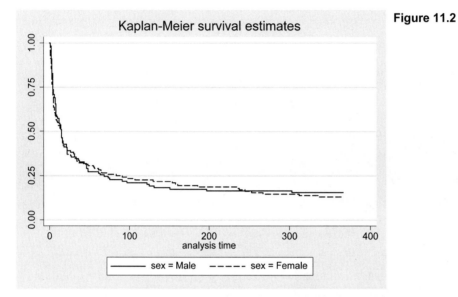

Figure 11.2

We can also formally test for the equality of survivor functions using a log-rank test. Unsurprisingly, this test finds no significant difference ($P = .6772$) between the smoking recidivism of men and women.

. **sts test *sex***

```
        failure _d:  smoking
  analysis time _t:  days
```

Log-rank test for equality of survivor functions

sex	Events observed	Events expected
Male	93	95.88
Female	108	105.12
Total	201	201.00

$$\text{chi2}(1) = 0.17$$
$$\text{Pr>chi2} = 0.6772$$

Cox Proportional Hazard Models

Regression methods allow us to take survival analysis further and examine the effects of multiple continuous or categorical predictors. One widely-used method known as Cox regression employs a proportional hazard model. The hazard rate for failure at time t is defined as the rate of failures at time t among those who have survived to time t:

$$h(t) \quad = \quad \frac{\text{probability of failing between times } t \text{ and } t + \Delta t}{(\Delta t)\,(\text{probability of failing after time } t)} \qquad\qquad [11.2]$$

We model this hazard rate as a function of the baseline hazard (h_0) at time t, and the effects of one or more x variables,

$$h(t) \quad = \quad h_0(t)\exp(\beta_1 x_1 + \beta_2 x_2 + \ldots + \beta_k x_k) \qquad\qquad [11.3a]$$

or, equivalently,

$$\ln[h(t)] \quad = \quad \ln[h_0(t)] + \beta_1 x_1 + \beta_2 x_2 + \ldots + \beta_k x_k \qquad\qquad [11.3b]$$

"Baseline hazard" means the hazard for an observation with all x variables equal to 0. Cox regression estimates this hazard nonparametrically and obtains maximum-likelihood estimates of the β parameters in [11.3]. Stata's **stcox** procedure ordinarily reports hazard ratios, which are estimates of $\exp(\beta)$. These indicate proportional changes relative to the baseline hazard rate.

Does age affect the onset of AIDS symptoms? Dataset *aids.dta* contains information that addresses this question. Note that with **stcox**, unlike most other Stata model-fitting commands, we list only the independent variable(s). The survival-analysis dependent variables, time variables, and censoring variables are understood automatically with **stset** data.

```
. stcox age, nolog

        failure _d:  aids
   analysis time _t:  time
               id:  case

Cox regression -- Breslow method for ties

No. of subjects =           51          Number of obs   =         51
No. of failures =           25
Time at risk    =         3164
                                         LR chi2(1)      =       5.00
Log likelihood  =    -86.576295          Prob > chi2     =     0.0254
```

_t	Haz. Ratio	Std. Err.	z	P>|z|	[95% Conf. Interval]	
age	1.084557	.0378623	2.33	0.020	1.01283	1.161363

We might interpret the estimated hazard ratio, 1.084557, with reference to two HIV-positive individuals whose ages are a and $a + 1$. The older person is 8.5% more likely to develop AIDS symptoms over a short period of time (that is, the ratio of their respective hazards is 1.084557).

This ratio differs significantly ($P = .020$) from 1. If we wanted to state our findings for a five-year difference in age, we could raise the hazard ratio to the fifth power:

```
. display exp(_b[age])^5
1.5005865
```

Thus, the hazard of AIDS onset is about 50% higher when the second person is five years older than the first. Alternatively, we could learn the same thing (and obtain the new confidence interval) by repeating the regression after creating a new version of *age* measured in five-year units. The **nolog noshow** options below suppress display of the iteration log and the **st**-dataset description.

```
. generate age5 = age/5
. label variable age5 "age in 5-year units"
. stcox age5, nolog noshow
```

```
Cox regression -- Breslow method for ties
```

No. of subjects =	51	Number of obs	=	51
No. of failures =	25			
Time at risk =	3164			
		LR chi2(1)	=	5.00
Log likelihood =	-86.576295	Prob > chi2	=	0.0254

_t	Haz. Ratio	Std. Err.	z	P>|z|	[95% Conf. Interval]	
age5	1.500587	.2619305	2.33	0.020	1.065815	2.112711

Like ordinary regression, Cox models can have more than one independent variable. Dataset *heart.dta* contains survival-time data from Selvin (1995) on 35 patients with very high cholesterol levels. Variable *time* gives the number of days each patient was under observation. *coronary* indicates whether a coronary event occurred at the end of this time period (*coronary* = 1) or not (*coronary* = 0). The data also include cholesterol levels and other factors thought to affect heart disease. File *heart.dta* was previously set up for survival-time analysis by an **stset** *time*, **failure(***coronary***)** command, so we can go directly to **st** analysis.

```
. describe patient - ab
```

variable name	storage type	display format	value label	variable label
patient	byte	%9.0g		Patient ID number
time	int	%9.0g		Time in days
coronary	byte	%9.0g		Coronary event (1) or none (0)
weight	int	%9.0g		Weight in pounds
sbp	int	%9.0g		Systolic blood pressure
chol	int	%9.0g		Cholesterol level
cigs	byte	%9.0g		Cigarettes smoked per day
ab	byte	%9.0g		Type A (1) or B (0) personality

```
. stdescribe
```

```
       failure _d:  coronary
  analysis time _t:  time
```

			per subject		
Category	total	mean	min	median	max
no. of subjects	35				
no. of records	35	1	1	1	1
(first) entry time		0	0	0	0
(final) exit time		2580.629	773	2875	3141
subjects with gap	0				
time on gap if gap	0				
time at risk	90322	2580.629	773	2875	3141
failures	8	.2285714	0	0	1

Cox regression finds that cholesterol level and cigarettes both significantly increase the hazard of a coronary event. Counterintuitively, weight appears to decrease the hazard. Systolic blood pressure and A/B personality do not have significant net effects.

```
. stcox weight sbp chol cigs ab, noshow nolog
```

```
Cox regression -- no ties
```

```
No. of subjects =        35          Number of obs   =        35
No. of failures =         8
Time at risk    =     90322
                                     LR chi2(5)      =     13.97
Log likelihood  = -17.263231         Prob > chi2     =    0.0158
```

_t	Haz. Ratio	Std. Err.	z	P>|z|	[95% Conf. Interval]	
weight	.9349336	.0305184	-2.06	0.039	.8769919	.9967034
sbp	1.012947	.0338061	0.39	0.700	.9488087	1.081421
chol	1.032142	.0139984	2.33	0.020	1.005067	1.059947
cigs	1.203335	.1071031	2.08	0.038	1.010707	1.432676
ab	3.04969	2.985616	1.14	0.255	.4476492	20.77655

After estimating the model, **stcox** can also generate new variables holding the estimated baseline cumulative hazard and survivor functions. Since "baseline" refers to a situation with all x variables equal to zero, however, we first need to center some variables so that 0 values make sense. A patient who weighs 0 pounds, or has 0 blood pressure, does not provide a useful comparison. Guided by the minimum values actually in our data, we might shift *weight* so that 0 indicates 120 pounds, *sbp* so that 0 indicates 105, and *chol* so that 0 indicates 340:

```
. summarize patient - ab
```

Variable	Obs	Mean	Std. Dev.	Min	Max
patient	35	18	10.24695	1	35
time	35	2580.629	616.0796	773	3141
coronary	35	.2285714	.426043	0	1
weight	35	170.0857	23.55516	120	225
sbp	35	129.7143	14.28403	104	154
chol	35	369.2857	51.32284	343	645
cigs	35	17.14286	13.07702	0	40
ab	35	.5142857	.5070926	0	1

```
. replace weight = weight - 120
(35 real changes made)
. replace sbp = sbp - 105
(35 real changes made)
. replace chol = chol - 340
(35 real changes made)
. summarize patient - ab
```

Variable	Obs	Mean	Std. Dev.	Min	Max
patient	35	18	10.24695	1	35
time	35	2580.629	616.0796	773	3141
coronary	35	.2285714	.426043	0	1
weight	35	50.08571	23.55516	0	105
sbp	35	24.71429	14.28403	-1	49
chol	35	29.28571	51.32284	3	305
cigs	35	17.14286	13.07702	0	40
ab	35	.5142857	.5070926	0	1

Zero values for all the x variables now make more substantive sense. To create new variables holding the baseline survivor and cumulative hazard function estimates, we repeat the regression with **basesurv( )** and **basechaz( )** options:

```
. stcox weight sbp chol cigs ab, noshow nolog basesurv(survivor)
      basechaz(hazard)
```

```
Cox regression -- no ties
```

No. of subjects =	35		Number of obs =	35
No. of failures =	8			
Time at risk =	90322			
			LR chi2(5) =	13.97
Log likelihood =	-17.263231		Prob > chi2 =	0.0158

_t	Haz. Ratio	Std. Err.	z	P>|z|	[95% Conf. Interval]	
weight	.9349336	.0305184	-2.06	0.039	.8769919	.9967034
sbp	1.012947	.0338061	0.39	0.700	.9488087	1.081421
chol	1.032142	.0139984	2.33	0.020	1.005067	1.059947
cigs	1.203335	.1071031	2.08	0.038	1.010707	1.432676
ab	3.04969	2.985616	1.14	0.255	.4476492	20.77655

Note that centering three x variables had no effect on the hazard ratios, standard errors, and so forth. The command created two new variables, arbitrarily named *survivor* and *hazard*. To graph the baseline survivor function, we plot *survivor* against *time* and connect data points in a stairstep fashion, as seen in Figure 11.3.

```
. graph twoway line survivor time, connect(stairstep) sort
```

Figure 11.3

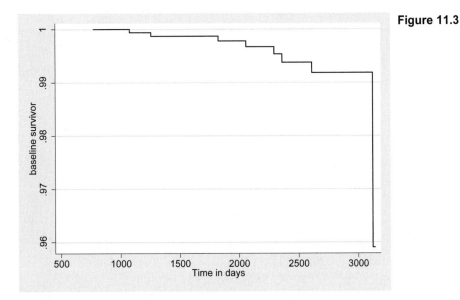

The baseline survivor function — which depicts survival probabilities for patients having "0" weight (120 pounds), "0" blood pressure (100), "0" cholesterol (340), 0 cigarettes per day, and a type B personality — declines with time. Although this decline looks precipitous at the right, notice that the probability really only falls from 1 to about .96. Given less favorable values of the predictor variables, the survival probabilities would fall much faster.

The same baseline survivor-function graph could have been obtained another way, without **stcox**. The alternative, shown in Figure 11.4 on the following page, employs an **sts graph** command with **adjustfor( )** option listing the predictor variables.

```
. sts graph, adjustfor(weight sbp chol cigs ab)
```

```
        failure _d:  coronary
  analysis time _t:  time
```

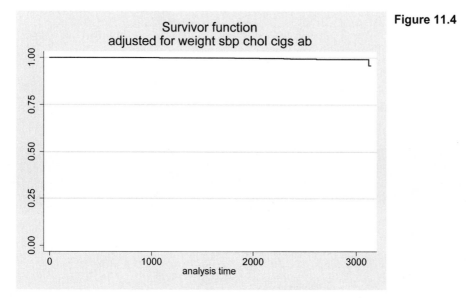

Figure 11.4

Figure 11.4, unlike Figure 11.3, follows the usual survivor-function convention of scaling the vertical axis from 0 to 1. Apart from this difference in scaling, Figures 11.3 and 11.4 depict the same curve.

Figure 11.5 graphs the estimated baseline cumulative hazard against time, using the variable (*hazard*) generated by our **stcox** command. This graph shows the baseline cumulative hazard increasing in 8 steps (because 8 patients "failed" or had coronary events), from near 0 to .033.

```
. graph twoway connected hazard time, connect(stairstep) sort
     msymbol(Oh)
```

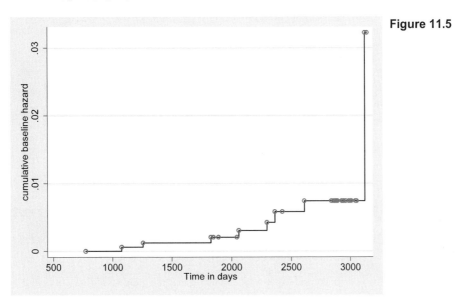

Figure 11.5

Exponential and Weibull Regression

Cox regression estimates the baseline survivor function empirically without reference to any theoretical distribution. Several alternative "parametric" approaches begin instead from assumptions that survival times do follow a known theoretical distribution. Possible distribution families include the exponential, Weibull, lognormal, log-logistic, Gompertz, or generalized gamma. Models based on any of these can be fit through the **streg** command. Such models have the same general form as Cox regression (equations [11.2] and [11.3]), but define the baseline hazard $h_0(t)$ differently. Two examples appear in this section.

If failures occur independently, with a constant hazard, then survival times follow an exponential distribution and could be analyzed by *exponential regression*. Constant hazard means that the individuals studied do not "age," in the sense that they are no more or less likely to fail late in the period of observation than they were at its start. Over the long term, this assumption seems unjustified for machines or living organisms, but it might approximately hold if the period of observation covers a relatively small fraction of their life spans. An exponential model implies that logarithms of the survivor function, $\ln(S(t))$, are linearly related to t.

A second common parametric approach, *Weibull regression*, is based on the more general Weibull distribution. This does not require failure rates to remain constant, but allows them to increase or decrease smoothly over time. The Weibull model implies that $\ln(-\ln(S(t)))$ is a linear function of $\ln(t)$.

Graphs provide a useful diagnostic for the appropriateness of exponential or Weibull models. For example, returning to *aids.dta*, we construct a graph (Figure 11.6) of ln(S(t)) versus time, after first generating Kaplan–Meier estimates of the survivor function S(t). The y-axis labels in Figure 11.6 are given a fixed two-digit, one-decimal display format (%2.1f) and oriented horizontally, to improve their readability.

```
. sts gen S = S
. generate logS = ln(S)
. graph twoway scatter logS time,
      ylabel(-.8(.1)0, format(%2.1f) angle(horizontal))
```

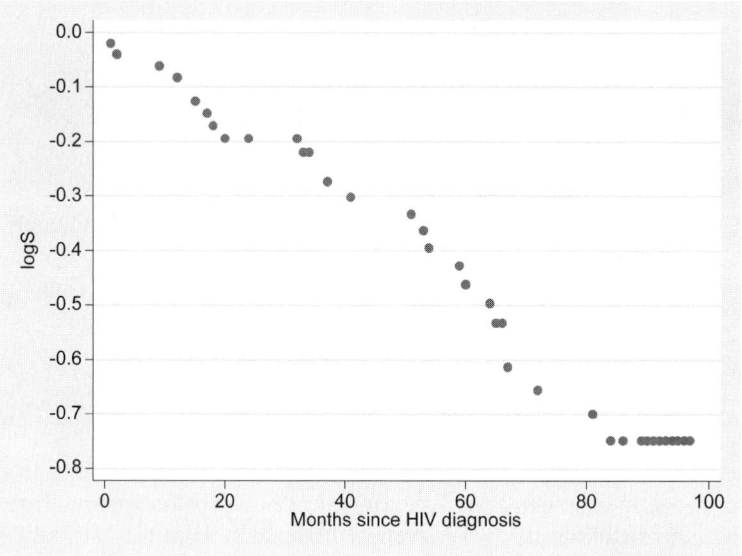

Figure 11.6

The pattern in Figure 11.6 appears somewhat linear, encouraging us to try an exponential regression:

```
.   streg age, dist(exponential) nolog noshow
```

Exponential regression -- log relative-hazard form

```
No. of subjects =          51              Number of obs   =          51
No. of failures =          25
Time at risk    =        3164
                                           LR chi2(1)      =        4.34
Log likelihood  =   -59.996976             Prob > chi2     =      0.0372
```

_t	Haz. Ratio	Std. Err.	z	P>|z|	[95% Conf. Interval]
age	1.074414	.0349626	2.21	0.027	1.008028 1.145172

The hazard ratio (1.074) and standard error (.035) estimated by this exponential regression do not greatly differ from their counterparts (1.085 and .038) in our earlier Cox regression. The similarity reflects the degree of correspondence between empirical hazard function and the constant hazard implied by an exponential distribution. According to this exponential model, the hazard of an HIV-positive individual developing AIDS increases about 7.4% with each year of age.

After **streg**, the **stcurve** command draws a graph of the models' cumulative hazard, survival, or hazard functions. By default, **stcurve** draws these curves holding all x variables in the model at their means. We can specify other x values by using the **at( )** option. The individuals in *aids.dta* ranged from 26 to 50 years old. We could graph the survival function at *age* = 26 by issuing a command such as

```
. stcurve, surviv at(age=26)
```

A more interesting graph uses the **at1( )** and **at2( )** options to show the survival curve at two different sets of x values, such as the low and high extremes of *age*:

```
. stcurve, survival at1(age=26) at2(age=50)
```

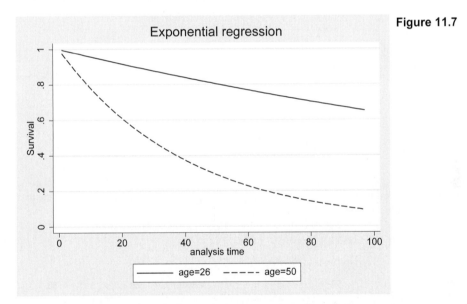

Figure 11.7

Figure 11.7 shows the predicted survival curve (for transition from HIV diagnosis to AIDS) falling more steeply among older patients. The significant *age* hazard ratio greater than 1 in our exponential regression table implied the same thing, but using **stcurve** with **at1( )** and **at2( )** values gives a strong visual interpretation of this effect. These options work in a similar manner with all three types of **stcurve** graphs:

stcurve, survival	Survival function.
stcurve, hazard	Hazard function.
stcurve, cumhaz	Cumulative hazard function.

Instead of the exponential distribution, **streg** can also fit survival models based on the Weibull distribution. A Weibull distribution might appear curvilinear in a plot of $\ln(S(t))$ versus t, but it should be linear in a plot of $\ln(-\ln(S(t)))$ versus $\ln(t)$, such as Figure 11.8. An exponential distribution, on the other hand, will appear linear in both plots and have a slope equal to 1 in the $\ln(-\ln(S(t)))$ versus $\ln(t)$ plot. In fact, the data points in Figure 11.8 are not far from a line with slope 1, suggesting that our previous exponential model is adequate.

```
. generate loglogS = ln(-ln(S))
. generate logtime = ln(time)
. graph twoway scatter loglogS logtime, ylabel(,angle(horizontal))
```

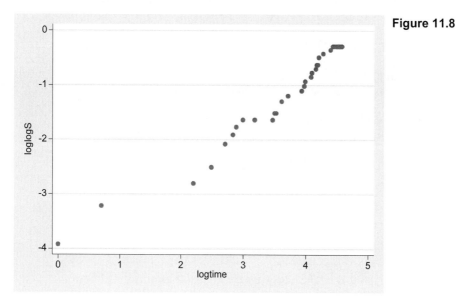

Figure 11.8

Although we do not need the additional complexity of a Weibull model with these data, results are given below for illustration.

```
. streg age, dist(weibull) noshow nolog
```

Weibull regression -- log relative-hazard form

No. of subjects =	51	
No. of failures =	25	
Time at risk =	3164	

Number of obs   =   51

LR chi2(1)   =   4.68
Log likelihood = -59.778257

Prob > chi2   =   0.0306

_t	Haz. Ratio	Std. Err.	z	P>|z|	[95% Conf. Interval]
age	1.079477	.0363509	2.27	0.023	1.010531 1.153127
/ln_p	.1232638	.1820858	0.68	0.498	-.2336179 .4801454
p	1.131183	.2059723			.7916643 1.616309
1/p	.8840305	.1609694			.6186934 1.263162

The Weibull regression obtains a hazard ratio estimate (1.079) intermediate between our previous Cox and exponential results. The most noticeable difference from those earlier models is the presence of three new lines at the bottom of the table. These refer to the Weibull distribution shape parameter p. A p value of 1 corresponds to an exponential model: the hazard does not change with time. $p > 1$ indicates that the hazard increases with time; $p < 1$ indicates that the hazard decreases. A 95% confidence interval for p ranges from .79 to 1.62, so we have no reason to reject an exponential ($p = 1$) model here. Different, but mathematically equivalent, parameterizations of the Weibull model focus on $\ln(p)$, p, or $1/p$, so Stata provides all three. **stcurve** draws survival, hazard, or cumulative hazard functions after **streg, dist(weibull)** just as it does after **streg, dist(exponential)** or other **streg** models.

Exponential or Weibull regression is preferable to Cox regression when survival times actually follow an exponential or Weibull distribution. When they do not, these models are misspecified and can yield misleading results. Cox regression, which makes no *a priori* assumptions about distribution shape, remains useful in a wider variety of situations.

In addition to exponential and Weibull models, **streg** can fit models based on the Gompertz, lognormal, log-logistic, or generalized gamma distributions. Type **help streg**, or consult the *Survival Analysis and Epidemiological Tables Reference Manual*, for syntax and a list of current options.

Poisson Regression

If events occur independently and with constant rate, then counts of events over a given period of time follow a Poisson distribution. Let r_j represent the incidence rate:

$$r_j \; = \; \frac{\text{count of events}}{\text{number of times event could have occurred}} \qquad [11.4]$$

The denominator in [11.4] is termed the "exposure" and is often measured in units such as person-years. We model the logarithm of incidence rate as a linear function of one or more predictor (x) variables:

$$\ln(r_t) \; = \; \beta_0 + \beta_1 x_1 + \beta_2 x_2 + \ldots + \beta_k x_k \qquad [11.5a]$$

Equivalently, the model describes logs of expected event counts:

$$\ln(\text{expected count}) \; = \; \ln(\text{exposure}) + \beta_0 + \beta_1 x_1 + \beta_2 x_2 + \ldots \beta_k x_k \qquad [11.5b]$$

Assuming that a Poisson process underlies the events of interest, Poisson regression finds maximum-likelihood estimates of the β parameters.

Data on radiation exposure and cancer deaths among workers at Oak Ridge National Laboratory provide an example. The 56 observations in dataset *oakridge.dta* represent 56 age/radiation-exposure categories (7 categories of age $\times$ 8 categories of radiation exposure). For each combination, we know the number of deaths and the number of person-years of exposure.

```
Contains data from c:\data\oakridge.dta
  obs:           56                      Radiation (Selvin 1995:474)
  vars:           4                      23 May 2008 12:08
  size:         840 (99.9% of memory free)
```

variable name	storage type	display format	value label	variable label
age	byte	%9.0g	ageg	Age group
rad	byte	%9.0g		Radiation exposure level
deaths	byte	%9.0g		Number of deaths
pyears	float	%9.0g		Person-years

. **summarize**

Variable	Obs	Mean	Std. Dev.	Min	Max
age	56	4	2.0181	1	7
rad	56	4.5	2.312024	1	8
deaths	56	1.839286	3.178203	0	16
pyears	56	3807.679	10455.91	23	71382

. **list in 1/6**

	age	rad	deaths	pyears
1.	< 45	1	0	29901
2.	45-49	1	1	6251
3.	50-54	1	4	5251
4.	55-59	1	3	4126
5.	60-64	1	3	2778
6.	65-69	1	1	1607

Does the death rate increase with exposure to radiation? Poisson regression finds a statistically significant effect:

. **poisson** *deaths rad,* **nolog exposure**(*pyears*) **irr**

```
Poisson regression                          Number of obs   =         56
                                            LR chi2(1)      =      14.87
                                            Prob > chi2     =     0.0001
Log likelihood =  -169.7364                 Pseudo R2       =     0.0420
```

deaths	IRR	Std. Err.	z	P>|z|	[95% Conf. Interval]
rad	1.236469	.0603551	4.35	0.000	1.123657 1.360606
pyears	(exposure)				

For the regression above, we specified the event count (*deaths*) as the dependent variable and radiation (*rad*) as the independent variable. The Poisson "exposure" variable is *pyears*, or person-years in each category of *rad*. The **irr** option calls for incidence rate ratios rather than regression coefficients in the results table — that is, we get estimates of $\exp(\beta)$ instead of β, the default. According to this incidence rate ratio, the death rate becomes 1.236 times higher (increases by 23.6%) with each increase in radiation category. Although that ratio is statistically significant, the fit is not impressive; the pseudo R^2 (equation [10.4]) is only .042.

To perform a goodness-of-fit test, comparing the Poisson model's predictions with the observed counts, use the postestimation command **estat gof** :

```
. poisgof
        Goodness-of-fit chi2  =   254.5475
        Prob > chi2(54)       =     0.0000
```

These goodness-of-fit test results ($\chi^2 = 254.5$, $P < .00005$) indicate that our model's predictions are significantly different from the actual counts — another sign that the model fits poorly.

We obtain better results when we include *age* as a second predictor. Pseudo R^2 then rises to .5966, and the goodness-of-fit test no longer leads us to reject our model.

```
. poisson deaths rad age, nolog exposure(pyears) irr
```

Poisson regression				Number of obs	=	56
				LR chi2(2)	=	211.41
				Prob > chi2	=	0.0000
Log likelihood = −71.4653				Pseudo R2	=	0.5966

deaths	IRR	Std. Err.	z	P>|z|	[95% Conf. Interval]	
rad	1.176673	.0593446	3.23	0.001	1.065924	1.298929
age	1.960034	.0997536	13.22	0.000	1.773955	2.165631
pyears	(exposure)					

```
. poisgof
        Goodness-of-fit chi2  =    58.00534
        Prob > chi2(53)       =     0.2960
```

For simplicity, to this point we have treated *rad* and *age* as if both were continuous variables, and we expect their effects on the log death rate to be linear. In fact, however, both independent variables are measured as ordered categories. *rad* = 1, for example, means 0 radiation exposure; *rad* = 2 means 0 to 19 milliseiverts; *rad* = 3 means 20 to 39 milliseiverts; and so forth. An alternative way to include radiation exposure categories in the regression, while watching for nonlinear effects, is as a set of dummy variables. On the following page, we use the **gen( )** option of **tabulate** to create 8 dummy variables, *r1* to *r8*, representing each of the 8 values of *rad*.

. **tabulate** *rad,* **gen(r)**
```
       Goodness-of-fit chi2  =   58.00534
       Prob > chi2(53)       =     0.2960
```

. tabulate rad, gen(r);

Radiation exposure level	Freq.	Percent	Cum.
1	7	12.50	12.50
2	7	12.50	25.00
3	7	12.50	37.50
4	7	12.50	50.00
5	7	12.50	62.50
6	7	12.50	75.00
7	7	12.50	87.50
8	7	12.50	100.00
Total	56	100.00	

. **describe**

```
Contains data from c:\data\oakridge.dta
  obs:            56                    Radiation (Selvin 1995:474)
  vars:           12                    23 May 2008 12:08
  size:        1,288  (99.9% of memory free)
```

variable name	storage type	display format	value label	variable label
age	byte	%9.0g	ageg	Age group
rad	byte	%9.0g		Radiation exposure level
deaths	byte	%9.0g		Number of deaths
pyears	float	%9.0g		Person-years
r1	byte	%8.0g		rad== 1.0000
r2	byte	%8.0g		rad== 2.0000
r3	byte	%8.0g		rad== 3.0000
r4	byte	%8.0g		rad== 4.0000
r5	byte	%8.0g		rad== 5.0000
r6	byte	%8.0g		rad== 6.0000
r7	byte	%8.0g		rad== 7.0000
r8	byte	%8.0g		rad== 8.0000

```
Sorted by:
   Note: dataset has changed since last saved
```

We now include seven of these dummies (omitting one to avoid multicollinearity) as regression predictors. The additional complexity of this dummy-variable model brings little improvement in fit. It does, however, add to our interpretation. The overall effect of radiation on death rate appears to come primarily from the two highest radiation levels (*r7* and *r8*, corresponding to 100 to 119 and 120 or more milliseiverts). At these levels, the incidence rates are about four times higher.

```
. poisson deaths r2-r8 age, nolog exposure(pyears) irr
```

```
Poisson regression                          Number of obs  =        56
                                            LR chi2(8)     =    215.44
                                            Prob > chi2    =    0.0000
Log likelihood = -69.451814                 Pseudo R2      =    0.6080
```

deaths	IRR	Std. Err.	z	P>|z|	[95% Conf. Interval]	
r2	1.473591	.426898	1.34	0.181	.8351884	2.599975
r3	1.630688	.6659257	1.20	0.231	.732428	3.630587
r4	2.375967	1.088835	1.89	0.059	.9677429	5.833389
r5	.7278113	.7518255	-0.31	0.758	.0961018	5.511957
r6	1.168477	1.20691	0.15	0.880	.1543195	8.847472
r7	4.433729	3.337738	1.98	0.048	1.013863	19.38915
r8	3.89188	1.640978	3.22	0.001	1.703168	8.893267
age	1.961907	.1000652	13.21	0.000	1.775267	2.168169
pyears	(exposure)					

Radiation levels 7 and 8 seem to have similar effects, so we might simplify the model by combining them. First, we test whether their coefficients are significantly different. They are not:

```
. test r7 = r8

 ( 1)   [deaths]r7 - [deaths]r8 = 0

          chi2(  1) =      0.03
        Prob > chi2 =      0.8676
```

Next, generate a new dummy variable *r78*, which equals 1 if either *r7* or *r8* equals 1:

```
. generate r78 = (r7 | r8)
```

Finally, substitute the new predictor for *r7* and *r8* in the regression:

```
. poisson deaths r2-r6 r78 age, irr ex(pyears) nolog
```

```
Poisson regression                          Number of obs  =        56
                                            LR chi2(7)     =    215.41
                                            Prob > chi2    =    0.0000
Log likelihood = -69.465332                 Pseudo R2      =    0.6079
```

deaths	IRR	Std. Err.	z	P>|z|	[95% Conf. Interval]	
r2	1.473602	.4269013	1.34	0.181	.8351949	2.599996
r3	1.630718	.6659381	1.20	0.231	.7324415	3.630655
r4	2.376065	1.08888	1.89	0.059	.9677823	5.833629
r5	.7278387	.7518538	-0.31	0.758	.0961055	5.512165
r6	1.168507	1.206942	0.15	0.880	.1543236	8.847704
r78	3.980326	1.580024	3.48	0.001	1.828214	8.665833
age	1.961722	.100043	13.21	0.000	1.775122	2.167937
pyears	(exposure)					

We could proceed to simplify the model further in this fashion. At each step, **test** helps to evaluate whether combining two dummy variables is justifiable.

Generalized Linear Models

Generalized linear models (GLM) have the form

$$g[E(y)] = \beta_0 + \beta_1 x_1 + \beta_2 x_2 + \ldots + \beta_k x_k, \qquad y \sim F \qquad [11.6]$$

where $g[\ ]$ is the *link function* and F the distribution family. This general formulation encompasses many specific models. For example, if $g[\ ]$ is the identity function and y follows a normal (Gaussian) distribution, we have a linear regression model:

$$E(y) = \beta_0 + \beta_1 x_1 + \beta_2 x_2 + \ldots + \beta_k x_k, \qquad y \sim \text{Normal} \qquad [11.7]$$

If $g[\ ]$ is the logit function and y follows a Bernoulli distribution, we have logit regression instead:

$$\text{logit}[E(y)] = \beta_0 + \beta_1 x_1 + \beta_2 x_2 + \ldots + \beta_k x_k, \qquad y \sim \text{Bernoulli} \qquad [11.8]$$

Because of its broad applications, GLM could have been introduced at several different points in this book. Its relevance to this chapter comes from the ability to fit event models. Poisson regression, for example, requires that $g[\ ]$ is the natural log function and that y follows a Poisson distribution:

$$\ln[E(y)] = \beta_0 + \beta_1 x_1 + \beta_2 x_2 + \ldots + \beta_k x_k, \qquad y \sim \text{Poisson} \qquad [11.9]$$

As might be expected with such a flexible method, Stata's **glm** command permits many different options. Users can specify not only the distribution family and link function, but also details of the variance estimation, fitting procedure, output, and offset. These options make **glm** a useful alternative even when applied to models for which a dedicated command (such as **regress**, **logistic**, or **poisson**) already exists.

We might represent a "generic" **glm** command as follows:

```
. glm y x1 x2 x3, family(familyname) link(linkname)
     lnoffset(exposure) eform vce(jknife)
```

where **family( )** specifies the y distribution family, **link( )** the link function, and **lnoffset( )** an "exposure" variable such as that needed for Poisson regression. The **eform** option asks for regression coefficients in exponentiated form, $\exp(\beta)$ rather than β. Standard errors are estimated through jackknife (**jknife**) calculations.

Possible distribution families are

family(gaussian)	Gaussian or normal (default)
family(igaussian)	Inverse Gaussian
family(binomial)	Bernoulli binomial
family(poisson)	Poisson
family(nbinomial)	Negative binomial
family(gamma)	Gamma

We can also specify a number or variable indicating the binomial denominator N (number of trials), or a number indicating the negative binomial variance and deviance functions, by declaring them in the **family( )** option:

> **family(binomial #)**
> **family(binomial** *varname***)**
> **family(nbinomial #)**

Possible link functions are

link(identity)	Identity (default)
link(log)	Log
link(logit)	Logit
link(probit)	Probit
link(cloglog)	Complementary log-log
link(opower #)	Odds power
link(power #)	Power
link(nbinomial)	Negative binomial
link(loglog)	Log-log
link(logc)	Log-complement

Coefficient variances or standard errors can be estimated in a variety of ways. A partial list of **glm** variance-estimating options is given below:

opg	Berndt, Hall, Hall, and Hausman "B-H-cubed" variance estimator.
oim	Observed information matrix variance estimator.
robust	Huber/White/sandwich estimator of variance.
unbiased	Unbiased sandwich estimator of variance
nwest	Heteroskedasticity and autocorrelation-consistent variance estimator.
jknife	Jackknife estimate of variance.
jknife1	One-step jackknife estimate of variance.
bstrap	Bootstrap estimate of variance. The default is 199 repetitions; specify some other number by adding the **bsrep(#)** option.

For a full list of options with some technical details, look up **glm** in the *Base Reference Manual*. A more in-depth treatment of GLM topics can be found in Hardin and Hilbe (2007).

Chapter 6 began with the simple regression of mean composite SAT scores (*csat*) on per-pupil expenditures (*expense*) of the 50 U.S. states and District of Columbia (*states.dta*):

```
. regress csat expense
```

Source	SS	df	MS		
Model	48708.3001	1	48708.3001	Number of obs =	51
Residual	175306.21	49	3577.67775	F(1, 49) =	13.61
				Prob > F =	0.0006
				R-squared =	0.2174
				Adj R-squared =	0.2015
Total	224014.51	50	4480.2902	Root MSE =	59.814

csat	Coef.	Std. Err.	t	P>|t|	[95% Conf. Interval]	
expense	-.0222756	.0060371	-3.69	0.001	-.0344077	-.0101436
_cons	1060.732	32.7009	32.44	0.000	995.0175	1126.447

We can fit the same model and obtain exactly the same estimates via a **glm** command,

```
. glm csat expense, link(identity) family(gaussian)

Iteration 0:    log likelihood = -279.99869

Generalized linear models                       No. of obs       =         51
Optimization     : ML                           Residual df      =         49
                                                Scale parameter =   3577.678
Deviance         =   175306.2097                (1/df) Deviance =   3577.678
Pearson          =   175306.2097                (1/df) Pearson  =   3577.678

Variance function: V(u) = 1                     [Gaussian]
Link function    : g(u) = u                     [Identity]

                                                AIC              =   11.05877
Log likelihood   = -279.9986936                 BIC              =   175113.6
```

csat	Coef.	OIM Std. Err.	z	P>|z|	[95% Conf. Interval]
expense	-.0222756	.0060371	-3.69	0.000	-.0341082 -.0104431
_cons	1060.732	32.7009	32.44	0.000	996.6399 1124.825

Because **link(identity)** and **family(gaussian)** are default options, we could have left them out of the previous **glm** command.

We could also fit the same OLS model but obtain bootstrap standard errors.

```
. glm csat expense, link(identity) family(gaussian) vce(bstrap)

Iteration 0:    log likelihood = -279.99869

Bootstrap iterations (199)
----+--- 1 ---+--- 2 ---+--- 3 ---+--- 4 ---+--- 5
.................................................. 50
.................................................. 100
.................................................. 150
..............................................

Generalized linear models                       No. of obs       =         51
Optimization     : ML                           Residual df      =         49
                                                Scale parameter =   3572.235
Deviance         =   175306.2097                (1/df) Deviance =   3577.678
Pearson          =   175306.2097                (1/df) Pearson  =   3577.678

Variance function: V(u) = 1                     [Gaussian]
Link function    : g(u) = u                     [Identity]

                                                AIC              =   11.05877
Log likelihood   = -279.9986936                 BIC              =   175113.6
```

csat	Coef.	Bootstrap Std. Err.	z	P>|z|	[95% Conf. Interval]
expense	-.0222756	.0038002	-5.86	0.000	-.0297238 -.0148275
_cons	1060.732	24.01884	44.16	0.000	1013.656 1107.808

The bootstrap standard errors reflect observed variation among coefficients estimated from 199 samples of $n = 51$ cases each, drawn by random sampling with replacement from the original $n = 51$ dataset. In this example, the bootstrap standard errors are less than the corresponding theoretical standard errors, and the resulting confidence intervals are narrower.

Similarly, we could use **glm** to repeat the first **logistic** regression of Chapter 10 (use dataset *shuttle0.dta*). In the example below, we ask for jackknife standard errors and odds ratio or exponential-form (**eform**) coefficients:

```
. glm any date, link(logit) family(bernoulli) eform vce(jknife)

Iteration 0:   log likelihood = -12.995268
Iteration 1:   log likelihood = -12.991098
Iteration 2:   log likelihood = -12.991096

Jackknife iterations (23)
————+—— 1 ——+—— 2 ——+— 3 ——+—— 4 ——+—— 5
. . . . . . . . . . . . . . . . . . . . .
```

```
Generalized linear models                  No. of obs       =          23
Optimization      : ML                     Residual df      =          21
                                           Scale parameter =           1
Deviance        =   25.98219269            (1/df) Deviance =    1.237247
Pearson         =   22.8885488             (1/df) Pearson  =    1.089931

Variance function: V(u) = u*(1-u/1)        [Binomial]
Link function    : g(u) = ln(u/(1-u))      [Logit]

                                           AIC              =    1.303574
Log likelihood   = -12.99109634            BIC              = -39.86319
```

any	Odds Ratio	Jackknife Std. Err.	z	P>|z|	[95% Conf. Interval]	
date	1.002093	.0015486	1.35	0.176	.9990623	1.005133

The final **poisson** regression of the present chapter (use dataset *oakridg0.dta*) corresponds to this **glm** model (results not shown):

```
. glm deaths r2-r6 r78 age, link(log) family(poisson)
     lnoffset(pyears) eform
```

Although **glm** can replicate the models fit by many specialized commands, and adds some new capabilities, the specialized commands have their own advantages including speed and customized options. A particular attraction of **glm** is its ability to fit models for which Stata has no specialized command.

12

Principal Components, Factor, and Cluster Analysis

Principal components and factor analysis provide methods for simplification, combining many correlated variables into a smaller number of underlying dimensions. Along the way to achieving simplification, the analyst must choose from a daunting variety of options. If the data really do reflect distinct underlying dimensions, different options might nonetheless converge on similar results. In the absence of distinct underlying dimensions, however, different options often lead to divergent results. Experimenting with these options can tell us how stable a particular finding is, or how much it depends on arbitrary choices about the analytical technique.

Stata accomplishes principal components and factor analysis with five basic commands:

pca   Principal components analysis.

factor   Extracts factors of several different types.

screeplot Constructs a scree graph (plot of the eigenvalues) from the recent **pca** or **factor**.

rotate   Performs orthogonal (uncorrelated factors) or oblique (correlated factors) rotation, after **factor**.

predict Generates factor scores (composite variables) and other case statistics after **pca**, **factor**, or **rotate**.

The composite variables generated by **predict** can subsequently be saved, listed, graphed, or analyzed like any other Stata variable.

Users who create composite variables by the older method of adding other variables together without doing factor analysis could assess their results by calculating an α (alpha) reliability coefficient:

alpha   Cronbach's α reliability

Instead of combining variables, cluster analysis combines observations by finding non-overlapping, empirically-based typologies or groups. Cluster analysis methods are even more diverse than those of factor analysis. Stata's **cluster** command provides tools for performing cluster analysis, graphing the results, and forming new variables to identify the resulting groups. Principal components, factor analysis, cluster analysis, and related commands are detailed in Stata's *Multivariate Statistics* reference manual.

Methods described in this chapter can be accessed through the following menus:

Statistics > Multivariate analysis

Graphics > Multivariate analysis graphs

Example Commands

. **pca** *x1-x20*

Obtains principal components of the variables *x1* through *x20*.

. **pca** *x1-x20*, **mineigen(1)**

Obtains principal components of the variables *x1* through *x20*. Retains components having eigenvalues greater than 1.

. **factor** *x1-x20*, **ml factor(5)**

Performs maximum likelihood factor analysis of the variables *x1* through *x20*. Retains only the first five factors.

. **screeplot**

Scree plot or graph of eigenvalues versus factor or component number, from the most recent **factor** command.

. **rotate, varimax factors(2)**

Performs orthogonal (varimax) rotation of the first two factors from the most recent **factor** command.

. **rotate, promax factors(3)**

Performs oblique (promax) rotation of the first three factors from the most recent **factor** command.

. **predict** *f1 f2 f3*

Generates three new factor score variables named *f1*, *f2*, and *f3*, based upon the most recent **factor** and **rotate** commands.

. **alpha** *x1-x10*

Calculates Cronbach's α reliability coefficient for a composite variable defined as the sum of *x1-x10*. The sense of items entering negatively is ordinarily reversed. Options can override this default, or form a composite variable by adding together either the original variables or their standardized values.

. **cluster centroidlinkage** *x y z w*, **L2 name(**L2cent**)**

Performs agglomerative cluster analysis with centroid linkage, using variables x, y, z, and w. Euclidean distance (**L2**) measures dissimilarity among observations. Results from this cluster analysis are saved with the name *L2cent*.

. **cluster dendrogram, ylabel(0(.5)3) cutnumber(20)**
   xlabel(, angle(vertical))

Draws a cluster analysis tree graph or dendrogram showing results from the previous cluster analysis. **cutnumber(20)** specifies that the graph begins with only 20 clusters remaining, after some previous fusion of the most-similar observations. Labels are printed in a compact vertical fashion below the graph.

. **cluster generate** *ctype* **= groups(3), name(***L2cent***)**

Creates a new variable *ctype* (values of 1, 2, or 3) that classifies each observation into one of the top three groups found by the cluster analysis named *L2cent*.

Principal Components

To illustrate principal components and factor analysis commands, we will use a small dataset, *planets.dta*, describing the nine classical planets of this solar system (from Beatty et al. 1981). The data include several variables in both raw and natural logarithm form. Logarithms are employed here to reduce skew and linearize relationships among the variables.

```
Contains data from C:\data\planets.dta
  obs:             9                          Solar system data
  vars:           12                          23 May 2008 15:01
  size:          477 (99.9% of memory free)
```

variable name	storage type	display format	value label	variable label
planet	str7	%9s		Planet
dsun	float	%9.0g		Mean dist. sun, km*10^6
radius	float	%9.0g		Equatorial radius in km
rings	byte	%8.0g	ringlbl	Has rings?
moons	byte	%8.0g		Number of known moons
mass	float	%9.0g		Mass in kilograms
density	float	%9.0g		Mean density, g/cm^3
logdsun	float	%9.0g		natural log dsun
lograd	float	%9.0g		natural log radius
logmoons	float	%9.0g		natural log (moons + 1)
logmass	float	%9.0g		natural log mass
logdense	float	%9.0g		natural log dense

```
Sorted by:  dsun
```

To extract initial factors or principal components, use the command **factor** followed by a variable list (variables in any order) and one of the following options:

pcf Principal components factoring
pf   Principal factoring (default)
ipf Principal factoring with iterated communalities
ml   Maximum-likelihood factoring

Principal components are calculated through the specialized command **pca**. Type **help pca** or **help factor** to see options for these commands.

To obtain principal components factors, type

```
.  factor rings logdsun - logdense, pcf
```

(obs=9)

Factor analysis/correlation   Number of obs   =   9
      Method: principal-component factors   Retained factors =   2
      Rotation: (unrotated)   Number of params =   11

Factor	Eigenvalue	Difference	Proportion	Cumulative
Factor1	4.62365	3.45469	0.7706	0.7706
Factor2	1.16896	1.05664	0.1948	0.9654
Factor3	0.11232	0.05395	0.0187	0.9842
Factor4	0.05837	0.02174	0.0097	0.9939
Factor5	0.03663	0.03657	0.0061	1.0000
Factor6	0.00006	.	0.0000	1.0000

LR test: independent vs. saturated: chi2(15) = 100.49 Prob>chi2 = 0.0000

Factor loadings (pattern matrix) and unique variances

Variable	Factor1	Factor2	Uniqueness
rings	0.9792	0.0772	0.0353
logdsun	0.6710	-0.7109	0.0443
lograd	0.9229	0.3736	0.0088
logmoons	0.9765	0.0003	0.0465
logmass	0.8338	0.5446	0.0082
logdense	-0.8451	0.4705	0.0644

Only the first two components have eigenvalues greater than 1, and these two components explain over 96% of the six variables' combined variance. The unimportant 3rd through 6th principal components might safely be disregarded in subsequent analysis.

Two **factor** options provide control over the number of factors extracted:

      factors(#)   where # specifies the number of factors

      mineigen(#)   where # specifies the minimum eigenvalue for retained factors

The principal components factoring (**pcf**) procedure automatically drops factors with eigenvalues below 1, so

```
. factor rings logdsun - logdense, pcf
```

is equivalent to

```
. factor rings logdsun - logdense, pcf mineigen(1)
```

In this example, we would also have obtained the same results by typing

```
. factor rings logdsun - logdense, pcf factors(2)
```

To see a scree graph (plot of eigenvalues versus component or factor number) after any **factor**, use the **screeplot** command. A horizontal line at eigenvalue = 1 in Figure 12.1 marks the usual cutoff for retaining principal components, and again emphasizes the unimportance in this example of components 3 through 6.

```
. screeplot, yline(1)
```

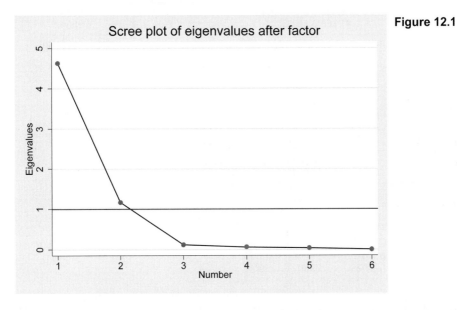

Scree plot of eigenvalues after factor

Figure 12.1

Rotation

Rotation further simplifies factor structure. After factoring, type **rotate** followed by an option to specify the type of rotation. Two common types are:

varimax   Varimax orthogonal rotation, for uncorrelated factors or components (default).

promax( )   Promax oblique rotation, allowing correlated factors or components. Choose a number (promax power) ≤ 4; the higher the number, the greater the degree of inter-factor correlation. **promax(3)** is the default.

Type **help rotate** for a complete list of rotation methods and other options. For example:

factors( )   As it does with **factor** , this option specifies how many factors to retain.

entropy   Minimum entropy orthogonal rotation.

Rotation can be performed following any factor analysis, whether it employed the **pcf**, **pf**, **ipf**, or **ml** options. In this section, we will follow through on our **pcf** example. For orthogonal (default) rotation of the first two components found in the planetary data, we simply type **rotate**.

```
. rotate
```

Factor analysis/correlation
    Method: principal-component factors
    Rotation: orthogonal varimax (Kaiser off)

Number of obs   =   9
Retained factors =   2
Number of params =   11

Factor	Variance	Difference	Proportion	Cumulative
Factor1	3.36900	0.94539	0.5615	0.5615
Factor2	2.42361	.	0.4039	0.9654

LR test: independent vs. saturated: chi2(15) = 100.49 Prob>chi2 = 0.0000

Rotated factor loadings (pattern matrix) and unique variances

Variable	Factor1	Factor2	Uniqueness
rings	0.8279	0.5285	0.0353
logdsun	0.1071	0.9717	0.0443
lograd	0.9616	0.2580	0.0088
logmoons	0.7794	0.5882	0.0465
logmass	0.9936	0.0678	0.0082
logdense	-0.3909	-0.8848	0.0644

Factor rotation matrix

	Factor1	Factor2
Factor1	0.7980	
Factor2	0.6026	-0.7980

The example above accepts all the defaults: varimax rotation and the same number of factors retained in the last **factor**. We could have asked for the same rotation explicitly, by adding options to the command: **rotate, varimax factors(2)**.

For oblique promax rotation (allowing correlated factors) of the most recent factoring, type

```
. rotate, promax
```

```
Factor analysis/correlation                    Number of obs   =        9
    Method: principal-component factors        Retained factors =       2
    Rotation: oblique promax (Kaiser off)      Number of params =      11
```

Factor	Variance	Proportion	Rotated factors are correlated
Factor1	4.12467	0.6874	
Factor2	3.32370	0.5539	

```
LR test: independent vs. saturated:  chi2(15) =  100.49 Prob>chi2 = 0.0000
```

Rotated factor loadings (pattern matrix) and unique variances

Variable	Factor1	Factor2	Uniqueness
rings	0.7626	0.3466	0.0353
logdsun	-0.1727	1.0520	0.0443
lograd	0.9926	0.0060	0.0088
logmoons	0.6907	0.4275	0.0465
logmass	1.0853	-0.2154	0.0082
logdense	-0.1692	-0.8719	0.0644

Factor rotation matrix

	Factor1	Factor2
Factor1	0.9250	0.7898
Factor2	0.3800	-0.6134

By default, this example used a promax power of 3. We could have specified the promax power and desired number of factors explicitly:

```
. rotate, promax(3) factors(2)
```

promax(4) would permit further simplification of the loading matrix, at the cost of stronger interfactor correlations and less total variance explained.

After promax rotation, *rings*, *lograd*, *logmoons*, and *logmass* load most heavily on factor 1. This appears to be a "large size/many satellites" dimension. *logdsun* and *logdense* load higher on factor 2, forming a "far out/low density" dimension. A post-factor-analysis graphing command, **loadingplot**, helps to visualize these results (Figure 12.2).

```
. loadingplot, factors(2) yline(0) xline(0)
```

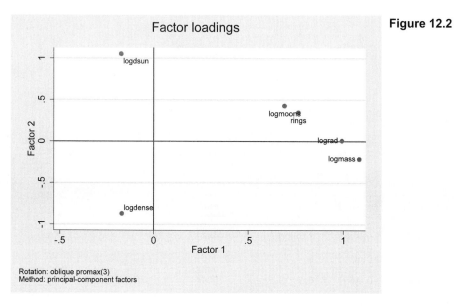

Figure 12.2

Factor Scores

Factor scores are linear composites, formed by standardizing each variable to zero mean and unit variance, and then weighting with factor score coefficients and summing for each factor. **predict** performs these calculations automatically, using the most recent **rotate** or **factor** results. In the **predict** command we supply names for the new variables, such as *f1* and *f2*.

```
. predict f1 f2
(regression scoring assumed)
```

Scoring coefficients (method = regression; based on promax(3) rotated factors)

Variable	Factor1	Factor2
rings	0.22099	0.12674
logdsun	-0.09689	0.48769
lograd	0.30608	-0.03840
logmoons	0.19543	0.16664
logmass	0.34386	-0.14338
logdense	-0.01609	-0.39127

```
. label variable f1 "Large size/many satellites"
. label variable f2 "Far out/low density"
```

```
. list planet f1 f2
```

	planet	f1	f2
1.	Mercury	-.9172388	-1.256881
2.	Venus	-.5160229	-1.188757
3.	Earth	-.3939372	-1.035242
4.	Mars	-.6799535	-.5970106
5.	Jupiter	1.342658	.3841085
6.	Saturn	1.184475	.9259058
7.	Uranus	.7682409	.9347457
8.	Neptune	.647119	.8161058
9.	Pluto	-1.43534	1.017025

Being standardized variables, the new factor scores *f1* and *f2* have means (approximately) equal to zero and standard deviations equal to one:

```
. summarize f1 f2
```

Variable	Obs	Mean	Std. Dev.	Min	Max
f1	9	-3.31e-09	1	-1.43534	1.342658
f2	9	9.93e-09	1	-1.256881	1.017025

Thus, the factor scores are measured in units of standard deviations from their means. Mercury, for example, is about .92 standard deviations below average on the large size/many satellites (*f1*) dimension because it is small and has no satellites. Mercury is 1.26 standard deviations below average on the far out/low density (*f2*) dimension because it is close to the sun and high density. Saturn, in contrast, is 1.18 and .93 standard deviations above average on these two dimensions.

Promax rotation permits correlations between factor scores:

```
. correlate f1 f2
(obs=9)
```

	f1	f2
f1	1.0000	
f2	0.4974	1.0000

Scores on factor 1 have a moderate positive correlation with scores on factor 2. That is, far out/low density planets are more likely to be larger, with many satellites.

Another post-factor-analysis graphing command, **scoreplot**, draws a scatterplot of the observations' factor scores. Used with principal components factors, such graphs can help to identify multivariate outliers, or clusters of observations that stand apart from the rest. Figure 12.3 reveals three distinct types of planets.

```
. scoreplot, mlabel(planet) yline(0) xline(0)
```

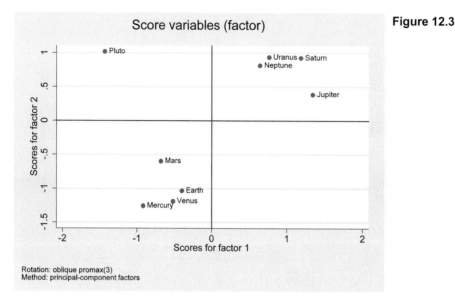

Figure 12.3

The inner, rocky planets (such as Mercury, low on "large size/many satellites" factor 1; low also on "far out/low density" factor 2) cluster together at the lower left. The outer gas giants have opposite characteristics, and cluster together at the upper right. Pluto, which physically resembles some outer-system moons, is unique among the nine classical planets for being high on the "far out/low density" dimension, and at the same time low on the "large size/many satellites" dimension. Our factor analysis thus identifies Pluto as a different kind of object that does not fit with either main group of planets. Recognizing Pluto's exceptionalism, the International Astronomical Union voted in 2006 to reclassify it among the dozens of known "dwarf planets," leaving us with only eight true planets.

If we employ varimax instead of promax rotation, we get uncorrelated factor scores:

```
. quietly factor rings logdsun - logdense, pcf
. quietly rotate
. quietly predict varimax1 varimax2

. correlate varimax1 varimax2
(obs=9)
```

	varimax1	varimax2
varimax1	1.0000	
varimax2	0.0000	1.0000

Once created by **predict**, factor scores can be treated like any other Stata variable — listed, correlated, graphed, and so forth. Factor scores often are used in social and behavioral science to combine many test or questionnaire items into composite variables or indexes. In physical-science fields such as climatology or remote sensing, factor scores obtained by principal components without rotation help to analyze large datasets. In these applications, principal components are called "empirical orthogonal functions." The first empirical orthogonal function, or EOF1, equals the factor score for the first unrotated principal component. EOF2 is the score for the second principal component, and so forth.

Principal Factoring

Principal factoring extracts principal components from a modified correlation matrix, in which the main diagonal consists of communality estimates instead of 1's. The **factor** options **pf** and **ipf** both perform principal factoring. They differ in how communalities are estimated:

    pf   Communality estimates equal R^2 from regressing each variable on all the others.

    ipf   Iterative estimation of communalities.

Whereas principal components analysis focuses on explaining the variables' variance, principal factoring explains intervariable correlations.

We apply principal factoring with iterated communalities (**ipf**) to the planetary data:

```
. factor rings logdsun - logdense, ipf
```

(obs=9)

```
Factor analysis/correlation                    Number of obs    =        9
    Method: iterated principal factors         Retained factors =        5
    Rotation: (unrotated)                       Number of params =       15

    Beware: solution is a Heywood case
            (i.e., invalid or boundary values of uniqueness)
```

Factor	Eigenvalue	Difference	Proportion	Cumulative
Factor1	4.59663	3.46817	0.7903	0.7903
Factor2	1.12846	1.05107	0.1940	0.9843
Factor3	0.07739	0.06438	0.0133	0.9976
Factor4	0.01301	0.01176	0.0022	0.9998
Factor5	0.00125	0.00137	0.0002	1.0000
Factor6	-0.00012	.	-0.0000	1.0000

LR test: independent vs. saturated: chi2(15) = 100.49 Prob>chi2 = 0.0000

Factor loadings (pattern matrix) and unique variances

Variable	Factor1	Factor2	Factor3	Factor4	Factor5
rings	0.9760	0.0665	0.1137	-0.0206	-0.0223
logdsun	0.6571	-0.6705	0.1411	0.0447	0.0082
lograd	0.9267	0.3700	-0.0450	0.0486	0.0166
logmoons	0.9674	-0.0107	0.0078	-0.0859	0.0160
logmass	0.8378	0.5458	0.0056	0.0282	-0.0071
logdense	-0.8460	0.4894	0.2059	-0.0061	0.0100

Variable	Uniqueness
rings	0.0292
logdsun	0.0966
lograd	-0.0004
logmoons	0.0564
logmass	-0.0007
logdense	0.0022

In this example, Stata gives an ominous warning: "Beware: solution is a Heywood case." Clicking on the linked Heywood case warning yields an explanation of the problem, which in this instance reflects our unusually small sample ($n = 9$). For simplicity, we will continue the analysis anyway, but in research such a warning would give reason to rethink our analysis.

Only the first two factors have eigenvalues above 1. With **pcf** or **pf** factoring, we can simply disregard minor factors. Using **ipf**, however, we must decide how many factors to retain, and then repeat the analysis asking for exactly that many factors. Here we will retain two factors:

```
. factor rings logdsun - logdense, ipf factor(2)
(obs=9)
```

Factor analysis/correlation	Number of obs	=	9
Method: iterated principal factors	Retained factors	=	2
Rotation: (unrotated)	Number of params	=	11

Beware: solution is a Heywood case
(i.e., invalid or boundary values of uniqueness)

Factor	Eigenvalue	Difference	Proportion	Cumulative
Factor1	4.57495	3.47412	0.8061	0.8061
Factor2	1.10083	1.07631	0.1940	1.0000
Factor3	0.02452	0.02013	0.0043	1.0043
Factor4	0.00439	0.00795	0.0008	1.0051
Factor5	-0.00356	0.02182	-0.0006	1.0045
Factor6	-0.02537	.	-0.0045	1.0000

LR test: independent vs. saturated: chi2(15) = 100.49 Prob>chi2 = 0.0000

Factor loadings (pattern matrix) and unique variances

Variable	Factor1	Factor2	Uniqueness
rings	0.9747	0.0537	0.0470
logdsun	0.6533	-0.6731	0.1202
lograd	0.9282	0.3605	0.0086
logmoons	0.9685	-0.0228	0.0614
logmass	0.8430	0.5462	-0.0089
logdense	-0.8294	0.4649	0.0960

After **ipf** factor analysis, we could create composite variables using **rotate** and **predict** as before. Due to the Heywood-case problem, **ipf** factor scores here are less plausible than our earlier **pcf** results. As a research strategy, it often helps to repeat factor analyses using several different methods, and compare these looking for stable conclusions.

Maximum-Likelihood Factoring

Maximum-likelihood factoring, unlike Stata's other **factor** options, provides formal hypothesis tests that help in determining the appropriate number of factors. To obtain a single maximum-likelihood factor for the planetary data, type

```
. factor rings logdsun - logdense, ml nolog factor(1)
```

(obs=9)

```
Factor analysis/correlation              Number of obs   =        9
    Method: maximum likelihood            Retained factors =       1
    Rotation: (unrotated)                 Number of params =       6
                                          Schwarz's BIC   =  97.8244
    Log likelihood = -42.32054            (Akaike's) AIC  =  96.6411
```

Factor	Eigenvalue	Difference	Proportion	Cumulative
Factor1	4.47258	.	1.0000	1.0000

```
LR test: independent vs. saturated:  chi2(15) =  100.49 Prob>chi2 = 0.0000
LR test:    1 factor vs. saturated:  chi2(9)  =   51.73 Prob>chi2 = 0.0000
```

Factor loadings (pattern matrix) and unique variances

Variable	Factor1	Uniqueness
rings	0.9873	0.0254
logdsun	0.5922	0.6493
lograd	0.9365	0.1229
logmoons	0.9589	0.0805
logmass	0.8692	0.2445
logdense	-0.7715	0.4049

The **ml** output includes two likelihood-ratio χ^2 tests:

LR test: independent vs. saturated
> This tests whether a no-factor (independent) model fits the observed correlation matrix significantly worse than a saturated or perfect-fit model. A low probability (here 0.0000, meaning $P < .00005$) indicates that a no-factor model is too simple.

LR test: 1 factor vs. saturated
> This tests whether the current 1-factor model fits significantly worse than a saturated model. The low P-value here suggests that one factor is too simple, as well.

Perhaps a 2-factor model will do better:

```
. factor rings logdsun - logdense, ml nolog factor(2)
(obs=9)
```

Factor analysis/correlation	Number of obs	=	9
Method: maximum likelihood	Retained factors	=	2
Rotation: (unrotated)	Number of params	=	11
	Schwarz's BIC	=	36.6881
Log likelihood = -6.259338	(Akaike's) AIC	=	34.5187

```
Beware: solution is a Heywood case
       (i.e., invalid or boundary values of uniqueness)
```

Factor	Eigenvalue	Difference	Proportion	Cumulative
Factor1	3.64200	1.67115	0.6489	0.6489
Factor2	1.97085	.	0.3511	1.0000

```
LR test: independent vs. saturated:   chi2(15) =  100.49 Prob>chi2 = 0.0000
LR test:   2 factors vs. saturated:   chi2(4)  =    6.72 Prob>chi2 = 0.1513
(tests formally not valid because a Heywood case was encountered)
```

Factor loadings (pattern matrix) and unique variances

Variable	Factor1	Factor2	Uniqueness
rings	0.8655	-0.4154	0.0783
logdsun	0.2092	-0.8559	0.2236
lograd	0.9844	-0.1753	0.0003
logmoons	0.8156	-0.4998	0.0850
logmass	0.9997	0.0264	0.0000
logdense	-0.4643	0.8857	0.0000

Now we find the following:

LR test: independent vs. saturated

The first test is unchanged; a no-factor model is too simple.

LR test: 2 factors vs. saturated

The 2-factor model is *not* significantly worse ($P = .1513$) than a perfect-fit model.

These tests suggest that two factors provide an adequate model.

Computational routines performing maximum-likelihood factor analysis often yield Heywood solutions, or unrealistic results such as negative variance or zero uniqueness. When this happens (as in the 2-factor **ml** example above), the χ^2 tests lack formal justification. Viewed descriptively, the tests can still provide informal guidance regarding the appropriate number of factors.

Cluster Analysis — 1

Cluster analysis encompasses a variety of methods that divide observations into groups or clusters, based on their dissimilarities across a number of variables. It is most often used as an exploratory approach, for developing empirical typologies, rather than as a means of testing pre-specified hypotheses. Indeed, there exists little formal theory to guide hypothesis testing for the common clustering methods. The number of choices available at each step in the analysis is daunting, and all the more so because they can lead to many different results. This section provides an entry point to beginning cluster analysis. We review some basic ideas and illustrate them through a simple example. The following section considers a somewhat larger example. Stata's *Multivariate Statistics Reference Manual* introduces and defines the full range of choices available. Everitt *et al.* (2001) cover topics in more detail, including helpful comparisons among the many cluster-analysis methods.

Clustering methods fall into two broad categories, *partition* and *hierarchical*. Partition methods break the observations into a pre-set number of nonoverlapping groups. We have two ways to do this:

cluster kmeans   Kmeans cluster analysis
   User specifies the number of clusters (K) to create. Stata then finds these through an iterative process, assigning observations to the group with the closest mean.

cluster kmedians   Kmedians cluster analysis
   Similar to Kmeans, but with medians.

Partition methods tend to be computationally simpler and faster than hierarchical methods. The necessity of declaring the exact number of clusters in advance is a disadvantage for exploratory work, however.

Hierarchical methods involve a process of smaller groups gradually fusing to form increasingly large ones. Stata takes an *agglomerative* approach in hierarchical cluster analysis: it starts out with each observation considered as its own separate "group." The closest two groups are merged, and this process continues until a specified stopping-point is reached, or all observations belong to one group. A graphical display called a *dendrogram* or *tree diagram* visualizes hierarchical clustering results. Several choices exist for the *linkage method*, which specifies what should be compared between groups that contain more than one observation:

cluster singlelinkage   Single linkage cluster analysis
   Computes the dissimilarity between two groups as the dissimilarity between the closest pair of observations between the two groups. Although simple, this method has low resistance to outliers or measurement errors. Observations tend to join clusters one at a time, forming unbalanced, drawn-out groups in which members have little in common, but are linked by intermediate observations — a problem called *chaining*.

cluster averagelinkage   Average linkage cluster analysis
Uses the average dissimilarity of observations between the two groups, yielding properties intermediate between single and complete linkage. Simulation studies report that this works well for many situations and is reasonably robust (see Everitt *et al.* 2001, and sources they cite). Commonly used in archaeology.

cluster completelinkage   Complete linkage cluster analysis
Uses the farthest pair of observations between the two groups. Less sensitive to outliers than single linkage, but with the opposite tendency towards clumping many observations into tight, spatially compact clusters.

cluster waveragelinkage   Weighted-average linkage cluster analysis
cluster medianlinkage   Median linkage cluster analysis.
Weighted-average linkage and median linkage are variations on average linkage and centroid linkage, respectively. In both cases, the difference is in how groups of unequal size are treated when merged. In average linkage and centroid linkage, the number of elements of each group are factored into the computation, giving correspondingly larger influence to the larger group (because each observation carries the same weight). In weighted-average linkage and median linkage, the two groups are given equal weighting regardless of how many observations there are in each group. Median linkage, like centroid linkage, is subject to reversals.

cluster centroidlinkage   Centroid linkage cluster analysis
Centroid linkage merges the groups whose means are closest (in contrast to average linkage which looks at the average distance between elements of the two groups). This method is subject to reversals — points where a fusion takes place at a lower level of dissimilarity than an earlier fusion. Reversals signal an unstable cluster structure, are difficult to interpret, and cannot be graphed by **cluster dendrogram**.

cluster wardslinkage   Ward's linkage cluster analysis
Joins the two groups that result in the minimum increase in the error sum of squares. Does well with groups that are multivariate normal and of similar size, but poorly when clusters have unequal numbers of observations.

All clustering methods begin with some definition of dissimilarity (or similarity). Dissimilarity measures reflect the differentness or distance between two observations, across a specified set of variables. Generally, such measures are designed so that two identical observations have a dissimilarity of 0, and two maximally different observations have a dissimilarity of 1. Similarity measures reverse this scaling, so that identical observations have a similarity of 1. Stata's **cluster** options offer many choices of dissimilarity or similarity measures. For purposes of calculation, Stata internally transforms similarity to dissimilarity:

dissimilarity = 1 − similarity

The default dissimilarity measure is the Euclidean distance, option **L2** (or **Euclidean**). This defines the distance between observations i and j as

$$\{\textstyle\sum_k (x_{ki} - x_{kj})^2\}^{1/2}$$

where x_{ki} is the value of variable x_k for observation i, x_{kj} the value of x_k for observation j, and summation occurs over all the x variables considered. Other choices available for measuring the (dis)similarities between observations based on continuous variables include the squared Euclidean distance (**L2squared**), the absolute-value distance (**L1**), maximum-value distance (**Linfinity**), and correlation coefficient similarity measure (**correlation**). Choices for binary variables include simple matching (**matching**), Jaccard binary similarity coefficient (**Jaccard**), and others. The **gower** option works with a mix of continuous and binary variables. Type **help measure option** for a complete list and detailed explanations of dissimilarity-measure options.

Earlier in this chapter, a principal components analysis of variables in *planets.dta* (Figure 12.3) identified three types of planets: inner rocky planets, outer gas giants, and in a class by itself, Pluto. Cluster analysis provides an alternative approach to the question of planet "types." Because variables such as number of moons (*moons*) and mass in kilograms (*mass*) are measured in incomparable units, with hugely different variances, we should standardize in some way to avoid results dominated by the highest-variance items. A common, although not automatic, choice is standardization to zero mean and unit standard deviation. This is accomplished through the **egen** command (and using variables in log form, for the same reasons discussed earlier). **summarize** confirms that the new z variables have (near) zero means, and standard deviations equal to one.

```
. egen zrings = std(rings)
. egen zlogdsun = std(logdsun)
. egen zlograd = std(lograd)
. egen zlogmoon = std(logmoons)
. egen zlogmass = std(logmass)
. egen zlogdens = std(logdense)
. summ zrings-zlogdens
```

Variable	Obs	Mean	Std. Dev.	Min	Max
zrings	9	-1.99e-08	1	-.8432741	1.054093
zlogdsun	9	-1.16e-08	1	-1.393821	1.288216
zlograd	9	-3.31e-09	1	-1.3471	1.372751
zlogmoon	9	0	1	-1.207296	1.175849
zlogmass	9	-4.14e-09	1	-1.74466	1.365167
zlogdens	9	-1.32e-08	1	-1.453143	1.128901

The "three types" conclusion suggested by our principal components analysis is robust, and could have been found through cluster analysis as well. For example, we might perform a hierarchical cluster analysis with average linkage, using Euclidean distance (**L2**) as our dissimilarity measure. The option **name(*L2avg*)** gives the results from this particular analysis a name, so that we can refer to them in later commands. The results-naming feature is convenient when we need to try a number of cluster analyses and compare their outcomes.

```
. cluster averagelinkage zrings zlogdsun zlograd zlogmoon zlogmass
      zlogdens, L2 name(L2avg)
```

Nothing seems to happen, although we might notice that our dataset now contains three new variables with names based on *L2avg*. These new *L2avg** variables are not directly of interest, but can be used unobtrusively by the **cluster dendrogram** command to draw a cluster analysis tree or dendrogram visualizing the most recent hierarchical cluster analysis results (Figure 12.4). The **label(*planet*)** option here causes planet names (values of *planet*) to appear as labels below the graph.

```
. cluster dendrogram, label(planet) ylabel(0(1)5)
```

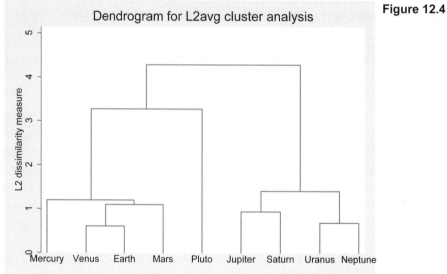

Figure 12.4

Dendrograms such as Figure 12.4 provide key interpretive tools for hierarchical cluster analysis. We can trace the agglomerative process from each observation comprising its own unique cluster, at bottom, to all fused into one cluster, at top. Venus and Earth, and also Uranus and Neptune, are the least dissimilar or most alike pairs. They are fused first, forming the first two multi-observation clusters at a height (dissimilarity) below 1. Jupiter and Saturn, then Venus–Earth and Mars, then Venus–Earth–Mars and Mercury, and finally Jupiter–Saturn and Uranus–Neptune are fused in quick succession, all with dissimilarities slightly above 1. At this point we have the same three groups suggested in Figure 12.3 by principal components: the inner rocky planets, the gas giants, and Pluto. The three clusters remain stable until, at much higher dissimilarity (above 3), Pluto fuses with the inner rocky planets. At a dissimilarity above 4, the final two clusters fuse.

So, how many types of planets are there? Figure 12.4 makes clear that the answer is "it depends." How much dissimilarity do we want to accept within each type? The long vertical lines between the three-cluster stage and the two-cluster stage in the upper part of the graph indicate that we have three fairly distinct types. We could reduce this to two types only by fusing an observation (Pluto) that is quite dissimilar to others in its group. We could expand it to five types only by drawing distinctions between several planets (e.g., Mercury and Venus–Earth–Mars) that by solar-system standards are not greatly dissimilar. Thus, the dendrogram makes a case for a three-type scheme.

The **cluster generate** command creates a new variable indicating the type or group to which each observation belongs. In this example, **groups(3)** calls for three groups. The **name(*L2avg*)** option specifies the particular results we named *L2avg*. This option is most useful when our session included multiple cluster analyses.

```
. cluster generate plantype = groups(3), name(L2avg)
. label variable plantype "Planet type"
. list planet plantype
```

	planet	plantype
1.	Mercury	1
2.	Venus	1
3.	Earth	1
4.	Mars	1
5.	Jupiter	3
6.	Saturn	3
7.	Uranus	3
8.	Neptune	3
9.	Pluto	2

The inner rocky planets have been coded as *plantype* = 1; the gas giants as *plantype* = 3; and Pluto is by itself as *plantype* = 2. The group designations as 1, 2, and 3 follow the left-to-right ordering of final clusters in the dendrogram (Figure 12.4). Once the data have been saved, our new typology could be used like any other categorical variable in subsequent analyses.

These planetary data have a strong pattern of natural groups, which is why such different techniques as cluster analysis and principal components point towards similar conclusions. We could have chosen other dissimilarity measures and linkage methods for this example, and still arrived at much the same place. Complex or weakly patterned data, on the other hand, often yield quite different results depending on the methods used. The clusters found by one method might not prove replicable under other methods, or with slightly different analytical decisions.

Cluster Analysis — 2

Discovering a simple, robust typology to describe nine planets was straightforward. For a more challenging example, consider the cross-national data in *nations.dta*. These data include living-conditions variables that provide a basis for classifying nations into types.

```
Contains data from c:\data\nations.dta
  obs:           109                       Data on 109 nations, ca. 1985
  vars:           15                       23 May 2008 15:02
  size:         4,578 (99.9% of memory free)
```

variable name	storage type	display format	value label	variable label
country	str8	%9s		Country
pop	float	%9.0g		1985 population in millions
birth	byte	%8.0g		Crude birth rate/1000 people
death	byte	%8.0g		Crude death rate/1000 people
chldmort	byte	%8.0g		Child (1-4 yr) mortality 1985
infmort	int	%8.0g		Infant (<1 yr) mortality 1985
life	byte	%8.0g		Life expectancy at birth 1985
food	int	%8.0g		Per capita daily calories 1985
energy	int	%8.0g		Per cap energy consumed, kg oil
gnpcap	int	%8.0g		Per capita GNP 1985
gnpgro	float	%9.0g		Annual GNP growth % 65-85
urban	byte	%8.0g		% population urban 1985
school1	int	%8.0g		Primary enrollment % age-group
school2	int	%8.0g		Secondary enroll % age-group
school3	byte	%8.0g		Higher ed. enroll % age-group

Working with these same data in Chapter 8, we saw that nonlinear transformations (logs or square roots) helped to normalize distributions and linearize relationships among some of these variables. Similar arguments for nonlinear transformations could apply to cluster analysis, but to keep our example simple, we will not pursue them here. Linear transformations to standardize the variables in some fashion remain essential, however. Otherwise, the variable *gnpcap*, which ranges from about $100 to $19,000 (standard deviation $4,400) would overwhelm other variables such as *life*, which ranges from 40 to 78 years (standard deviation 11 years). In the previous section, we standardized planetary data by subtracting each variable's mean, then dividing by its standard deviation, so that the resulting z-scores all had standard deviations of one. In this section we take a different approach, range standardization, which also works well for cluster analysis.

Range standardization involves dividing each variable by its range. There is no command to do this directly in Stata, but we can improvise one easily enough. To do so, we make use of results that Stata stores unobtrusively after **summarize**. Recall that we can see a complete list of stored results after **summarize** by typing the command **return list**. (After modeling procedures such as **regress** or **factor**, use the command **ereturn list** instead.) In this example, we view the results stored after **summarize** *pop*, then use the maximum and minimum values (stored as scalars that Stata names r(max) and r(min)) to calculate a new, range-standardized version of population.

```
. summarize pop
```

Variable	Obs	Mean	Std. Dev.	Min	Max
pop	109	38.9211	125.3888	1	1040.3

```
. return list
```

scalars:
```
                r(N) =  109
            r(sum_w) =  109
             r(mean) =  38.92110117750431
              r(Var) =  15722.35690184565
               r(sd) =  125.3888228744717
              r(min) =  1
              r(max) =  1040.300048828125
              r(sum) =  4242.400028347969
```

```
. generate rpop = pop/(r(max) - r(min))
. label variable rpop "Range-standardized population"
```

Similar commands create range-standardized versions of other living-conditions variables:

```
. quietly summ birth, detail
. generate rbirth = birth/(r(max) - r(min))
. label variable rbirth "Range-standardized bith rate"
. quietly summ infmort, detail
```

and so forth, defining the 8 new variables listed below.

```
. describe rpop-rschool2
```

variable name	storage type	display format	value label	variable label
rpop	float	%9.0g		Range-standardized population
rbirth	float	%9.0g		Range-standardized bith rate
rinf	float	%9.0g		Range-standardized infant mortality
rlife	float	%9.0g		Range-standardized life expectancy
rfood	float	%9.0g		Range-standardized food per capita
renergy	float	%9.0g		Range-standardized energy per capita
rgnpcap	float	%9.0g		Range-standardized GNP per capita
rschool2	float	%9.0g		Range-standardized secondary school %

If our **generate** commands were done correctly, these new range-standardized variables should all have ranges equal to 1. **tabstat** confirms that they do.

```
. tabstat rpop - rschool2, statistics(range)
```

stats	rpop	rbirth	rinf	rlife	rfood	renergy	rgnpcap
range	1	1	1	1	.9999999	1	1

stats	rschool2
range	1

Once the variables of interest have been standardized, we can proceed with cluster analysis. As we divide more than 100 nations into "types," we have no reason to assume that each type will include a similar number of nations. Average linkage (used in our planetary example), along with some other methods, gives each observation the same weight. This tends to make larger clusters more influential as agglomeration proceeds. Weighted average and median linkage methods, on the other hand, give equal weight to each cluster regardless of how many observations it contains. Such methods consequently tend to work better for detecting clusters of unequal size. Median linkage, like centroid linkage, is subject to reversals (which will occur with these data), so the following example applies weighted average linkage. Absolute-value distance (**L1**) provides our dissimilarity measure.

```
. cluster waveragelinkage rpop - rschool2, L1 name(L1wav)
```

The full cluster analysis proves unmanageably large for a dendrogram:

```
. cluster dendrogram
too many leaves; consider using the cutvalue() or cutnumber() options
r(198);
```

Following the error-message advice, Figure 12.5 employs a **cutnumber(100)** option to form a dendrogram that starts with only 100 groups, after the first few fusions have taken place.

```
. cluster dendrogram, ylabel(0(.5)3) cutnumber(100)
```

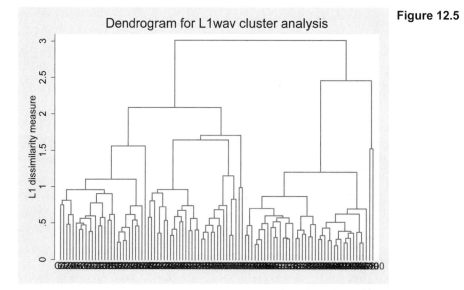

Figure 12.5

The bottom labels in Figure 12.5 are unreadable, but we can trace the general flow of this clustering process. Most of the fusion takes place at dissimilarities below 1. Two nations at the far right are unusual; they resist fusion until about 1.5 and then form a stable two-nation group quite different from all the rest. This is one of four clusters remaining above dissimilarities of 2. The first and second of these four final clusters (reading left to right) appear heterogeneous, formed through successive fusion of a number of somewhat distinct major subgroups. The third cluster, in contrast, appears more homogeneous. It combines many nations that fused into two subgroups at dissimilarities below 1, and then fused into one group at slightly above 1.

Figure 12.6 gives another view of this analysis, this time using the **cutvalue(1)** option to show only clusters with dissimilarities above 1. The **showcount** option calls for bottom labels (n=18, n=9, etc.) that indicate the number of observations in each group. We see that groups 3, 10, and 11 each consists of a single observation, unlike any others.

```
. cluster dendrogram, ylabel(0(.5)3) cutvalue(1) showcount
```

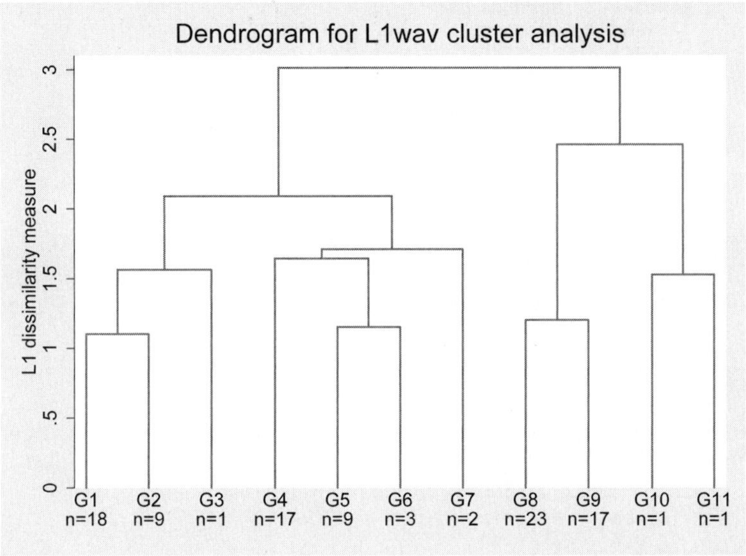

Figure 12.6

Dendrogram for L1wav cluster analysis

As Figure 12.6 shows, there are 11 groups remaining at dissimilarities above 1. For purposes of illustration, we will consider only the top four groups, which have dissimilarities above 2. **cluster generate** creates a categorical variable for the final four groups from the cluster analysis we named *L1wav*.

```
. cluster generate ctype = groups(4), name(L1wav)
. label variable ctype "Country type"
```

We could next examine which countries belong to which groups by typing

```
. sort cype
. by ctype:  list country
```

The commands below reshape this information to create the more compact list seen on the following page. The columns are country types.

```
. by ctype (country), sort: gen id = _n
. keep country id ctype
. drop if missing(ctype)
. reshape wide country, i(id) j(ctype)
. list
```

	id	country1	country2	country3	country4
1.	1	Algeria	Argentin	Banglade	China
2.	2	Brazil	Australi	Benin	India
3.	3	Burma	Austria	Bolivia	
4.	4	Chile	Belgium	Botswana	
5.	5	Colombia	Canada	BurkFaso	
6.	6	CostaRic	Denmark	Burundi	
7.	7	DomRep	Finland	Cameroon	
8.	8	Ecuador	France	CenAfrRe	
9.	9	Egypt	Greece	ElSalvad	
10.	10	Indonesi	HongKong	Ethiopia	
11.	11	Jamaica	Hungary	Ghana	
12.	12	Jordan	Ireland	Guatemal	
13.	13	Malaysia	Israel	Guinea	
14.	14	Mauritiu	Italy	Haiti	
15.	15	Mexico	Japan	Honduras	
16.	16	Morocco	Kuwait	IvoryCoa	
17.	17	Panama	Netherla	Kenya	
18.	18	Paraguay	NewZeala	Liberia	
19.	19	Peru	Norway	Madagasc	
20.	20	Philippi	Poland	Malawi	
21.	21	SauArabi	Portugal	Mauritan	
22.	22	SriLanka	S_Korea	Mozambiq	
23.	23	Syria	Singapor	Nepal	
24.	24	Thailand	Spain	Nicaragu	
25.	25	Tunisia	Sweden	Niger	
26.	26	Turkey	TrinToba	Nigeria	
27.	27	Uruguay	U_K	Pakistan	
28.	28	Venezuel	U_S_A	PapuaNG	
29.	29		UnArEmir	Rwanda	
30.	30		W_German	Senegal	
31.	31		Yugoslav	SierraLe	
32.	32			Somalia	
33.	33			Sudan	
34.	34			Tanzania	
35.	35			Togo	
36.	36			YemenAR	
37.	37			YemenPDR	
38.	38			Zaire	
39.	39			Zambia	
40.	40			Zimbabwe	

The two-nation cluster seen at far right in Figure 12.5 turns out to be type 4, China and India. The broad, homogeneous third cluster in Figure 12.5, type 3, contains a large group of the poorest nations, mainly in Africa. The relatively diverse type 2 contains nations with higher living conditions including the U.S., Europe, and Japan. Type 1, also diverse, contains nations with intermediate conditions. Whether this or some other typology is meaningful remains a substantive question, not a statistical one, and depends on the uses for which a typology is needed. Choosing different options in the steps of our cluster analysis would have returned different results. By experimenting with a variety of reasonable choices, we could gain a sense of which findings are most stable.

Time Series Analysis

Stata's time series capabilities are covered in the 450-page *Time-Series Reference Manual*. This chapter provides a brief introduction, beginning with two basic analytical tools: time plots and smoothing. We then move on to illustrate the use of correlograms, ARIMA and ARMAX modeling, and diagnostic tests for stationarity and white noise. Further applications, notably periodograms and the flexible ARCH family of models, are left to the reader's explorations.

A thorough, technical treatment of time series topics can be found in Hamilton (1994). Other sources include Box, Jenkins, and Reinsel (1994), Chatfield (2004), Diggle (1990), Enders (2004), and Shumway (1988).

Menus for time series operations come under the following headings:

Statistics > Time series

Statistics > Multivariate time series

Statistics > Cross-sectional time series

Graphics > Time-series graphs

Example Commands

. **ac *y*, lags(8) level(95) generate(*newvar*)**

Graphs autocorrelations of variable *y*, with 95% confidence intervals (default), for lags 1 through 8. Stores the autocorrelations as the first 8 values of *newvar*.

. **arch D.*y*, arch(1/3) ar(1) ma(1)**

Fits an ARCH (autoregressive conditional heteroskedasticity) model for first differences of *y*, including first- through third-order ARCH terms, and first-order AR and MA disturbances.

. **arima *y*, arima(1,1,1)**

Fits a simple ARIMA(1,1,1) model, with first differencing and first-order AR and MA terms. Possible options could specify alternative estimation strategies, linear constraints, and robust estimates of variance.

. `arima y, arima(1,0,1/2) sarima(1,0,1,12)`

Fits ARIMA$(1,0,2)\times(1,0,1)_{12}$ model with first-order AR terms, first and second-order MA terms, and also a multiplicative seasonal component with period 12.

. `arima y x1 x2 x3, arima(1,0,1)`

Performs ARMAX (autoregressive moving average with exogenous variables) regression of y on three predictors. Errors are modeled as first-order autoregressive and moving average process.

. `arima D.y x1 L1.x1 x2, ar(1) ma(1 12)`

Fits ARMAX model in which first differences of y are regressed on $x1$, lag-1 values of $x1$, and $x2$, including AR(1), MA(1), and MA(12) disturbances.

. `corrgram y, lags(8)`

Obtains autocorrelations, partial autocorrelations, and Q tests for lags 1 through 8.

. `dfuller y`

Performs Dickey–Fuller unit root test for stationarity.

. `estat dwatson`

After **regress** with time series data, calculates a Durbin–Watson statistic testing first-order autocorrelation.

. `egen newvar = ma(y), nomiss t(7)`

Generates *newvar* equal to the span-7 moving average of y, replacing the start and end values with shorter, uncentered averages.

. `generate date = mdy(month,day,year)`

Creates variable *date*, equal to days since January 1, 1960, from the three variables *month*, *day*, and *year*.

. `generate date = date(str_date, "mdy")`

Creates variable *date* from the string variable *str_date*, where *str_date* contains dates in month, day, year form such as "11/19/2001", "4/18/98", or "June 12, 1948". Type **help dates** for many other date functions and options.

. `generate newvar = L3.y`

Generates *newvar* equal to lag-3 values of y.

. `pac y, lags(8) yline(0) ciopts(bstyle(outline))`

Graphs partial autocorrelations with confidence intervals and residual variance for lags 1 through 8. Draws a horizontal line at 0; shows the confidence interval as an outline, instead of a shaded area (default).

. `pergram y, generate(newvar)`

Draws the sample periodogram (spectral density function) of variable y and creates *newvar* equal to the raw periodogram values.

. `smooth 73 `*`y`*`, generate(`*`newvar`*`)`

Generates *newvar* equal to span-7 running medians of *y*, re-smoothing by span-3 running medians. Compound smoothers such as "3RSSH" or "4253h,twice" are possible. Type **help smooth**, or **help tssmooth**, for other smoothing and filters.

. `tsset `*`date`*`, format(%td)`

Defines the dataset as a time series. Time is indicated by variable *date*, which is formatted as daily. For "panel" data with parallel time series for a number of different units, such as cities, **tsset** *city year* identifies both panel and time variables. Most of the commands in this chapter require that the data be **tsset**.

. `tssmooth ma `*`newvar`*` = `*`y`*`, window(2 1 2)`

Applies a moving-average filter to *y*, generating *newvar*. The **window(2 1 2)** option finds a span-5 moving average by including 2 lagged values, the current observation, and 2 leading values in the calculation of each smoothed point. Type **help tssmooth** for a list of other possible filters including weighted moving averages, exponential or double exponential, Holt–Winters, and nonlinear.

. `tssmooth nl `*`newvar`*` = `*`y`*`, smoother(4253h,twice)`

Applies a nonlinear smoothing filter to *y*, generating *newvar*. The **smoother(4253h,twice)** option iteratively finds running medians of span 4, 2, 5, and 3, then applies Hanning, and then repeats on the residuals. **tssmooth nl**, unlike other **tssmooth** procedures, cannot work around missing values.

. `wntestq `*`y`*`, lags(15)`

Box–Pierce portmanteau Q test for white noise (also provided by **corrgram**).

. `xcorr `*`x`*` `*`y`*`, lags(8) xline(0)`

Graphs cross-correlations between input (x) and output (y) variable for lags 1–8. **xcorr** *x y,* **table** gives a text version that includes the actual correlations. If we add a **generate(**newvar**)** option to the **xcorr** command, it will store the correlations as a variable.

Smoothing

Many time series exhibit high-frequency variations that make it difficult to discern underlying patterns. Smoothing such series breaks the data into two parts, one that varies gradually, and a second "rough" part containing the leftover rapid changes:

$$\text{data} = \text{smooth} + \text{rough}$$

To illustrate smoothing methods, we examine data on daily water consumption for the town of Milford, New Hampshire over seven months from January through July 1983 (*MILwater.dta*; source Hamilton 1985b). Milford's usual patterns of water use were interrupted by alarming news midway through this period.

```
Contains data from c:\data\MILwater.dta
  obs:            212                          Milford daily water use, 1/1/83
                                                 - 7/31/83
  vars:             4                          24 May 2008 07:10
  size:         2,968 (99.9% of memory free)
```

variable name	storage type	display format	value label	variable label
month	byte	%9.0g		Month
day	byte	%9.0g		Date
year	int	%9.0g		Year
water	int	%9.0g		Water use in 1000 gallons

```
Sorted by:
```

Before further analysis, we need to convert the month, day, and year information into a single numerical index of time. Stata's **mdy( )** function does this, creating an elapsed-date variable (named *date* here) indicating the number of days since January 1, 1960.

```
. generate date = mdy(month,day,year)
. list in 1/5
```

	month	day	year	water	date
1.	1	1	1983	520	8401
2.	1	2	1983	600	8402
3.	1	3	1983	610	8403
4.	1	4	1983	590	8404
5.	1	5	1983	620	8405

The January 1, 1960 reference date is arbitrary but fixed. We can provide more understandable formatting for *date*, and also set up our data for later analyses, by using the **tsset** (**t**ime **s**eries **set**) command to identify *date* as the time index variable and to specify the **%td** (**d**aily) display option for this variable.

```
. tsset date, format(%td)
        time variable:  date, 01jan1983 to 31jul1983
                delta:  1 day
. list in 1/5
```

	month	day	year	water	date
1.	1	1	1983	520	01jan1983
2.	1	2	1983	600	02jan1983
3.	1	3	1983	610	03jan1983
4.	1	4	1983	590	04jan1983
5.	1	5	1983	620	05jan1983

Dates in the new *date* format, such as "05jan1983", are more readable than the underlying numeric values such as "8405" (days since January 1, 1960). If desired, we could use **%td** formatting to produce other formats, such as "05 Jan 1983" or "01/05/83". Stata offers a number of variable-definition, display-format, and dataset-format features that are important with time series. Many of these involve ways to input, convert, and display dates. Full descriptions of date functions are found in the *Data Management Reference Manual* and the *User's Guide*, or they can be explored within Stata by typing **help dates**.

Figure 13.1 uses **twoway line** to draw a simple time plot of *water* against *date*. The graph shows a pattern of day-to-day variation, as well as an upward trend in water use when summer arrives. Value labels for our date-format variable are labeled automatically (01jan1983, etc.) on the *x* (or *t*) axis, but Stata's default choices here lead to crowded, unsatisfactory results.

```
. graph twoway line water date
```

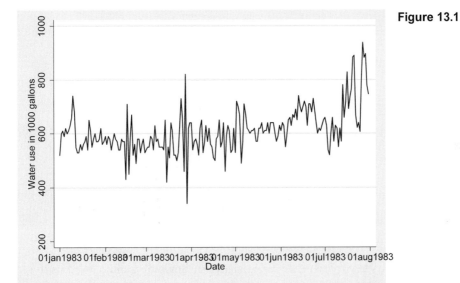

Figure 13.1

A better way to draw time plots when the *x* axis is a date variable is to use the special time series command **twoway tsline**. This command allows us to describe the *x* axis in terms of dates, without having to reference the numerical elapsed dates underneath. For example, we could draw a time plot similar to Figure 13.1 but with a less crowded time axis, as seen in Figure 13.2 on the next page. Note that the **tsline** command does not accept an *x* variable, only one or more *y* variables. With **tsset** data, the time dimension has already been defined. Options **tlabel( )** and **ttick( )** work in a **twoway tsline** plotjust as **xlabel( )** and **xtick( )** would in any other **twoway** plot, except that they understand date notations such as 01jan1983. In Figure 13.2 we have also suppressed the *x*-axis (time or *t*-axis) title with the option **ttitle("")**, because the word "Date" there seems superfluous below an axis labeled 01jan1983, 01mar1983, and so forth.

```
. graph twoway tsline water, ylabel(300(100)900) ttitle("")
     tlabel(01jan1983 01mar1983 01may1983 01jul1983, grid)
     ttick(01feb1983 01apr1983 01jun1983 01aug1983)
```

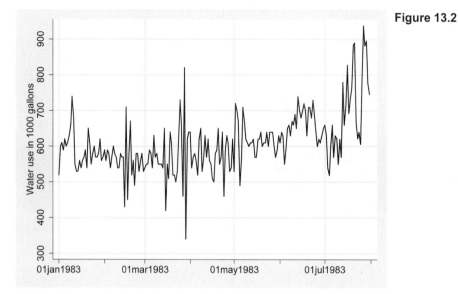

Figure 13.2

Visual inspection plays a key role in time series analysis. Smoothing often helps us to see underlying patterns in jagged series. The simplest smoothing method is to calculate a "moving average" at each data point based on present, earlier, and later values of *y*. For example, a "moving average of span 3" refers to the mean of y_{t-1}, y_t, and y_{t+1}. We could use Stata's explicit subscripting to **generate** such a variable:

```
. generate water3a = (water[_n-1] + water[_n] + water[_n+1])/3
```

A better way would be to apply the **ma** (moving average) function of **egen** :

```
. egen water3b = ma(water), nomiss t(3)
```

The **nomiss** option in this **egen** command requests shorter, uncentered moving averages in the tails; otherwise, the first and last values of *water3* would be missing. The **t(3)** option calls for moving averages of span 3. Any odd-number span ≥3 could be used.

For time series (**tsset**) data, powerful smoothing tools are available through the **tssmooth** commands. All but **tssmooth nl** can handle missing values.

tssmooth ma	moving-average filters, unweighted or weighted
tssmooth exponential	single exponential filters
tssmooth dexponential	double exponential filters
tssmooth hwinters	nonseasonal Holt–Winters smoothing
tssmooth shwinters	seasonal Holt–Winters smoothing
tssmooth nl	nonlinear filters

tssmooth ma, for example, can calculate moving averages of span 3, identical to those from our previous **egen** command:

```
. tssmooth ma water3c = water, window(1 1 1)
The smoother applied was
     (1/3)*[x(t-1) + 1*x(t) + x(t+1)]; x(t)= water
```

Type **help tssmooth ma**, **help tssmooth exponential**, etc. for the syntax of each command.

Figure 13.3 graphs a 5-day moving average of Milford water use (*water5*), together with the raw data (*water*). The **twoway tsline** command overlays a time plot of smoothed *water5* values with a second timeplot of raw *water* values (thinner line). *T*-axis labels mark the dates, as they did in Figure 13.2. In Figure 13.3, however, we specified a simpler date display format giving just the month and day, **format(%tdmd)**. The shorter format leaves room to label the beginning of each month in this graph, as opposed to every other month as done in Figure 13.2.

```
. tssmooth ma water5 = water, window(2 1 2)
The smoother applied was
     (1/5)*[x(t-2) + x(t-1) + 1*x(t) + x(t+1) + x(t+2)]; x(t)= water

. graph twoway tsline water5, clwidth(thick)

     ||  tsline water, clwidth(thin) clpattern(solid)

     ||  , ylabel(300(100)900) ytitle("Water use in 1000 gallons")
     ttitle("") tlabel(01jan1983 01feb1983 01mar1983 01apr1983
     01may1983 01jun1983 01jul1983 01aug1983, grid format(%tdmd))
     legend(order(2 1) position(4) ring(0) rows(2)
        label(1 "5-day average") label(2 "daily water use"))
```

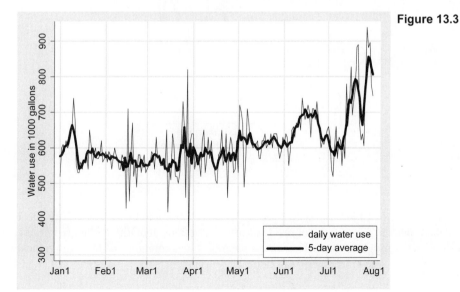

Figure 13.3

Moving averages share a drawback of other mean-based statistics: they have little resistance to outliers. Because outliers form prominent spikes in the water time series, we might want to try a more resistant smoothing method. The **tssmooth nl** command performs outlier-resistant nonlinear smoothing, employing methods and terminology described in Velleman and Hoaglin (1981) and Velleman (1982). For example,

```
. tssmooth nl water5r = water, smoother(5)
```

creates a new variable named *water5r*, holding the values of *water* after smoothing by running medians of span 5. Compound smoothers using running medians of different spans, in combination with "hanning" (¼, ½, and ¼ -weighted moving averages of span 3) and other techniques, can be specified in Velleman's original notation. One compound smoother that seems particularly useful is called "4253h, twice." Applying this to *water*, we calculate smoothed variable *water4r*:

```
. tssmooth nl water4r = water, smoother(4253h,twice)
```

Figure 13.4 graphs these new smoothed values, *water4r*. Compare Figure 13.4 with 13.3 to see how 4253h, twice smoothing performs relative to a span-5 moving average. Although both smoothers have similar spans, 4253h, twice does more to reduce the jagged variations.

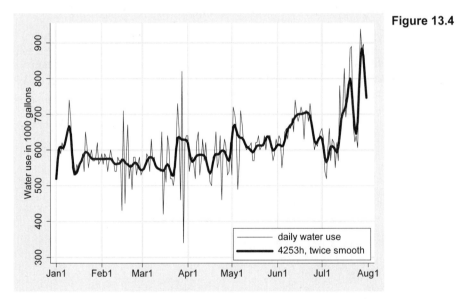

Figure 13.4

Sometimes our goal in smoothing is to look for patterns in smoothed plots. With these particular data, however, the "rough" or residuals after smoothing actually hold more interest. We can calculate the rough as the difference between data and smooth, and then graph these residuals in their own time plot, as seen in Figure 13.5.

```
. generate rough = water - water4r
. label variable rough "Residuals from 4253h, twice"
. graph twoway tsline rough, ttitle("")
     tlabel(01jan1983 01feb1983 01mar1983 01apr1983
     01may1983 01jun1983 01jul1983 01aug1983, grid format(%tdmd))
```

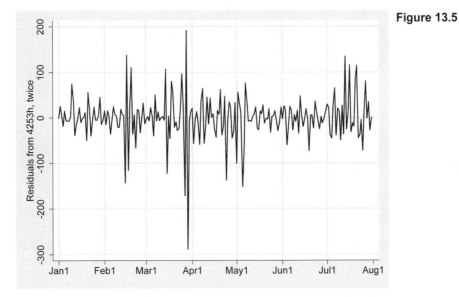

Figure 13.5

The wildest fluctuations in Figure 13.5 occur around March 27–29. Water use abruptly dropped, rose again, and then dropped lower and bounced even higher before settling towards more usual levels. On these days, local newspapers carried stories that hazardous chemical wastes had been discovered in one of the wells that supplied the town's water. Initial reports alarmed people, and consumption dropped precipitously. Over the next few days, water use vacillated between high and low peaks, in response to new developments or in compensation for delayed use. Things calmed down after the questionable well was taken offline.

The smoothing techniques described in this section tend to make the most sense when the observations are equally spaced in time. For time series with uneven spacing, lowess regression (see Chapter 8) can provide a practical alternative.

Further Time Plot Examples

Dataset *atlantic.dta* contains time series of climate, ocean, and fisheries variables for the northern Atlantic from 1950–2000 (the original data sources include Buch 2000, and others cited in Hamilton, Brown, and Rasmussen 2003). The variables include sea temperatures on Fylla Bank off west Greenland; air temperatures in Nuuk, Greenland's capital city; two climate indexes called the North Atlantic Oscillation (NAO) and the Arctic Oscillation (AO); and catches of cod and shrimp in west Greenland waters.

```
Contains data from c:\data\atlantic.dta
  obs:            51                          Greenland climate & fisheries
  vars:            8                          24 May 2008 10:13
  size:         1,938 (99.9% of memory free)

              storage  display   value
variable name  type    format    label    variable label

year           int     %ty                Year
fylltemp       float   %9.0g              Fylla Bank temperature at 0-40m
fyllsal        float   %9.0g              Fylla Bank salinity at 0-40m
nuuktemp       float   %9.0g              Nuuk air temperature
wNAO           float   %9.0g              Winter (Dec-Mar)
                                             Lisbon-Stykkisholmur NAO
wAO            float   %9.0g              Winter (Dec-Mar) AO index
tcod1          float   %9.0g              Division 1 cod catch, 1000t
tshrimp1       float   %9.0g              Division 1 shrimp catch, 1000t

Sorted by:  year
```

Before analyzing the time series, we use **tsset** to define these as yearly time series data (unlike our previous daily example). Variable *year* contains the time information.

```
. tsset year, yearly
        time variable:  year, 1950 to 2000
                delta:  1 year
```

Two special qualifiers exist for time series data: **tin** (times **in**) and **twithin** (times **within**). To list Fylla temperatures and NAO values for the years 1950 through 1955, type

```
. list year fylltemp wNAO if tin(1950,1955)
```

```
     year   fylltemp   wNAO

1.   1950      2.1      1.4
2.   1951      1.9    -1.26
3.   1952      1.6      .83
4.   1953      2.1      .18
5.   1954      2.3      .13

6.   1955      1.2    -2.52
```

The **twithin** qualifier works similarly, but excludes the two endpoints.

```
. list year fylltemp wNAO if twithin(1950,1955)
```

```
     year   fylltemp   wNAO

2.   1951      1.9    -1.26
3.   1952      1.6      .83
4.   1953      2.1      .18
5.   1954      2.3      .13
```

We use **tssmooth nl** to define a new variable, *fyll4*, containing 4253h, twice smoothed values of *fylltemp* (oceanographic data from Buch 2000).

```
. tssmooth nl fyll4 = fylltemp, smoother(4253h, twice)
```

Figure 13.6 graphs raw (*fylltemp*) and smoothed (*fyll4*) Fylla Bank temperatures. Raw temperatures are shown as spike-plot deviations from the mean (1.67 °C), so this graph emphasizes both decadal cycles and annual variations. With years rather than calender dates defining the horizontal axis, **twoway line** and other **twoway** plot types work well here; we do not need the date capabilities of **tsline**.

```
. graph twoway spike fylltemp year, base(1.67) yline(1.67)

     || line fyll4 year, clpattern(solid)

     ||  , ytitle("Fylla Bank temperature, degrees C")
     ylabel(0(1)3) xtitle("") xtick(1955(10)1995) legend(off)
```

Figure 13.6

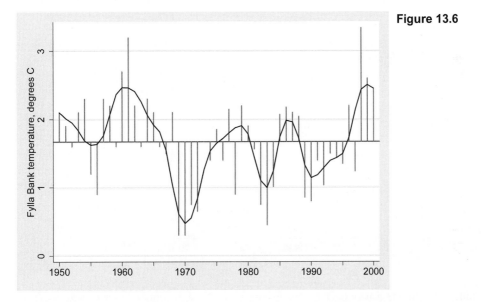

The smoothed values of Figure 13.6 exhibit irregular periods of generally warmer and cooler water. Of course, "warmer" is a relative term around Greenland; these summer sea temperatures rise no higher than 3.34 °C (37 °F).

Fylla Bank temperatures are influenced by winds and currents associated with a large-scale atmospheric pattern called the North Atlantic Oscillation, or NAO. Figure 13.7 on the next page graphs smoothed temperatures together with smoothed values of the NAO (a new variable named *wNAO4*).

```
. tssmooth nl wNAO4 = wNAO, smoother(4253h, twice)
```

For the overlaid graph in Figure 13.7, temperature defines the left axis scale, **yaxis(1)**, and NAO the right, **yaxis(2)**. Further *y*-axis options specify whether they refer to axis 1 or 2. For example, a horizontal line drawn by **yline(0, axis(2))** marks the zero point of the NAO index. On both axes, numerical labels are written horizontally. The legend appears at the 5 o'clock position inside the plot space, **position(5) ring(0)**.

```
. graph twoway line fyll4 year, yaxis(1)
    ylabel(0(1)3, angle(horizontal) nogrid axis(1))
    ytitle("Fylla Bank temperature, degrees C", axis(1))

    ||  line wNAO4 year, yaxis(2) ytitle("Winter NAO index",
            axis(2))
    ylabel(-3(1)3, angle(horizontal) axis(2)) yline(0, axis(2))

    ||  , xtitle("") xlabel(1950(10)2000, grid) xtick(1955(5)1995)
    legend(label(1 "Fylla temperature") label(2 "NAO index")
            cols(1) position(5) ring(0))
```

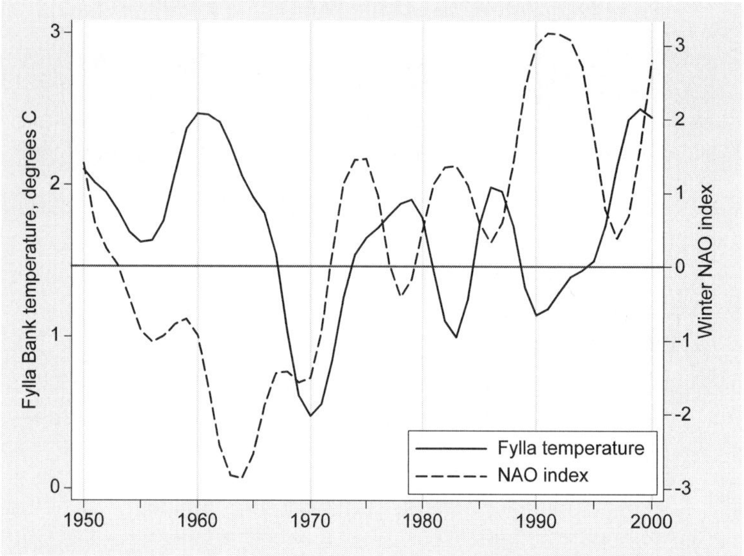

Figure 13.7

Overlaid plots provide a way to visually examine how several time series vary together. In Figure 13.7, we see evidence of a negative correlation: high-NAO periods correspond to low temperatures. The physical mechanism responsible involves northerly winds that bring Arctic air and water to west Greenland during high-NAO phases. The negative NAO– temperature correlation became stronger during the later part of this time series, roughly 1973 to 1997.

Lags, Leads, and Differences

Time series analysis often involves lagged variables, or values from previous times. Lags can be specified by explicit subscripting. For example, the following command creates variable *wNAO_1*, equal to the previous year's NAO value:

```
. generate wNAO_1 = wNAO[_n-1]
(1 missing value generated)
```

An alternative way to achieve the same thing, using **tsset** data, is with Stata's **L.** (lag) operator:

```
. generate wNAO_1 = L.wNAO
```

Lag operators are often simpler than an explicit-subscripting approach. More importantly, the lag operators also respect panel data. To generate lag 2 values, use

```
. generate wNAO_2 = L2.wNAO
(2 missing values generated)
. list year wNAO wNAO_1 wNAO_2 if tin(1950,1954)
```

	year	wNAO	wNAO_1	wNAO_2
1.	1950	1.4	.	.
2.	1951	-1.26	1.4	.
3.	1952	.83	-1.26	1.4
4.	1953	.18	.83	-1.26
5.	1954	.13	.18	.83

We could have obtained this same list without generating any new variables, by instead typing

```
. list year wNAO L.wNAO L2.wNAO if tin(1950,1954)
```

The **L.** operator is one of several that simplify working with time series datasets. Other time series operators are **F.** (lead), **D.** (difference), and **S.** (seasonal difference). These operators can be typed either in upper or lower case — for example, **F2.**wNAO or **f2.**wNAO.

Time Series Operators
L.   Lag y_{t-1} (**L1.** means the same thing)
L2.   2-period lag y_{t-2} (similarly, **L3.**, etc. **L(1/4).** means **L1.** through **L4.**)
F.   Lead y_{t+1} (**F1.** means the same thing)
F2.   2-period lead y_{t+2} (similarly, **F3.**, etc.)
D.   Difference $y_t - y_{t-1}$ (**D1.** means the same thing)
D2.   Second difference $(y_t - y_{t-1}) - (y_{t-1} - y_{t-2})$ (similarly, **D3.**, etc.)
S.   Seasonal difference $y_t - y_{t-1}$, (which is the same as **D.**)
S2.   Second seasonal difference $(y_t - y_{t-2})$ (similarly, **S3.**, etc.)

In the case of seasonal differences, **S12.** does not mean "12th difference," but rather a first difference at lag 12. For example, if we had monthly temperatures instead of yearly, we might want to calculate **S12.**temp, which would be the differences between December 2000 temperature and December 1999 temperature, November 2000 temperatures and November 1999 temperature, and so forth.

Lag operators can appear directly in most analytical commands involving **tsset** data. We could regress 1973–97 *fylltemp* on *wNAO*, including as additional predictors *wNAO* values from one, two, and three years previously, without first creating any new lagged variables.

```
. regress fylltemp wNAO L1.wNAO L2.wNAO L3.wNAO if tin(1973,1997)
```

Source	SS	df	MS		
Model	3.1884913	4	.797122826		
Residual	3.48929123	20	.174464562		
Total	6.67778254	24	.278240939		

```
Number of obs =      25
F( 4,    20) =    4.57
Prob > F      =  0.0088
R-squared     =  0.4775
Adj R-squared =  0.3730
Root MSE      =  .41769
```

fylltemp	Coef.	Std. Err.	t	P>|t|	[95% Conf. Interval]
wNAO					
--.	-.1688424	.0412995	-4.09	0.001	-.2549917 -.0826931
L1.	.0043805	.0421436	0.10	0.918	-.0835294 .0922905
L2.	-.0472993	.050851	-0.93	0.363	-.1533725 .058774
L3.	.0264682	.0495416	0.53	0.599	-.0768738 .1298102
_cons	1.727913	.1213588	14.24	0.000	1.474763 1.981063

Equivalently, we could have typed

```
. regress fylltemp L(0/3).wNAO if tin(1973,1997)
```

The estimated model is

$$\text{predicted } fylltemp_t = 1.728 - .169wNAO_t + .004wNAO_{t-1} - .047wNAO_{t-2} + .026wNAO_{t-3}$$

Coefficients on the lagged terms are not statistically significant; it appears that current (unlagged) values of $wNAO_t$ provide the most parsimonious prediction. Indeed, if we re-estimate this model without the lagged terms, the adjusted R^2 rises from .37 to .43. Either model is very rough, however. A Durbin–Watson test for autocorrelated errors is inconclusive (relative to the critical bounds for this statistic), but that is not reassuring given the small sample size.

```
. estat dwatson
```

```
Durbin-Watson d-statistic(  5,    25) =  1.423806
```

Autocorrelated errors, commonly encountered with time series, invalidate the usual OLS confidence intervals and tests. More suitable regression methods for time series, involving the **arima** command, appear later in this chapter.

Correlograms

Autocorrelation coefficients estimate the correlation between a variable and itself at particular lags. For example, first-order autocorrelation is the correlation between y_t and y_{t-1}. Second order refers to $\text{Cor}[y_t, y_{t-2}]$, and so forth. A correlogram graphs correlation versus lags.

Stata's **corrgram** command provides simple correlograms and related information. The maximum number of lags it shows can be limited by the data, by **matsize**, or to some arbitrary lower number that is set by specifying the **lags( )** option:

```
. corrgram fylltemp, lags(9)
```

LAG	AC	PAC	Q	Prob>Q	-1 0 1 [Autocorrelation]	-1 0 1 [Partial Autocor]
1	0.4038	0.4141	8.8151	0.0030		
2	0.1996	0.0565	11.012	0.0041		
3	0.0788	0.0045	11.361	0.0099		
4	0.0071	-0.0556	11.364	0.0228		
5	-0.1623	-0.2232	12.912	0.0242		
6	-0.0733	0.0880	13.234	0.0395		
7	0.0490	0.1367	13.382	0.0633		
8	-0.1029	-0.2510	14.047	0.0805		
9	-0.2228	-0.2779	17.243	0.0450		

Lags appear at the left side of the table, followed by columns of autocorrelations (AC) and partial autocorrelations (PAC). For example, the correlation between *fylltemp*$_t$ and *fylltemp*$_{t-2}$ is .1996, and the partial autocorrelation (adjusted for lag 1) is .0565. The Q statistics (Box–Pierce portmanteau) test a series of null hypotheses that all autocorrelations up to and including each lag are zero. Because the P-values seen here are mostly below .05, we should reject the null hypotheses and conclude that *fylltemp* shows significant autocorrelation. If none of the Q statistic probabilities had been below .05, we might conclude instead that the series was "white noise" with no significant autocorrelation.

At the right in this output are character-based plots of the autocorrelations and partial autocorrelations. Inspection of such plots plays a role in the specification of time series models. More refined graphical autocorrelation plots can be obtained through the **ac** command:

```
. ac fylltemp, lags(9)
```

The resulting correlogram, Figure 13.8, includes a shaded area marking pointwise 95% confidence intervals. Correlations outside of these intervals are individually significant.

Figure 13.8

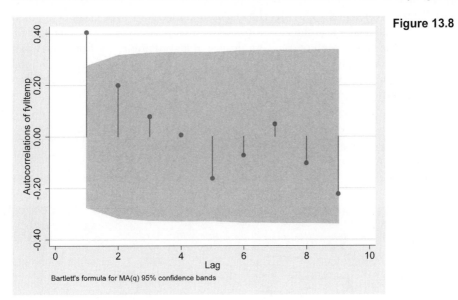

A similar command, **pac**, produces graphs of partial autocorrelations, with approximate confidence intervals (estimating the standard error as $1/\sqrt{n}$). The default plot produced by both **ac** and **pac** has the look shown in Figure 13.8 on the previous page. For the **pac** graph in Figure 13.9, however, we chose different options: drawing a baseline at zero correlation, and indicating the confidence interval as an outline instead of a shaded area.

```
. pac fylltemp, yline(0) lags(9) ciopts(bstyle(outline))
```

Figure 13.9

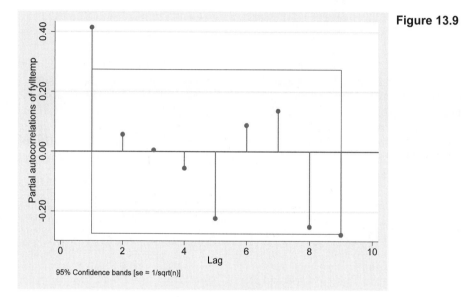

Cross-correlograms help to explore relationships between two time series. Figure 13.10 (next page) shows the cross-correlogram of *wNAO* and *fylltemp* over 1973–97. The cross-correlation is substantial and negative at 0 lag, and closer to zero at other positive or negative lags (with minor positive peaks at ±4). Recall the nonsignificance of lagged predictors from our earlier OLS regression. Because the physical mechanism connecting the North Atlantic Oscillation and sea temperatures around Greenland involves winds, it makes sense that the strongest correlation is within the same year.

```
. xcorr wNAO fylltemp if tin(1973,1997), lags(9) xlabel(-9(1)9, grid)
```

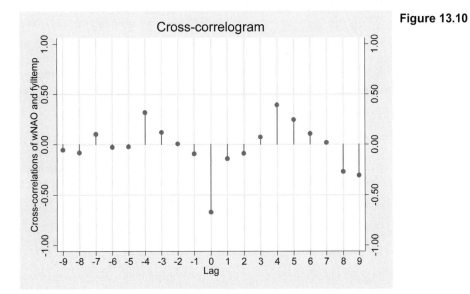

Figure 13.10

If we list our input or independent variable first in the **xcorr** command, and the output or dependent variable second — as done for Figure 13.10 — then positive lags denote correlations between input at time t and output at time $t+1$, $t+2$, etc. Thus, we see a modest positive correlation of .394 between winter NAO index and Fylla temperature four years later.

The actual cross-correlation coefficients, and a text version of the cross-correlogram, can be obtained with the **table** option:

```
. xcorr wNAO fylltemp if tin(1973,1997), lags(9) table
```

```
                  -1       0       1
  LAG     CORR  [Cross-correlation]

  -9   -0.0541           |
  -8   -0.0786           |
  -7    0.1040           |
  -6   -0.0261           |
  -5   -0.0230           |
  -4    0.3185           |—
  -3    0.1212           |
  -2    0.0053           |
  -1   -0.0909           |
   0   -0.6740      —————|
   1   -0.1386          —|
   2   -0.0865           |
   3    0.0757           |
   4    0.3940           |—
   5    0.2464           |—
   6    0.1100           |
   7    0.0183           |
   8   -0.2699         ——|
   9   -0.3042         ——|
```

In the next section, we explore more formal models for this time series.

ARIMA Models

Autoregressive integrated moving average (ARIMA) models can be estimated through the **arima** command. This command encompasses autoregressive (AR), moving average (MA), or ARIMA models. It also can estimate structural models that include one or more predictor variables and ARIMA disturbances. These are termed ARMAX models, for autoregressive moving average with exogenous variables. The general form of such ARMAX models, in matrix notation, is

$$y_t = \mathbf{x}_t \boldsymbol{\beta} + \mu_t \qquad [13.1]$$

where y_t is the vector of dependent-variable values at time t, $\mathbf{x_t}$ is a matrix of exogenous predictor-variable values (usually including a constant). μ_t is a vector of "everything else" disturbances. Those disturbances can be autoregressive or moving-average, of any order. For example, ARMA(1,1) disturbances are

$$\mu_t = \rho\mu_{t-1} + \theta\epsilon_{t-1} + \epsilon_t \qquad [13.2]$$

where ρ is a first-order autoregression parameter, θ is a first-order moving average parameter, and ϵ_t represents random, uncorrelated "white-noise" (normal i.i.d.) errors. **arima** fits simple models as a special case of [13.1] and [13.2], with a constant (β_0) replacing the structural term $\mathbf{x_t}\boldsymbol{\beta}$. Therefore, a simple ARMA(1,1) model becomes

$$\begin{aligned} y_t &= \beta_0 + \mu_t \\ &= \beta_0 + \rho\mu_{t-1} + \theta\epsilon_{t-1} + \epsilon_t \end{aligned} \qquad [13.3]$$

Some sources present an alternative version. In the ARMA(1,1) case, they show y_t as a function of the previous y value (y_{t-1}) and the present (ϵ_t) and lagged (ϵ_{t-1}) disturbances:

$$y_t = \alpha + \rho y_{t-1} + \theta\epsilon_{t-1} + \epsilon_t \qquad [13.4]$$

Because in the simple structural model $y_t = \beta_0 + \mu_t$, equation [13.3] (used by Stata) is equivalent to [13.4], apart from rescaling the constant $\alpha = (1-\rho)\beta_0$.

Using **arima**, an ARMA(1,1) model (equation [13.3]) can be specified in either of two ways:

```
. arima y, ar(1) ma(1)
```

or

```
. arima y, arima(1,0,1)
```

The **i** in **arima** stands for "integrated," referring to models that also involve differencing. To fit an ARIMA(2,1,1) model, use

```
. arima y, arima(1/2,1,1)
```

or equivalently,

```
. arima D.y, ar(1 2) ma(1)
```

Either command specifies a model in which first differences of the dependent variable $(y_t - y_{t-1})$ are a function of first differences one and two lags previous $(y_{t-1} - y_{t-2}$ and $y_{t-2} - y_{t-3})$, and also of present and previous errors $(\epsilon_t$ and $\epsilon_{t-1})$.

To estimate a structural model in which y_t depends on two predictor variables x (present and lagged values, x_t and x_{t-1}) and w (present values only, w_t), with ARIMA(1,0,3) errors and also multiplicative seasonal ARIMA$(1,0,1)_{12}$ errors looking back 12 periods (as might be appropriate for monthly data, over a number of years), a suitable command could have the form

```
. arima y x L.x w, arima(1,0,1/3) sarima(1,0,1,12)
```

In econometric notation, this corresponds to an ARIMA$(1,0,3)\times(1,0,1)_{12}$ model.

A time series y is considered "stationary" if its mean and variance do not change with time, and if the covariance between y_t and y_{t+u} depends only on the lag u, and not on the particular values of t. ARIMA modeling assumes that our series is stationary, or can be made stationary through appropriate differencing or transformation. We can check this assumption informally by inspecting time plots for trends in level or variance. Formal statistical tests for "unit roots" (a nonstationary AR(1) process in which $\rho_1 = 1$, also known as a "random walk") also help. Stata offers three unit root tests, **pperron** (Phillips–Perron), **dfuller** (augmented Dickey–Fuller), and **dfgls** (augmented Dickey–Fuller using GLS, generally a more powerful test than **dfuller**).

Applied to Fylla Bank temperatures, a **pperron** test rejects the null hypothesis of a unit root ($P < .01$).

```
. pperron fylltemp, lag(3)
```

```
Phillips-Perron test for unit root          Number of obs   =        50
                                            Newey-West lags =         3

                  ———————— Interpolated Dickey-Fuller ————————
          Test       1% Critical      5% Critical      10% Critical
        Statistic       Value            Value            Value
Z(rho)   -29.871       -18.900          -13.300          -10.700
Z(t)      -4.440        -3.580           -2.930           -2.600

MacKinnon approximate p-value for Z(t) = 0.0003
```

Test results sometimes prove sensitive to the number of lags selected, but in this example other lags larger or smaller than 3 lead to similar conclusions. A Dickey–Fuller GLS test evaluating the null hypothesis that *fylltemp* has a unit root (versus the alternative hypothesis that it is stationary with a possibly nonzero mean, but no linear time trend) rejects this null hypothesis as well ($P < .05$). Both tests thus confirm the visual impression of stationarity given by Figure 13.6.

```
. dfgls fylltemp, notrend maxlag(3)
```

DF-GLS for fylltemp   Number of obs =   47

[lags]	DF-GLS mu Test Statistic	1% Critical Value	5% Critical Value	10% Critical Value
3	-2.304	-2.620	-2.211	-1.913
2	-2.479	-2.620	-2.238	-1.938
1	-3.008	-2.620	-2.261	-1.959

```
Opt Lag (Ng-Perron seq t) = 0 [use maxlag(0)]
Min SC   = -.6735952 at lag  1 with RMSE   .6578912
Min MAIC = -.2683716 at lag  2 with RMSE   .6569351
```

For a stationary series, correlograms provide guidance about selecting a preliminary ARIMA model:

AR(p)   An autoregressive process of order p has autocorrelations that damp out gradually with increasing lag. Partial autocorrelations cut off after lag p.

MA(q)   A moving average process of order q has autocorrelations that cut off after lag q. Partial autocorrelations damp out gradually with increasing lag.

ARMA(p,q)   A mixed autoregressive–moving average process has autocorrelations and partial autocorrelations that damp out gradually with increasing lag.

Correlogram spikes at seasonal lags (for example, at 12, 24, 36 in monthly data) indicate a seasonal pattern. Identification of seasonal models follows similar guidelines, but applied to autocorrelations and partial autocorrelations at seasonal lags.

Figures 13.8 and 13.9 weakly suggest an AR(1) process, so we will try this as a simple model for *fylltemp*.

```
. arima fylltemp, arima(1,0,0) nolog
```

ARIMA regression

Sample: 1950 - 2000   Number of obs   =   51
                                        Wald chi2(1)   =   7.53
Log likelihood = -48.66274   Prob > chi2   =   0.0061

fylltemp	Coef.	OPG Std. Err.	z	P>|z|	[95% Conf. Interval]	
fylltemp						
_cons	1.68923	.1513096	11.16	0.000	1.392669	1.985792
ARMA						
ar L1.	.4095759	.1492491	2.74	0.006	.1170531	.7020987
/sigma	.627151	.0601859	10.42	0.000	.5091889	.7451131

After we fit an **arima** model, its coefficients and other results are saved temporarily in Stata's usual way. For example, to see the recent model's AR(1) coefficient and standard error, type

```
. display [ARMA]_b[L1.ar]
.4095759

. display [ARMA]_se[L1.ar]
.14924909
```

The AR(1) coefficient in this example is statistically distinguishable from zero ($t = 2.74$, $P = .006$), which gives one indication of model adequacy. A second test is whether the residuals appear to be uncorrelated "white noise." We can obtain residuals (also predicted values, and other case statistics) after **arima** through **predict**:

```
. predict fyllres, resid
. corrgram fyllres, lags(15)
```

					-1 0 1	-1 0 1
LAG	AC	PAC	Q	Prob>Q	[Autocorrelation]	[Partial Autocor]
1	-0.0173	-0.0176	.0162	0.8987		
2	0.0467	0.0465	.13631	0.9341		
3	0.0386	0.0497	.22029	0.9742		
4	0.0413	0.0496	.31851	0.9886		
5	-0.1834	-0.2450	2.2955	0.8069		
6	-0.0498	-0.0602	2.4442	0.8747		
7	0.1532	0.2156	3.8852	0.7929		
8	-0.0567	-0.0726	4.087	0.8492		
9	-0.2055	-0.3232	6.8055	0.6574		
10	-0.1156	-0.2418	7.6865	0.6594		
11	0.1397	0.2794	9.0051	0.6214		
12	-0.0028	0.1606	9.0057	0.7024		
13	0.1091	0.0647	9.8519	0.7060		
14	0.1014	-0.0547	10.603	0.7169		
15	-0.0673	-0.2837	10.943	0.7566		

corrgram 's portmanteau Q tests find no significant autocorrelation among residuals out to lag 15. We could obtain exactly the same result by requesting a **wntestq** (white noise test Q statistic) for 15 lags.

```
. wntestq fyllres, lags(15)
```

Portmanteau test for white noise

Portmanteau (Q) statistic =		10.9435
Prob > chi2(15)	=	0.7566

By these criteria, our AR(1) or ARIMA(1,0,0) model appears adequate. More complicated versions, with MA or higher-order AR terms, do not offer much improvement in fit.

A similar AR(1) model fits *fylltemp* over just the years 1973–1997. During this period, however, information about the winter North Atlantic Oscillation (*wNAO*) significantly improves the predictions. For a simple ARMAX (autoregressive moving average with exogenous variables) model, we include *wNAO* as a predictor but keep an AR(1) term to account for autocorrelation of errors.

```
. arima fylltemp wNAO if tin(1973,1997), ar(1) nolog
```

ARIMA regression

Sample: 1973 - 1997

Number of obs	=	25
Wald chi2(2)	=	12.73
Prob > chi2	=	0.0017

Log likelihood = -10.3481

fylltemp	Coef.	OPG Std. Err.	z	P>|z|	[95% Conf. Interval]	
fylltemp						
wNAO	-.1736227	.0531688	-3.27	0.001	-.2778317	-.0694138
_cons	1.703462	.1348599	12.63	0.000	1.439141	1.967782
ARMA						
ar						
L1.	.2965222	.237438	1.25	0.212	-.1688478	.7618921
/sigma	.36536	.0654008	5.59	0.000	.2371767	.4935432

```
. predict fyllhat
(option xb assumed; predicted values)
. label variable fyllhat "predicted temperature"
. predict fyllres2 if tin(1973,1997), resid
. corrgram fyllres2, lags(9)
```

LAG	AC	PAC	Q	Prob>Q	-1 0 1 -1 0 1 [Autocorrelation] [Partial Autocor]
1	0.0501	0.0558	.07046	0.7907	
2	-0.0121	-0.0127	.07479	0.9633	
3	-0.5363	-0.5932	8.9001	0.0306	
4	-0.0222	0.1126	8.9159	0.0632	
5	-0.0684	-0.1821	9.0736	0.1062	
6	0.1720	-0.0747	10.125	0.1195	
7	0.0709	0.0655	10.314	0.1715	
8	0.0812	0.0491	10.576	0.2269	
9	-0.1148	-0.1731	11.132	0.2668	

wNAO exhibits a significant, negative effect in this model, and the residuals pass a portmanteau test for white noise ($Q_9 = 11.444$, $P = .2668$). The AR(1) coefficient is not statistically significant, but is nevertheless useful because if we dropped this term, the residuals would no longer pass the test for white noise.

Figure 13.11 graphs our model's predicted values, *fyllhat*, together with the observed temperature series *fylltemp*. The model succeeds reasonably well in fitting the main warming/cooling episodes and a few of the minor variations. To have the *y*-axis labels displayed with the same number of decimal places (0.5, 1.0, 1.5,... instead of .5, 1, 1.5,...) in this graph, we specify their format as **%2.1f**.

```
. graph twoway line fylltemp year if tin(1973, 1997)

    || line fyllhat year if tin(1973, 1997)

    ||  , ylabel(.5(.5)2.5, angle(horizontal) format(%2.1f))
    ytitle("Degrees C") xlabel(1975(5)1995, grid) xtitle("")
    legend(label(1 "observed temperature")
        label(2 "model prediction") position(5) ring(0) col(1))
```

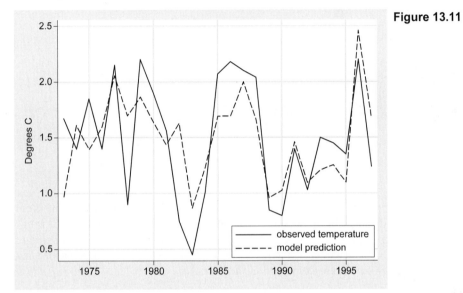

Figure 13.11

The final section of this chapter presents a more elaborate illustration of ARMAX modeling.

ARMAX Models

Figure 3.55 in Chapter 3 plotted daily snow depth measurements in the city of Boston, and also at a location in New Hampshire's White Mountains, during the winter of 1999–2000. The graph also depicted the actual and model-predicted number of skier/snowboarder visits to one ski resort in the White Mountains, near where the snow depth was measured. Figure 13.12, on the following page, reproduces this same graph.

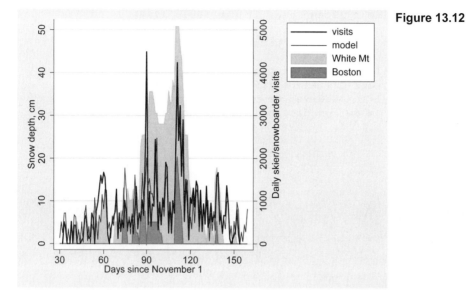

Figure 13.12

The data and analysis behind Figure 13.12 derive from a study of ski areas, weather, and climate, described in the *International Journal of Climatology* (Hamilton, Brown, and Keim 2007). For the original study, the authors analyzed daily weather and skier attendance at two ski areas over a period of nine winters. Dataset *whitemt2.dta* contains a subset of their data, covering one ski area over one winter season, 1999–2000. The data are indexed by elapsed date (variable *edate*) through a previous command of the form

```
. tsset edate, daily
```

whitemt2.dta also contains a second time indicator, a day-of-season variable counting the days since November 1 each season, which was needed for drawing graphs such as Chapter 3's Figure 3.54. Other variables include dummy indicators for each day of the week (*day1*, for example, equals 1 if the day is Sunday), and daily snowfall and snow depth in centimeters at Boston and the White Mountains location. *visits* gives the daily number of skiers and snowboarders visiting the ski area.

```
Contains data from C:\data\whitemt2.dta
  obs:            182                    Winter 1999-2000 at NH ski area
                                         (Hamilton, Brown, Keim 2007)
  vars:            16                    25 May 2008 17:07
  size:         8,190 (99.9% of memory free)
```

variable name	storage type	display format	value label	variable label
edate	int	%td		Elapsed date
season	int	%9.0g	season	Winter season
dayseason	int	%9.0g		Days since November 1
visits	float	%9.0g		Daily skier/snowboarder visits
day1	byte	%8.0g		dayweek==Sun
day2	byte	%8.0g		dayweek==Mon
day3	byte	%8.0g		dayweek==Tue
day4	byte	%8.0g		dayweek==Wed
day5	byte	%8.0g		dayweek==Thu
day6	byte	%8.0g		dayweek==Fri
day7	byte	%8.0g		dayweek==Sat
bosfall	float	%9.0g		Snow fall in Boston, cm
mtfall	float	%9.0g		Snow fall in mountains, cm
mtdepth	float	%9.0g		Snow depth in mountains, cm
bosdepth	float	%9.0g		Snow depth in Boston, cm
model	float	%9.0g		ARMAX predicted visits

```
Sorted by: edate
```

The dataset includes predicted values from a time series model, which were graphed without explanation in Chapter 3. These predictions came from an ARMAX (autoregressive moving average with exogenous variables) model estimated by **arima**. The original study fit models based on multi-year data from two ski areas, but the one-year, one-area subset here allows for a compact illustration of the method— and leads to a similar conclusion regarding its "backyard effect" hypothesis.

On the following page, we see an ARMAX regression of daily ski area *visits* on 12 predictors:

> Dummy indicators for each day of the week (*day1*, *day3*, etc.). Monday, often a low-attendance day at this resort, is the omitted category.

> Snow depth in the mountains the previous day (L1.*mtdepth*), and snowfall in the mountains both one and two days previously (L1.*mtfall* and L2.*mtfall*).

> Similarly, the previous day's snow depth in Boston (L1.*bosdepth*), and snowfall in Boston at lags 1 and 2 (L1.*bosfall* and L2.*bosfall*).

Through experimentation with multiyear data, the study's authors found that it worked well to model disturbances as regular and multiplicative "seasonal" (reflecting 7-day weekly, not annual cycles) first-order autoregressive and moving-average processes — **arima(1,0,1) sarima(1,0,1,7)**. In such models, disturbances at time t (μ_t; see equation [13.1]) have the form

$$\mu_t = \rho_1 \mu_{t-1} + \rho_{7,1} \mu_{t-7} - \rho_1 \rho_{7,1} \mu_{t-8} + \epsilon_t + \theta_1 \epsilon_{t-1} + \theta_{7,1} \epsilon_{t-7} + \theta_1 \theta_{7,1} \epsilon_{t-8} \qquad [13.5]$$

where the ρ are autoregression parameters (present disturbances μ_t regressed on lagged disturbances) and the θ are moving-average parameters (present white-noise errors ϵ_t regressed on lagged errors).

```
. arima attend  day1 day3 day4 day5 day6 day7  L1.mtdepth
     L1.mtfall  L2.mtfall  L1.bosdepth  L1.bosfall  L2.bosfall,
     arima(1,0,1)  sarima(1,0,1,7)

(setting optimization to BHHH)
Iteration 0:   log likelihood = -928.83082  (not concave)
Iteration 1:   log likelihood = -927.72396
Iteration 2:   log likelihood = -926.73491
Iteration 3:   log likelihood = -926.57525
Iteration 4:   log likelihood = -925.99813
(switching optimization to BFGS)
Iteration 5:   log likelihood = -925.71423
Iteration 6:   log likelihood = -925.67082
Iteration 7:   log likelihood = -925.58537
Iteration 8:   log likelihood = -925.48515
Iteration 9:   log likelihood = -925.43945
Iteration 10:  log likelihood = -925.40022
Iteration 11:  log likelihood = -925.37681
Iteration 12:  log likelihood = -925.37219
Iteration 13:  log likelihood = -925.37119
Iteration 14:  log likelihood = -925.37094
(switching optimization to BHHH)
Iteration 15:  log likelihood = -925.37091
Iteration 16:  log likelihood = -925.3709
```

ARIMA regression

Sample: **03dec1999 - 03apr2000**   Number of obs   =   123
                                           Wald chi2(**16**) =   227.08
Log likelihood = **-925.3709**   Prob > chi2   =   0.0000

visits	Coef.	OPG Std. Err.	z	P>|z|	[95% Conf. Interval]	
visits						
day1	829.5041	166.5313	4.98	0.000	503.1088	1155.9
day3	349.5057	171.7344	2.04	0.042	12.91237	686.099
day4	131.6191	199.9863	0.66	0.510	-260.3468	523.585
day5	508.6724	233.1467	2.18	0.029	51.71321	965.6316
day6	245.0255	221.6643	1.11	0.269	-189.4285	679.4795
day7	790.0761	190.256	4.15	0.000	417.1811	1162.971
mtdepth						
L1.	17.84307	7.687006	2.32	0.020	2.776816	32.90932
mtfall						
L1.	-.8688855	19.33456	-0.04	0.964	-38.76393	37.02616
L2.	-35.05528	16.06164	-2.18	0.029	-66.53552	-3.575034
bosdepth						
L1.	112.5427	25.19787	4.47	0.000	63.15577	161.9296
bosfall						
L1.	21.16441	27.46685	0.77	0.441	-32.66962	74.99845
L2.	-33.90994	24.36751	-1.39	0.164	-81.66938	13.8495
_cons	52.40108	186.0097	0.28	0.778	-312.1713	416.9734
ARMA						
ar						
L1.	.5022812	.2482918	2.02	0.043	.0156382	.9889242
ma						
L1.	-.150929	.278841	-0.54	0.588	-.6974474	.3955893
ARMA7						
ar						
L1.	-.7203389	.2019019	-3.57	0.000	-1.116059	-.3246185
ma						
L1.	.8788344	.1828954	4.81	0.000	.520366	1.237303
/sigma	444.9262	30.20298	14.73	0.000	385.7295	504.123

```
. predict e1, resid
. corrgram e1, lags(14)
```

					-1 0 1	-1 0 1
LAG	AC	PAC	Q	Prob>Q	[Autocorrelation]	[Partial Autocor]
1	-0.0038	-0.0038	.00182	0.9659		
2	0.0685	0.0689	.5984	0.7414		
3	-0.1508	-0.1518	3.5111	0.3193		
4	0.0707	0.0685	4.1576	0.3851		
5	-0.0123	0.0060	4.1773	0.5242		
6	0.1076	0.0787	5.6992	0.4577		
7	0.0181	0.0388	5.7428	0.5701		
8	-0.1798	-0.2127	10.063	0.2606		
9	0.0367	0.0781	10.245	0.3311		
10	0.0490	0.0694	10.571	0.3919		
11	0.0428	-0.0276	10.823	0.4582		
12	0.1173	0.1654	12.728	0.3891		
13	-0.0033	-0.0228	12.73	0.4689		
14	0.0818	0.1151	13.672	0.4744		

corrgram's Q tests indicate that residuals from this model are not significantly different from white noise, out to at least lag 14: $Q_{14} = 13.672$, $P = .4744$. Several coefficients in the model are statistically indistinguishable from zero, however, according to their z tests. For example, the coefficients on *day4* is nonsignificant ($z = 0.66$, $P = .510$), indicating that skier visits do not go up much on Wednesday. We also do not gain much benefit from having both 1- and 2-day lagged measures of snowfall; perhaps 1-day lags would suffice. Model simplification to improve parsimony, without sacrificing too much fit, involves an iterative process of dropping nonsignificant effects, re-estimating the model, and testing residuals again for autocorrelation.

A simplified version of this model, similar to the reduced model for multiyear data described by Hamilton et al. (2007, Table 1), appears on the following page. The simpler version drops Wednesday (*day4*) and lag-2 snow depth (L2.*mtdepth* and L2.*bosdepth*) effects. It also employs robust estimation of the variance-covariance matrix (hence the standard errors), as a precaution against heteroskedastic or nonnormal errors.

Substituting day-of-week names for *day1*, *day3*, etc. to improve readability, and showing disturbance terms μ as differences between observed and predicted y ($y - \hat{y}$), we could write out the regression model as equation [13.6]. Coefficient estimates are taken from the **arima** table on the next page. Seasonal ARMA disturbances follow the form of equation [13.5].

$$
\begin{aligned}
y_t =\ & 70.4 + 760sunday + 319tuesday + 440thursday + 187friday + 736saturday \\
& + 18mtdepth_{t-1} + 10mtfall_{t-1} + 81bosdepth_{t-1} + 29bosfall_{t-1} \\
& + .43(y_{t-1} - \hat{y}_{t-1}) + .39(y_{t-7} - \hat{y}_{t-7}) - (.43)(.39)(y_{t-8} - \hat{y}_{t-8}) \\
& - .08\epsilon_{t-1} - .21\epsilon_{t-7} + (.08)(.21)\epsilon_{t-8} + \epsilon_t
\end{aligned}
\qquad [13.6]
$$

```
. arima  visits  day1 day3 day5 day6 day7
     L1.mtdepth  L1.mtfall  L1.bosdepth  L1.bosfall,
     nolog arima(1,0,1) sarima(1,0,1,7) vce(robust)
```

ARIMA regression

Sample: **03dec1999 - 03apr2000**   Number of obs   =   123
                                                 Wald chi2(13)   =   165.61
Log pseudolikelihood = -929.3802   Prob > chi2   =   0.0000

visits	Coef.	Semi-robust Std. Err.	z	P>|z|	[95% Conf. Interval]	
visits						
day1	760.3589	184.9516	4.11	0.000	397.8604	1122.857
day3	318.7005	117.8276	2.70	0.007	87.76266	549.6382
day5	439.6547	126.6586	3.47	0.001	191.4085	687.9009
day6	187.4986	138.4221	1.35	0.176	-83.8037	458.801
day7	735.9067	171.2135	4.30	0.000	400.3345	1071.479
mtdepth						
L1.	18.12641	6.764066	2.68	0.007	4.869087	31.38374
mtfall						
L1.	10.36519	15.34777	0.68	0.499	-19.71589	40.44627
bosdepth						
L1.	80.54236	26.13901	3.08	0.002	29.31084	131.7739
bosfall						
L1.	29.13017	15.44579	1.89	0.059	-1.143021	59.40337
_cons	70.38917	106.8985	0.66	0.510	-139.1281	279.9064
ARMA						
ar						
L1.	.4318007	.1659101	2.60	0.009	.1066228	.7569785
ma						
L1.	-.083972	.2139253	-0.39	0.695	-.5032579	.335314
ARMA7						
ar						
L1.	.3857269	.2652439	1.45	0.146	-.1341416	.9055953
ma						
L1.	-.206814	.3163991	-0.65	0.513	-.8269448	.4133168
/sigma	461.9048	58.43922	7.90	0.000	347.3661	576.4436

The most interesting conclusion from this one-season model, consistent with findings from the larger study, is the existence of a "backyard effect." Snow depth in Boston has a statistically significant impact on the number of skiers visiting a ski area 225 kilometers (140 miles) to its north the next day, even controlling for snow depth in the mountains. In fact, a centimeter of snow in Boston is worth more than a centimeter of snow in the mountains, judging by their respective coefficients (about 81 vs. 18 skier visits per centimeter). The previous day's snowfall in Boston also appears to have a larger impact than snowfall in the mountains (29 vs. 10), although neither of those coefficients is statistically significant here.

Following up on the simplified model, we can obtain predicted values, and for graphical purposes replace several negative predictions (fewer than zero skiers) with zeroes. Note that this *yhat* is identical to the variable *model*, graphed earlier in Figures 3.55 and 13.12.

```
. predict yhat
(option xb assumed; predicted values)
(1 missing value generated)
. replace yhat=0 if yhat<0
(6 real changes made)
. correlate yhat model
(obs=181)
```

	yhat	model
yhat	1.0000	
model	1.0000	1.0000

predict also calculates residuals. This results in 59 missing values representing days in November and April, before the opening or after the closing of that year's ski season. A portmanteau test finds no significant residual autocorrelation ($Q_{14} = 16.861$, $P = .2636$), so the assumption of white-noise errors appears plausible.

```
. predict e2, resid
(59 missing values generated)
. corrgram e2, lag(14)
```

					-1 0 1	-1 0 1
LAG	AC	PAC	Q	Prob>Q	[Autocorrelation]	[Partial Autocor]
1	-0.0034	-0.0033	.00141	0.9700		
2	0.0836	0.0846	.88907	0.6411		
3	-0.1817	-0.1847	5.1189	0.1633		
4	0.0374	0.0330	5.3	0.2579		
5	-0.0358	-0.0075	5.4673	0.3615		
6	0.1419	0.1084	8.1139	0.2299		
7	0.0076	0.0211	8.1217	0.3220		
8	-0.1848	-0.2344	12.687	0.1231		
9	0.0314	0.0954	12.82	0.1709		
10	0.0817	0.1227	13.729	0.1857		
11	0.0708	-0.0113	14.417	0.2108		
12	0.1315	0.1535	16.812	0.1568		
13	0.0101	0.0038	16.826	0.2074		
14	-0.0157	0.0284	16.861	0.2636		

Figure 13.13 graphs the model's predictions together with the actual skier visits. It resembles Figure 13.12 because both show the same data. In Figure 13.13, we take advantage of the time-axis labeling features of **tsline** to draw a more reader-friendly version of this plot.

```
. graph twoway tsline visits yhat if dayseason>29 & dayseason<160,
      legend(rows(2) position(2) ring(0) label(1 "actual")
        label(2 "model"))
      ylabel(0(1000)4000) ttitle("Date, 1999-2000 ski season")
      tlabel(01dec1999 01jan2000 01feb2000 01mar2000 01apr2000,
        format(%tdmd))
```

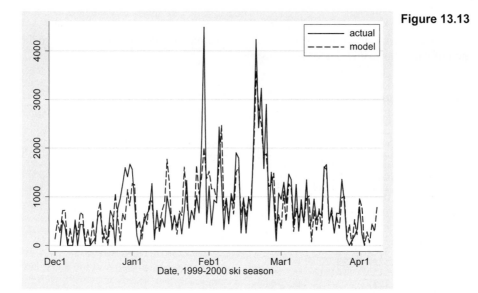

Figure 13.13

14

Survey Data Analysis

Survey research places strong emphasis on drawing valid inferences about populations, despite situations where simple random sampling is impossible. In place of the simple random-sampling ideal, complex sampling strategies are employed to achieve more representative samples, and post-sampling adjustments applied where needed. Because standard statistical methods assume simple random sampling, we also need specialized methods designed for survey data that can take information about the sampling process into account.

Stata has developed a reputation for strength in survey data analysis, based on a unified approach encompassing a wide selection of analytical methods. All of these methods work from an underlying definition of the sample structure, which can include probability weights adjusting for design or selection bias. The sampling design could also involve complexities such as stratification, multi-stage clustering, finite populations at one or more levels, with-replacement sampling, replication weights, or poststratification. Once the basic design elements have been specified (via the **svyset** command), a dataset is saved with this information. Subsequent analyses using **svy:** commands will automatically apply weights and other survey design information as appropriate.

Most Stata survey procedures can be called using menus, and especially the many submenus of Statistics > Survey data analysis . Typing **help survey** brings up a starting place with links to further information. The *Survey Data Reference Manual* provides examples and technical details for the full set of Stata survey-related commands. Other good references include a book on sampling by Levy and Lemeshow (1999), and on the analysis of survey data, with biostatistical examples, by Korn and Graubard (1999). Sul and Forthofer (2006) provide a brief overview of some main issues in survey data analysis. See Moore (2008) for a well-illustrated discussion of some pitfalls in survey question wording.

Example Commands

. `svyset _n [pweight = censuswt]`

Declares dataset as survey type, with probability weights (proportional to the inverse of the probabilities of selection) given by variable *censuswt*. The _n specifies individual observations (the default) as primary sampling units or PSUs.

. `svyset _n [pweight = censuswt], strata(district)`
Declares dataset as survey type, from one-stage stratified sampling: the population was partitioned into strata, and individuals sampled independently within each strata. In this example, variable *district* identifies the strata, and *censuswt* the probability weights.

. `svyset school [pweight = finalwt], fpc(nschools)`
    `|| _n, fpc(nstudents)`
Declares dataset as survey type from two-stage cluster sampling. In the first stage, schools were selected randomly, so schools are the PSUs. In the second stage, students were selected at random within the selected schools. Finite population corrections (FPCs) are specified for each stage: *nschools* is the total number of schools in the population, and *nstudents* is the number of students in each school.

. `svy: tabulate vote, percent miss ci`
Obtains a table of weighted percentages and confidence intervals for variable *vote*, including the missing values, based on **svyset** data.

. `svy: tabulate vote gender, column pearson lr wald`
Obtains a weighted cross-tabulation of *vote* by *gender*, with proportions based on each column (*gender*). Pearson χ^2, likelihood-ratio χ^2, and Wald test statistics are reported.

. `svy: regress y x1 x2 x3`
Performs survey regression of *y* on three predictors, *x1*, *x2* and *x3*. Other regression-type survey estimation procedures include the following, along with many others. Type **help svy estimation** for a complete listing.

`svy: cnreg`	Censored-normal regression
`svy: glm`	Generalized linear models
`svy: intreg`	Interval regression
`svy: nl`	Nonlinear least squares
`svy: tobit`	Tobit regression
`svy: truncreg`	Truncated regression
`svy: stcox`	Cox proportional hazards model
`svy: streg`	Parametric survival models
`svy: cloglog`	Complementary log-log regression
`svy: logit`	Logit regression
`svy: ologit`	Ordered logit regression
`svy: mlogit`	Multinomial logit regression
`svy: probit`	Probit regression
`svy: nbreg`	Negative binomial regression
`svy: zinb`	Zero-inflated negative binomial regression
`svy: heckman`	Heckman selection model

. `svy, subpop(voted): regress y x1 x2 x3`
Performs survey regression analysis using only a subpopulation defined by 1 values of the {0,1} variable *voted*. Selecting subsets of the data in the usual way with **if** or **in** statements is not appropriate for most survey analysis; use a **svy, subpop( ):** option instead.

. **svy jackknife: mprobit** *y x1 x2 x3*

> Performs the multinomial probit regression of multiple-category variable *y* on three predictors, resulting in a table with jackknife estimates of coefficients, standard errors, *t* tests, and confidence intervals.

. **estat effects, deff deft meff meft**

> After estimation, produces a table of design and misspecification effects for point estimates. Type **help svy_estat** for a full list of **estat** postestimation capabilities.

Probability Weights

Although phrases such as "weighted data" or "weighted analysis" appear often in research writing, they can mean many different things. Stata distinguishes four general types of weights, each appropriate for different situations and purposes.

aweight   Analytical weights, used in weighted least squares (WLS) regression and similar procedures.

fweight   Frequency weights, counting the number of duplicated observations. Frequency weights must be integers.

iweight   Importance weights, however you define "importance."

pweight   Probability or sampling weights, proportional to the inverse of the probability that an observation is included due to sampling strategy.

Survey researchers often use probability weights to adjust for biases in their sampling methods. The biases could arise from intentional features of the sampling design, or from accidental properties of the data-collection process. Either way, initial sampling yields a dataset not well representing the population of interest. Probability weighting attempts to compensate for departures from simple random sampling, and give us a more realistic picture of sampling variability and population characteristics. This section illustrates the use of probability weights to adjust for known bias in the sampling design.

Since 2001, the Granite State Poll at the University of New Hampshire has conducted statewide telephone surveys several times each year. Each survey contacts a new sample of about 500 people, asking a variety of opinion questions along with respondent background characteristics. The poll's political findings attain national importance every four years during the run-up to New Hampshire's presidential primary elections. Dataset *Granite_08_01s.dta* contains questions from a Granite State Poll of 555 people, conducted in January 2008.

. **use** *Granite_08_01s.dta,* **clear**
(Granite State Poll, January 2008)

. **describe, short**

```
Contains data from C:\data\Granite_08_01s.dta
  obs:           555                          Granite State Poll, January 2008
  vars:           46                          13 Apr 2008 17:22
  size:       37,740 (99.9% of memory free)
Sorted by:
```

The pollsters telephone a random sample of New Hampshire household phone numbers which could yield, in theory, a representative sample of households. For voting research and many other purposes, however, we hope to generalize not about the population of all households, but about the population of all adults or all voters. Some households contain only one adult, whereas others have more. Among respondents to this particular poll, about 28% said that they lived in a one-adult household. (Responses in this example are limited to one, two, or three or more adults, as a practical compromise for weighting purposes.)

```
. tab adults, miss
```

Adults in household	Freq.	Percent	Cum.
1	154	27.75	27.75
2	306	55.14	82.88
3+	85	15.32	98.20
.	10	1.80	100.00
Total	555	100.00	

Although about 28% of our sample lived in one-adult households, it would be a mistake to guess that a similar percentage of New Hampshire adults do so. People from households with one adult were at least three times more likely to enter our sample, compared with those from households with three or more. To select a resident randomly, once a household has been called, the telephone interviewers ask to speak to the adult in the household who had the most recent birthday, or would call back later if that individual was not present. The non-missing responses above suggest that our 545 phone calls reached households with at least $(1 \times 154) + (2 \times 306) + (3 \times 85) = 1{,}021$ adults. Among this sample of pseudopeople, those living in one-adult households make up only 154/1,021 or 15%, much less than the 28% in our table.

Survey weights provide a way to adjust for such known sampling biases, and achieve more realistic results. That could matter, in this example, not only for describing household size, but for anything else (such as voting behavior) that might be correlated with household size. Single-adult households probably include a larger proportion of elderly people living alone. Two-adult households will include many young families. Multi-adult households will often be older families with adult children, or else young adults with roommates.

Probability weights are proportional to the inverse of the probability of selection. For our example, the conditional probability of selecting a particular person from a one-adult household (given that we phoned there) equals one. The probability of selecting a person from a two-person household equals 1/2, and from a three-person household 1/3. If we used the inverse of these probabilities, 1, 2 and 3, as weights then our sample would contain 1,021 pseudopeople — giving us the correct proportions, but leading to incorrect confidence intervals and other confusion. To maintain the true sample size, we can multiply these inverse probabilities by the ratio of real people to pseudopeople, 545/1,021. This results in weights of .53 (one-adult household), 1.07 (two adults) or 1.60 (three or more), with missing values given a neutral weight of 1. With a calculator you can confirm that the ratio of these weights remains 1:2:3.

```
. generate adultwt = adults*(545/1021)
. replace adultwt = 1 if adultwt >= .
. tab adults, summ(adultwt) miss
```

```
    Adults in │         Summary of adultwt
   household │    Mean   Std. Dev.        Freq.
────────────┼────────────────────────────────────
           1 │ .53379041          .          154
           2 │ 1.0675808          .          306
          3+ │ 1.6013712          .           85
           . │        1          0           10
────────────┼────────────────────────────────────
       Total │        1   .3441352          555
```

In this first step, we created a new variable named *adultwt*, which contains the probability weights to adjust for one known sampling bias, while preserving the original sample size. Any Stata procedure that allows **pweights** could now apply the probability weights as a correction for this bias. For example, people's opinions about the threat posed by global warming (*warmthr*), discussed later in this chapter, might be regressed on gender, age and education using a command such as

```
. logit warmthr gender age educ [pweight=adultwt]
```

Poststratification Weights

The previous section gave an example of weights based on the sampling design, which was known before data collection began. A second type of weights might be defined after we start to analyze our sample, and see that despite our best efforts, it appears unrepresentative in some respect. For instance, the sample might have a gender or age distribution noticeably different from those of our target population, making further analysis suspect. Poststratification refers to probability weights calculated so that the proportions of particular groups or strata in our sample more closely resemble the population.

The Granite State Poll sample, for example, is 59.82% female.

```
. tab gender
```

```
      Sex of │
  respondent │     Freq.       Percent        Cum.
────────────┼────────────────────────────────────
        Male │       223         40.18        40.18
      Female │       332         59.82       100.00
────────────┼────────────────────────────────────
       Total │       555        100.00
```

But according to the Census Bureau's 2006 estimates, the adult population of New Hampshire is only 51.6% female. If we guessed from the survey that the state's population was something close to 59.82% female, we would be far off the mark. Moreover, we could easily draw mistaken conclusions about other things correlated with gender, such as voting. This apparent response bias undermines our ability to draw inferences about a larger population.

There are many ways to approach poststratification. (For an alternative to the by-hand approach shown below, Stata's **svyset** command offers a **poststrata** option, illustrated in the *Survey Reference Manual*; type **help svyset** for the basic syntax.) If we know the true population percentages of key variables, as we do regarding gender, then weights to adjust for response bias can be calculated from population percentage divided by sample percentage. *Gender* is coded 0 for males, who make up 48.4% of the adult population of New Hampshire (according to Census estimates), but only 40.18% of this sample. It is coded 1 for females, who are 51.6% of the population and 59.82% of the sample. There are no missing values of *gender* in these data. Consequently, we calculate weights slightly below one for females (*genwt* = .86) and above one for males (*genwt* = 1.20).

```
. generate genwt = 48.4/40.18 if gender==0
. replace genwt = 51.6/59.82 if gender==1
. tab gender, summ(genwt)
```

Sex of respondent	Summary of genwt Mean	Std. Dev.	Freq.
Male	1.2045794	0	223
Female	.86258775	.	332
Total	1.0000006	.16781684	555

More elaborate poststratification weights could follow a similar approach. For example, suppose that for a different study (not the Granite State Poll) we wish to approximate a population age-race-sex distribution.

1. Obtain a table of age-race-sex percentages from Census or other data on the population of interest, such as adults living in a particular state. If we employ five groups for age (18–29, 30–39, etc.) and two for race (white, nonwhite), this results in 20 numbers such as the percentage of that state's adult population consisting of white males 18–29, the percentage of white females 18–29, and so forth.

2. Obtain a similar table of age-race-sex percentages from the sample, for example by creating and tabulating a new variable named *ARS* denoting age-race-sex combinations:

```
. egen ARS = group(agegroup race sex), lname(ars)
. tab ARS
```

3. Define a new set of weights, using **generate ... if** commands. For example, suppose we know that 8.6% of the adult population in a state consists of white males age 18–29, and 8.2% are white females in this age group. In our unweighted sample, however, we see only 2.6% white males 18–29, and 5.1% white females — so the young adults, particularly young males, are under-represented. We could create a new age-race-sex weight variable named *ARSwt* equal to 1 (a neutral weight) if we do not know a respondent's age-race-sex combination, and otherwise equal to the population percentage divided by corresponding sample percentage for their age-race-sex group. The first few commands could be

```
. generate ARSwt = 1 if ARS >= .
. label variable ARSwt "Age-race-sex weights"
. replace ARSwt = 8.6/2.6 if ARS == 1
. replace ARSwt = 8.2/5.1 if ARS == 2
```

Poststratification adjustments work best in connection with carefully-designed surveys, and should not be misunderstood as a cure for haphazard sampling. Such adjustments have been applied most extensively in areas such as voter opinion polls and social science surveys, where great effort goes into securing the most representative samples to begin with. These also are areas where independent evidence such as vote outcomes or replications by other researchers provides clear feedback on how well the adjustments succeed.

A single dataset might include weight variables calculated from more than one source, such as design weights and poststratification weights. To combine these into one overall weight variable, we multiply and then make an adjustment so that the final sum of weights equals the sample size.

```
. generate finalwt = adultwt*ARSwt
. replace finalwt = 1 if finalwt >= .
. quietly summ finalwt
. replace finalwt = finalwt*(r(N)/r(sum))
```

Any number of weight variables can exist in the same dataset. They affect analyses of other variables only when we apply **pweights** or special survey methods, as described below.

Declare Survey Data

Although **pweight** options permit probability weighting for many analyses, a more general approach (which also supports complex sampling designs) involves declaring the survey structure of the dataset through a **svyset** command. In the Granite State Poll example, we earlier defined *adultwt* to adjust for the number of adults in a household. We now use **svyset** to declare this as a survey-type dataset, with probability weights given by *adultwt*.

```
. use Granite_08_01s.dta, clear
(Granite State Poll, January 2008)

. svyset _n [pweight = adultwt]

      pweight: adultwt
          VCE: linearized
  Single unit: missing
     Strata 1: <one>
         SU 1: <observations>
        FPC 1: <zero>
```

The dataset appears basically unchanged, and analytical results will be no different — except those specifically designed for survey data, which are prefixed by **svy:** For example, the unweighted distribution of household size is

```
. tab adults
```

Adults in household	Freq.	Percent	Cum.
1	154	28.26	28.26
2	306	56.15	84.40
3+	85	15.60	100.00
Total	545	100.00	

To see the survey-weighted distribution of household size, type

```
. svy: tab adults, percent
(running tabulate on estimation sample)
```

```
Number of strata   =          1        Number of obs      =        545
Number of PSUs     =        545        Population size    =        545
                                       Design df          =        544
```

```
Adults in
household  │  percentages
───────────┼──────────────
        1  │      15.08
        2  │      59.94
       3+  │      24.98
───────────┼──────────────
    Total  │        100
───────────┴──────────────
```

  Key: percentages = **cell percentages**

The weighted results are quite different, as they should be. The **svy: tab** results show 15.08% in one-adult households — matching our initial calculation of 545/1,021 = .1508. Note also that the survey table shows the correct number of nonmissing observations for this variable. It is always good to check the consequences of our weighting commands by comparing weighted with unweighted tables, before going further.

Researchers using the Granite State Poll routinely calculate design weights based not only on the number of adults in a household, but also on the number of telephone lines. Households with more than one phone line have a higher chance of being called, and so deserve proportionately lower weights. Poststratification weights are calculated too, based on sex and geographical region of New Hampshire (although not on age and race, as in our hypothetical example above). Finally, these four weights are multiplied to form an overall weight variable named *censuswt*. For the remainder of this chapter, we declare *censuswt* as the appropriate probability weights in these Granite State Poll data. The **svydescribe** command describes how survey data are currently defined.

```
. svyset _n [pweight = censuswt]

      pweight: censuswt
          VCE: linearized
  Single unit: missing
     Strata 1: <one>
        SU 1: <observations>
       FPC 1: <zero>
```

```
. svydescribe

Survey: Describing stage 1 sampling units

        pweight: censuswt
            VCE: linearized
    Single unit: missing
       Strata 1: <one>
           SU 1: <observations>
          FPC 1: <zero>
```

			#Obs per Unit		
Stratum	#Units	#Obs	min	mean	max
1	555	555	1	1.0	1
1	555	555	1	1.0	1

A **svyset** command can declare much more information than just the probability weights used in our examples. **svyset** options allow for complex designs including stratified and multistage cluster sampling, finite population corrections, alternative methods of variance estimation, and a different approach to poststratification. Type **help svyset** to see the syntax and a complete list of options. The *Survey Data Reference Manual* gives more examples and technical details.

Survey-Weighted Tables and Graphs

In the heart of a cold winter, the Granite State Poll of January 2008 asked several questions about global warming. The first of these inquired,

> *Next, thinking about the issue of global warming, sometimes called the "greenhouse effect," how well do you feel you understand this issue — would you say very well, fairly well, not very well, or not at all?*

The weighted responses are shown on the following page. This **svy: tab** command applies weights according to the specification declared earlier by **svyset**. A **miss** option includes missing values in the table. Stata permits value labels for extended missing values (.a, .b etc.). For variable *warmund*, missing value .a denotes respondents who answered "DK/not sure." **ci** requests confidence intervals for the weighted percentages, shown as lower and upper bounds (lb and ub). Based on this sample, we are 95% confident that between 28.64% and 37.37% of New Hampshire adults believe they understand global warming very well.

```
. svy: tab warmund, percent miss ci
(running tabulate on estimation sample)
```

```
Number of strata   =        1          Number of obs    =        555
Number of PSUs     =      555          Population size  =   554.9165
                                       Design df        =        554
```

Understand global warming/greenhouse	percentages	lb	ub
Not at a	2.012	1.115	3.607
Not very	13.69	10.83	17.15
Fairly w	50.51	45.94	55.06
Very wel	32.85	28.64	37.37
DK/not s	.9411	.3566	2.46
Total	100		

```
Key:  percentages  =  cell percentages
      lb           =  lower 95% confidence bounds for cell percentages
      ub           =  upper 95% confidence bounds for cell percentages
```

Stata's native graph types are not ideal for viewing weighted, categorical-variable percentage distributions such as those in the table above. Fortunately, a user-written program named **catplot** does this job quite well. You can obtain the ado-files for **catplot** easily from the Web by typing

```
. findit catplot
```

and following the links to install them on your computer. (The **findit** command works for hundreds of other user-written programs as well.) Once it is installed, typing **help catplot** will show the command's syntax and options. Figure 14.1 contains a **catplot** bar graph of *warmund*, corresponding to our table above. Although **catplot** does not accept pweights, using aweights with this command will have the same visual effect. The **percent** option specifies that bar heights show percentages; **missing** asks that missing values be shown also.

```
. catplot bar warmund [aweight = censuswt], percent missing
```

Figure 14.1

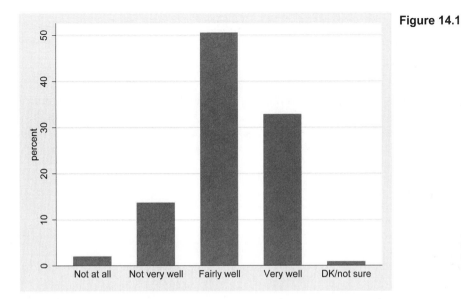

Bar graphs with value labels often are easier to read in a horizontal-bar format, particularly when we have many bars. Figure 14.2 shows a horizontal version including a title and better axis label, suitable for a report or presentation on the survey results. We also label the bar heights, so that weighted percentages can be read directly from the graph.

```
. catplot hbar warmund [aweight = censuswt],
      blabel(bar, format(%3.0f)) percent missing
      ytitle("Weighted percent")
      title("How well do you understand global warming?")
```

Figure 14.2

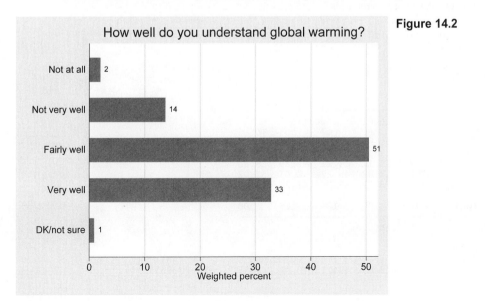

Figures 14.2 emphasizes the fact that a large majority of respondents claim to understand global warming fairly or very well. Only a few admit that they don't know, or do not understand it at all.

How does self-reported understanding relate to other variables in the survey, such as respondent background characteristics or their opinions about global warming itself? Before going further, it seems reasonable to create a new version of *warmund*, in which the few "don't know/not sure" responses are combined with the slightly larger group who said they understand this topic "not at all." The following commands leave the original variable *warmund* unchanged, but create an alternative version named *warmund2*.

```
. gen warmund2 = warmund
. replace warmund2 = 1 if warmund >= .
. label values warmund2 warmund
. tab warmund warmund2, miss
```

Understand global warming/green house	warmund2 Not at al	Not very	Fairly we	Very well	Total
Not at all	12	0	0	0	12
Not very well	0	75	0	0	75
Fairly well	0	0	288	0	288
Very well	0	0	0	175	175
DK/not sure	5	0	0	0	5
Total	17	75	288	175	555

Not surprisingly, more educated respondents tend to be more confident about their understanding of global warming. We can see this in a weighted cross-tabulation. The **column percent** option requests column percentages, because the column variable, *educ*, is our independent variable in this table.

```
. svy: tab warmund2 educ, column percent
(running tabulate on estimation sample)
```

Number of strata = 1	Number of obs	= 549
Number of PSUs = 549	Population size	= 547.08399
	Design df	= 548

warmund2	hig	Level of education tec	col	pos	Total
Not at a	4.229	.9411	2.533	.7452	2.217
Not very	23	13.11	13.21	4.829	13.88
Fairly w	47.21	57.42	48.27	52.67	51.02
Very wel	25.56	28.53	35.99	41.75	32.88
Total	100	100	100	100	100

Key: column percentages

Pearson:
    Uncorrected   chi2(9)   =   27.0999
    Design-based F(8.85, 4847.37)=   2.5953   P = 0.0058

A Pearson χ^2 test, corrected for the survey design and converted to an *F* statistic with noninteger degrees of freedom, indicates a statistically significant relationship (*P* = .0058) between understanding and level of education. The weighted percentage saying they understand global warming very well rises with education, from 26% among those with high school or less, up to 42% among those with postgraduate education. Few postgraduates say they understand this topic not at all, or not very well.

Figure 14.3 visualizes the results in a set of horizontal bar graphs. Weighted percentages within values of *educ* correspond to the column percentages in our cross-tabulation.

```
. catplot hbar warmund2 [aweight = censuswt],
    percent(educ) blabel(bar, format(%3.0f))
    ytitle("Weighted percent") by(educ, note("")
    title("How well do you understand global warming?"))
```

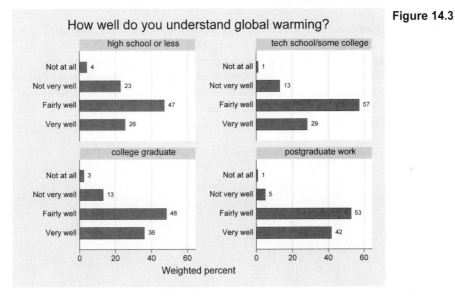

Figure 14.3

So how does understanding, as measured above, relate to people's opinions about the threat posed by global warming? The poll asked,

> *Do you think that global warming will pose a serious threat to you or your way of life in your lifetime, or not?*

Weighted responses on variable *warmthr* are shown on the next page. About 54% said warming would not pose a threat in their lifetime. Forty-one percent said yes it would, and 6% were unsure.

```
. svy: tab warmthr, percent miss
(running tabulate on estimation sample)
```

Number of strata	=	1		Number of obs	=	555
Number of PSUs	=	555		Population size	=	554.9165
				Design df	=	554

Global warming threat in your lifetime	percentages
No	53.66
Yes	40.77
DK/not s	5.57
Total	100

Key: percentages = **cell percentages**

Perhaps surprisingly, opinions about global warming are not much related to how well people say they understand the issue. A weighted cross-tabulation finds no significant relationship (P = .6871) between *warmund2* and *warmthr* (apart from those who don't know about either). In the table below we ask for **row** percentages, because the row variable (*warmund2*) now forms our independent variable. Majorities at each level of understanding believe that warming will not pose a threat, and these majorities are similar among those who say they understand warming not very well (59.92%), fairly well (55.34%), or very well (56.83%).

```
. svy: tab warmund2 warmthr, row percent
(running tabulate on estimation sample)
```

Number of strata	=	1		Number of obs	=	525
Number of PSUs	=	525		Population size	=	524.00984
				Design df	=	524

	Global warming threat in your lifetime		
warmund2	No	Yes	Total
Not at a	73.81	26.19	100
Not very	59.92	40.08	100
Fairly w	55.34	44.66	100
Very wel	56.84	43.16	100
Total	56.83	43.17	100

Key: **row percentages**

```
Pearson:
   Uncorrected   chi2(3)         =    1.7389
   Design-based  F(2.99, 1565.50)=    0.4919      P = 0.6871
```

Among scientists who study climate change, and understand its issues better than anyone, the possible consequences are a main focus of research. Among the general public, on the other hand, opinions about the consequences tend to resemble political views. The next step in understanding these Granite State Poll responses is to bring politics and background factors into the picture, through the use of regression-type models.

Regression-Type Modeling

Survey versions of many regression-type modeling methods exist for use with **svyset** data. These include linear, censored-normal, tobit, Poisson, and negative binomial regression; generalized linear modeling (GLM); Cox and parametric survival analysis models; instrumental-variable or endogenous-regressor models; and others. Type **help svy_estimation** to see a complete list. In general, survey estimation commands have the same syntax as their non-survey counterparts, but are prefixed by **svy:** For example, to regress *y* on three predictors, applying weights and design elements as specified when we **svyset** our data, simply type a command such as **svy: regress y x1 x2 x3**.

Regression methods for categorical dependent variables hold particular interest in the analysis of survey data like the Granite State Poll, which contains mostly categorical variables. Such methods include logit and probit models for dichotomous, polytomous, or ordered-category *y* variables. For example, to regress *warmund2* (understand global warming), which has four ordered categories, on gender, age, education, and political party identification (a 7-point scale from strong Democrat (1) to strong Republican (7)), we might fit an ordered logit model using the **svy: ologit** command.

```
. svy: ologit warmund2 gender age educ party
(running ologit on estimation sample)

Survey: Ordered logistic regression
```

Number of strata	=	1		Number of obs	=		525
Number of PSUs	=	525		Population size	=		524.91573
				Design df	=		524
				F(4, 521)	=		6.97
				Prob > F	=		0.0000

warmund2	Coef.	Linearized Std. Err.	t	P>|t|	[95% Conf. Interval]	
gender	-.7199688	.1904727	-3.78	0.000	-1.094153	-.3457849
age	.0070643	.0063311	1.12	0.265	-.0053732	.0195018
educ	.2982768	.0914617	3.26	0.001	.1186002	.4779535
party	.0578297	.0463835	1.25	0.213	-.0332908	.1489503
/cut1	-2.928684	.536781	-5.46	0.000	-3.983191	-1.874177
/cut2	-.8618262	.50222	-1.72	0.087	-1.848438	.1247858
/cut3	1.703624	.510058	3.34	0.001	.7016139	2.705633

Gender and education have significant effects. Gender is coded 0 for males and 1 for females, so the negative coefficient above indicates that females were less confident about their understanding of global warming. The positive coefficient on education is consistent with our earlier tabular analysis: self-reported understanding increases with the level of respondent's education. Political party identification, on a scale from 1 for strong Democrat to 7 for strong Republican, exhibits no significant effect on understanding of global warming.

The poll question asking whether global warming would pose a threat in respondents' lifetimes permitted yes or no answers, so a binary logit model is appropriate. In specifying this model we might consider not only how the perceived threat (*warmthr*) is related to gender, age, education, and party, as done for perceived understanding (*warmund2*) above, but also how the perceived threat is related to perceived understanding.

Analysts studying survey opinions about global warming often find that the effects of education or knowledge are modified by ideology. For example, a Pew Research Center study reported that "Among Republicans, higher education is linked to greater skepticism about global warming .... But among Democrats, the pattern is the reverse" (Pew 2007:2). More generally, ideology filters knowledge about global warming, so that people tend to accept new information that conforms to their pre-existing beliefs (Wood and Vedlitz 2007; Hamilton 2008). These and other studies suggest that education or knowledge, together with political orientation, might have interacting effects on perceived threats from global warming. To test this hypothesis we define two interaction terms: political party×education, and political party×understanding. Both interaction terms are defined below using new, centered versions of *party*, *educ*, and *warmund2* — calculated by subtracting the mean from each variable, so the resulting variables *party0*, *educ0*, and *warmund0* all have means of approximately zero, and are measured in deviations from those means. Centering can reduce problems of multicollinearity in models containing interaction effects, and also makes the main effects of the component variables easier to interpret.

```
. summ party educ warmund2

    Variable |       Obs        Mean    Std. Dev.       Min        Max
-------------+--------------------------------------------------------
       party |       541    3.689464    2.043884         1          7
        educ |       549    2.500911      1.0492         1          4
    warmund2 |       555    3.118919    .7482246         1          4

. generate party0 = party - 3.7
. label variable party0 "Political party (centered)"
. generate educ0 = educ - 2.5
. label variable educ0 "Education (centered)"
. generate warmund0 = warmund2 - 3.1
. label variable warmund0 "Understand global warming (centered)"
. generate partyed = party0 * educ0
. label variable partyed "Political party * education"
. generate partyund = party0 * warmund0
. label variable partyund "Political party * understanding"
```

The interaction terms *partyed* and *partyund* can be included, along with the centered versions of their component variables and other background factors, as predictors in a binary logit model for *warmthr*.

```
. svy: logit warmthr gender age educ0 party0 warmund0 partyed
        partyund
(running logit on estimation sample)

Survey: Logistic regression

Number of strata   =        1            Number of obs    =        499
Number of PSUs     =      499            Population size  = 498.64706
                                         Design df        =        498
                                         F(  7,    492)   =       8.57
                                         Prob > F         =     0.0000
```

warmthr	Coef.	Linearized Std. Err.	t	P>|t|	[95% Conf. Interval]	
gender	1.034997	.2302258	4.50	0.000	.5826629	1.48733
age	-.029915	.0074932	-3.99	0.000	-.0446372	-.0151927
educ0	.1617984	.1122198	1.44	0.150	-.0586842	.382281
party0	-.2829968	.0581085	-4.87	0.000	-.3971648	-.1688289
warmund0	.1608663	.1550788	1.04	0.300	-.1438231	.4655558
partyed	-.0440177	.0550783	-0.80	0.425	-.1522322	.0641968
partyund	-.1866498	.0807215	-2.31	0.021	-.3452464	-.0280532
_cons	.7015043	.4165917	1.68	0.093	-.1169897	1.519998

Predictors of the perceived threat from global warming turn out to be quite different from the predictors of understanding. Gender has a positive effect, indicating that women more often thought that global warming posed a threat in their lifetime. Older respondents were less likely to see a threat, perhaps because their personal horizons extend less far into the 21st century. Political party identification exerts a strong negative effect: respondents who identified themselves as Republican were less likely than Democrats to think that global warming posed a threat. Consistent with our earlier tabular analysis, the logit model does not find a simple relationship between understanding and perceived threat. The interaction term *partyund*, or political party×understanding, however, does have a significant effect. Moreover, this effect is negative, as expected. The political party×education interaction effect is not statistically significant, but it too is negative.

The positive, nonsignificant coefficient on understanding global warming (*warmund0*) is .161. Because *educ*, *party*, and *warmund2* all were centered before generating the interaction terms, we can interpret this main effect as the effect of a one-unit increase in understanding global warming (*warmund2* or *warmund0*), for respondents of average political party identification (that is, for respondents with *party* = 3.7, hence *party0* ≈ 0, so *partyund* = *party0*×*warmund0* would also equal 0). Similarly, the main effect or coefficient on *educ0* can be interpreted as the effect of a one-unit increase in education, for respondents of average political party identification.

The negative, significant interaction effect (coefficient on *partyund*) is −.187. Thus, for respondents who identify themselves as strong Democrats (*party* = 1, or as a mean deviation 1−3.7 = −2.7), the log odds that global warming is perceived as a threat increase by .667 with each one-unit increase in their self-reported understanding:

.162 + (−2.7×−.187) = .667

On the other hand, for respondents who identify themselves as strong Republicans (*party* = 7, or as a mean deviation 7−3.7 = 3.3), the log odds that warming is perceived as a threat decrease by .455 with each additional unit of self-reported understanding:

$.162 + (3.3\times-.187) = -.455$

In other words, Democrats who believe they understand global warming better also are more likely to believe that it poses a threat in their lifetimes. Conversely, Republicans who believe they understand global warming better are *less* likely to believe that it poses a threat. Ideology thus plays a role in filtering what people believe about science. In general terms, this Granite State Poll analysis replicates conclusions reached from surveys using a variety of different global-warming related questions, samples, and analytical methods.

Conditional Effect Plots for Interactions

One task that often faces survey researchers is reporting their findings to a wider audience, who might not want to hear about logit coefficients or odds ratios. Conditional effect plots, introduced in Chapters 8 and 10, help to visualize model results such as interaction effects. To graph the party×understanding effect found above, we could calculate predicted probabilities as a function of understanding (*warmund2*), separately for strong Democrats (*party* = 1) and strong Republicans (*party* = 7), while holding other predictors in the model constant at their means or other selected values. Chapter 10 showed how to draw such plots using the regression equations. In this section we take a different and sometimes easier approach: create some fake observations, with the desired predictor-variable values and missing dependent-variable values. These fake observations will not affect the regression estimation, but after estimation, we can use **predict** to calculate predicted values, and graph them.

The original survey dataset contains n = 555 observations. The understand-global-warming question (*warmund2*) has four possible answers, coded 1 through 4. In order to graph the party×understanding interaction, we use either the Data Editor or explicit commands (as shown below) to create 8 new fake observations. These new observations, numbers 556 through 563, are denoted by the indicator variable *fake* = 1. The fake observations are all assigned average values for *gender* (.6), *age* (54), and *educ0* (0).

```
. display _N
555
. set obs 563
obs was 555, now 563
. gen fake = 1 in 556/563
(555 missing values generated)
. replace gender = .6 in 556/563
gender was byte now float
(8 real changes made)
. replace age = 54 in 556/563
(8 real changes made)
. replace educ0 = 0 in 556/563
(8 real changes made)
```

We make the first four fake observations (numbers 556 through 559) "strong Democrats," and the last four (numbers 560 through 563) "strong Republicans."

```
. replace party = 1 in 556/559
(4 real changes made)
. replace party = 7 in 560/563
(4 real changes made)
```

Finally, to complete our eight fake observations, each one of the four Democrats gets one of the four possible values of *warmund2*, and so does each of the four Republicans. We calculate centered values *party0* and *warmund0*, then the interaction terms *partyed* and *partyund*, from these.

```
. replace warmund2 = 1 in 556
(1 real change made)
. replace warmund2 = 2 in 557
(1 real change made)
. replace warmund2 = 3 in 558
(1 real change made)
. replace warmund2 = 4 in 559
(1 real change made)
. replace warmund2 = 1 in 560
(1 real change made)
. replace warmund2 = 2 in 561
(1 real change made)
. replace warmund2 = 3 in 562
(1 real change made)
. replace warmund2 = 4 in 563
(1 real change made)
. replace party0 = party - 3.7 in 556/563
(8 real changes made)
. replace warmund0 = warmund2 - 3.1 in 556/563
(8 real changes made)
. replace partyed = party0*educ0 in 556/563
(8 real changes made)
. replace partyund = party0*warmund0 in 556/563
(8 real changes made)
```

Now, after repeating the earlier logit regression analysis, we can follow up with **predict** to calculate predicted probabilities (here named *phat*) that *warmthr* = 1 for the fake observations as well as the real ones. The results can be checked in a Data Editor window.

```
. quietly svy: logit warmthr gender age educ0 party0 warmund0
       partyed partyund
. predict phat
(option pr assumed; Pr(warmthr))
(30 missing values generated)
. edit warmthr gender age educ0 party0 warmund0
       partyed partyund phat if fake == 1
```

	warmthr	gender	age	educ0	party0	warmund0	partyed	partyund	phat	
556	.	.6	54	0	-2.7	-2.1	0	5.67	.2839527	
557	.	.6	54	0	-2.7	-1.1	0	2.97	.4353347	
558	.	.6	54	0	-2.7	-.1	0	.27	.599817	
559	.	.6	54	0	-2.7	.9	0	-2.43	.7445063	
560	.	.6	54	0	3.3	-2.1	0	-6.929999	.432629	
561	.	.6	54	0	3.3	-1.1	0	-3.63	.3260269	
562	.	.6	54	0	3.3	-.1	0	-.33	.2348206	
563	.	.6	54	0	3.3	.9	0	2.97	.1629601	

Figure 14.4 graphs these fake-observation probabilities against understanding, with axis titles and legend suitable for a research report.

```
. graph twoway connect phat warmund2 if party==1 & fake==1

    || connect phat warmund2 if party==7 & fake==1,
    lpattern(dash) msymbol(Th)

    || , legend(ring(0) position(10) rows(2)
    label(1 "Strong Democrat") label(2 "Strong Republican"))
    ytitle("Probability of seeing warming as a threat")
    xlabel(1 "Not at all" 2 "Not very well" 3 "Fairly well"
    4 "Very well") xscale(range(1 4.2))
    xtitle("How well do you understand global warming?")
```

Figure 14.4

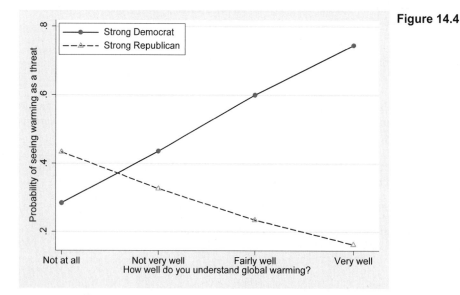

Figure 14.4 helps to visualize why our earlier cross-tabulation found no simple relationship between perceived understanding and threat. A relationship does exist, but it runs in opposite

directions depending on respondents' political orientation. In a simple cross-tabulation, these contrary patterns cancel each other out.

Other Tools for Survey Analysis

If we wish to see odds ratios instead of logit coefficients, the option **or** works just as it would with a non-survey **logit** command. For example,

```
. svy: logit warmthr gender age educ0 party0 warmund0
     partyed partyund, or
```

Jackknife and bootstrap standard errors are common options available with **logit** and other estimation commands, but they must be performed differently in the case of complex survey designs. For example, to obtain valid jackknife estimates of standard errors for the model shown above, use a **svy jackknife** command.

```
. svy jackknife: logit warmthr gender age educ0 party0 warmund0
     partyed partyund
```

(running logit on estimation sample)

```
Jackknife replications (499)
———+——— 1 ———+——— 2 ———+——— 3 ———+——— 4 ———+——— 5
..................................................   50
..................................................  100
..................................................  150
..................................................  200
..................................................  250
..................................................  300
..................................................  350
..................................................  400
..................................................  450
.................................................
```

Survey: Logistic regression

| Number of strata | = | 1 |
| Number of PSUs | = | 499 |

Number of obs	=	499
Population size	=	498.64706
Replications	=	499
Design df	=	498
F(7, 492)	=	7.93
Prob > F	=	0.0000

warmthr	Coef.	Jackknife Std. Err.	t	P>|t|	[95% Conf. Interval]	
gender	1.034997	.2362575	4.38	0.000	.5708121	1.499181
age	-.029915	.0077198	-3.88	0.000	-.0450824	-.0147476
educ0	.1617984	.1157079	1.40	0.163	-.0655375	.3891343
party0	-.2829968	.0604028	-4.69	0.000	-.4016725	-.1643211
warmund0	.1608663	.1612597	1.00	0.319	-.155967	.4776996
partyed	-.0440177	.0574733	-0.77	0.444	-.1569378	.0689024
partyund	-.1866498	.0854706	-2.18	0.029	-.3545772	-.0187224
_cons	.7015043	.4293683	1.63	0.103	-.1420923	1.545101

Jackknife estimation involves resampling the data *n* times, each time leaving out one observation (or in survey data, one primary sampling unit). This procedure yields alternative design-based standard errors for hypothesis tests and confidence intervals, with less reliance on theoretical assumptions. In the example above, the $n = 499$ jackknife replications lead to slightly different standard errors and *t* statistics than our initial analysis, but support the initial conclusions that the coefficient on *partyed* is not significantly different from zero, while the coefficient on *partyund* is.

Jackknife estimates for the standard error of a particular odds ratio, which equal the exponential (*e* to power) of the corresponding logit coefficient, could be obtained by commands such as:

```
. svy jackknife partyund = exp(_b[partyund]): logit warmthr gender age
    educ0 party0 warmund0 partyed partyund
```

See **help svy_jackknife** or the *Survey Reference Manual* for more on this general-purpose tool.

Postestimation statistics for survey data are obtained via **estat** commands, or equivalently through menu selections from Statistics > Survey data analysis > DEFF, MEFF, and other statistics . For example, **estat svyset** reports the survey design characteristics associated with the recent estimation (in this case, our logit model with jackknife variance estimation);

```
. estat svyset
      pweight: censuswt
          VCE: jackknife
          MSE: off
  Single unit: missing
```

Other **estat** commands can display design and misspecification effects (DEFF and MEFF) for each estimated parameter, or for linear combinations of parameters; tables of singleton and certainty strata; and the covariance or correlation matrix of parameter estimates. See **help svy_postestimation** for a full list of commands and options. The *Survey Data Reference Manual* supplies technical details, examples, and references regarding **estat** and Stata's other survey analysis capabilities.

15

Multilevel and Mixed-Effects Modeling

Mixed-effects modeling is basically regression analysis allowing two kinds of effects: *fixed effects*, meaning intercepts and slopes meant to describe the population as a whole, just as in ordinary regression; and also *random effects*, meaning intercepts and slopes that can vary across subgroups of the sample. All of the regression-type methods shown so far in this book involve fixed effects only. Mixed-effects modeling opens a new range of possibilities for multilevel models, growth curve analysis, and panel data or cross-sectional time series.

For example, a simple multilevel analysis might use data on grade point averages and hours of studying reported by college students in twenty different majors, and test hypotheses not only that grades improved with studying overall (a fixed effect), but also that the means and rates of such improvement differ by major (random effects). A more complex analysis might consider similar data but from multiple colleges, and allow also for differences between majors nested within colleges (two levels of random effects, so these models are sometimes termed "two-level"). This chapter presents examples to illustrate some basic ideas.

Three Stata commands provide the most general tools for multilevel and mixed-effects modeling. **xtmixed** fits linear models, like a mixed-effects counterpart to **regress**. Similarly, **xtmelogit** fits mixed-effects logit regression models for binary outcomes, like a generalization of **logit** or **logistic**; and **xtmepoisson** fits mixed-effects Poisson models for count outcomes, like a generalization of **poisson**. Both **xtmelogit** and **xtmepoisson** are new with Stata version 10. Stata also offers a number of more specialized procedures for conceptually related tasks. Examples include tobit, probit, and negative binomial models with random intercepts; type **help xt** to see a complete list with links to details about each command. Many of these commands were first developed for use with panel or cross-sectional time series data, hence their common **xt** designation.

The **xtmixed**, **xtmelogit**, and **xtmepoisson** procedures can be called either through typed commands or through menus,

Statistics > Multilevel mixed-effects models

Menus for other **xt** procedures are grouped separately under

Statistics > Longitudinal/panel data

The *Longitudinal/Panel Data Reference Manual* contains examples, technical details and references for mixed-effects and other **xt** methods. Luke (2004) provides a compact introduction to multilevel modeling. More extended treatments include books by Bickel (2007), McCulloch and Searle (2001), Raudenbush and Bryk (2002), and Verbeke and Molenberghs (2000). One particularly valuable resource for Stata users, *Multilevel and Longitudinal Modeling Using Stata* (Rabe-Hesketh and Skrondal 2008), describes not only the official Stata **xt** methods but also an unofficial program named **gllamm** (generalized linear latent and mixed models) that adds mixed-effect generalized linear modeling capabilities to Stata. Within Stata, type **findit gllamm** for information on how to obtain and install the ado-files for this program.

Example Commands

. **xtmixed** *crime year* || *city: year*

Performs mixed-effects regression of *crime* on *year*, with random intercept and slope for each value of *city*. Thus, we obtain trends in crime rates, which are a combination of the overall trend (fixed effects), and variations on that trend (random effects) for each city.

. **xtmixed** *SAT parentcoll prepcourse* || *city:* || *school: grades*

Fits a hierarchical or multilevel mixed-effects model predicting students's SAT scores as a function of (1) fixed or whole-sample effects of whether the individual students' parent(s) graduated from college, and whether the student took a preparation course; (2) random intercepts representing the effect of the city in which they attend school; and (3) a random intercept and slope for the effect of individual students' grades, which could be different from one school to the next. Individual students (observations) are nested within schools, which are nested within cities. Note the order of mixed-effects parts in the command.

. **xtmixed** *y x1 x2 x3* || *state: x4*
. **estimates store** *A*
. **xtmixed** *y x1 x2 x3* || *state:*
. **estimates store** *B*
. **lrtest** *A B*

Conducts likelihood-ratio χ^2 test of null hypothesis of no difference in fit between the more complex model *A* (model names arbitrary), which includes a random slope on *x4*, and the simpler model *B* ("*B* nested within *A*") that does not include a random slope on *x4*. This amounts to a test of whether the random slope on *x4* is statistically significant. The order of the two models specified in the **lrtest** command does not matter; if we had typed instead **lrtest** *B A*, Stata would still have correctly inferred that *B* is nested within *A*.

. **xtmixed** *y x1 x2 x3* || *state: x4 x5,* **ml nocons cov(unstructured)**

Performs mixed-effects regression of *y* on fixed-effects predictors *x1*, *x2* and *x3*; also on random effects of *x4* and *x5* for each value of *state*. Obtains estimates by maximum likelihood. The model should have no random intercept, and an unstructured covariance matrix in which random-effect variances and covariances all are estimated distinctly.

. **estat recov**

After **xtmixed**, displays the estimated variance-covariance matrix of the random effects.

. **predict** *re**, **reffects**

After **xtmixed** estimation, obtains best linear unbiased predictions (BLUPs) of all random effects in the model. The random effects are stored as variables named *re1*, *re2* and so forth, with appropriate variable labels.

. **predict** *yhat*, **fitted**

After **xtmixed** estimation, obtains predicted values of *y*. To obtain predictions from the fixed-effects portion of the model only, type **predict yhat, xb**. Other **predict** options find standard errors of the fixed portion (**stdp**), and residuals (**resid**) or standardized residuals (**rstan**). To see a full list of xtmixed postestimation commands, with links to their syntax and options, type **help xtmixed_postestimation**.

. **xtmelogit** *y x1 x2* **||** *state***:**

Performs mixed-effects logit regression of {0, 1} variable *y* on *x1* and *x2*, with random intercepts for each level of *state*.

. **predict** *phat*

After **xtmelogit** estimation, obtains predicted probabilities from the complete (fixed plus random) model. Type **help xtmelogit postestimation** to see other postestimation commands, as well as a complete list of options for **predict**, including Pearson residuals (**pearson**) and deviance residuals (**deviance**).

. **xtmepoisson** *accidents x1 x2 x3*, **exposure(***persondays***)** **||** *season***:**
    || *port***: ,** **irr**

Estimates mixed-effects Poisson model for *accidents*, a count of accidents on fishing vessels. Fixed-effect predictors, characteristics of individual vessels, are *x1*, *x2* and *x3*. Exposure is measured by the number of person-days at sea for that vessel. We include random intercepts for each *season* or year, and for *port* city nested within seasons. Report fixed-effect coefficients as incident rate ratios (**irr**).

. **gllamm** *warming sex race educ age class*,
    i(*region***) family(binomial) link(ologit) adapt**

Performs generalized linear latent and mixed modeling — in this example a mixed-effects ordered logit regression of ordinal variable *warming*, opinion about global warming, on fixed-effect predictors *sex*, *race*, *educ*, *age*, and *class*. Includes random intercepts for each value of *region*. Estimation by adaptive quadrature. The **family( )** and **link( )** options can specify other models including multinomial logit, probit, ordered probit and complementary log-log. **gllamm** is not an official program supplied with Stata, but a useful ado-file available at no cost online. Type **findit gllamm** for information on how to download and install the necessary files. Rabe-Hesketh and Skrondal (2008) provide details and many examples using **gllamm**.

Regression with Random Intercepts

To illustrate the **xtmixed** command, we begin with data on votes in the 2004 presidential election, from 3,054 U.S. counties (*election_2004i.dta*, based on data from Robinson 2005). In this election, Republican George W. Bush (receiving 50.7% of the popular vote) defeated Democrat John Kerry (48.3%) and Independent Ralph Nader (0.4%). One striking feature of this election was its geographical pattern: Kerry won states on the West coast, the Northeast, and around the Great Lakes, while Bush won everywhere else. Within states, Bush's support tended to be stronger in rural areas, whereas Kerry's votes were more concentrated in the cities (see for example Hamilton 2006a). Dataset *election_2004i* contains a categorical variable for census divisions (*cendiv*), which divide the U.S. into 9 geographical areas. It also gives the total number of votes cast (*votes*), the percent for Bush (*bush*), logarithm of population density (*logdens*) as an indicator of "rural-ness," and other variables for the percent of county population belonging to ethnic minorities (*minority*), or adults having college degrees (*colled*).

```
. use election_2004i.dta, clear
(US counties -- 2004 election (Robinson 2005))

. describe

Contains data from C:\data\election_2004i.dta
  obs:          3,054                          US counties -- 2004 election
                                                 (Robinson 2005)
  vars:            11                          14 Apr 2008 12:50
  size:       244,320 (99.5% of memory free)

              storage   display    value
variable name   type    format     label      variable label

fips            long     %9.0g                 FIPS code
state           str20    %20s                  State name
state2          str2     %9s                   State 2-letter abbreviation
region          byte     %9.0g      region     Region (4)
cendiv          byte     %15.0g     division   Census division (9)
county          str24    %24s                  County name
votes           float    %9.0g                 Total # of votes cast, 2004
bush            float    %9.0g                 % votes for GW Bush, 2004
logdens         float    %9.0g                 log10(people per square mile)
minority        float    %9.0g                 % population minority
colled          float    %9.0g                 % adults >25 w/4+ years college

Sorted by:  fips
```

The percent voting for Bush declined with population density, as shown by the scatterplot and regression line in Figure 15.1. Each data point represents one of the 3,054 counties.

```
. graph twoway scatter bush logdens, msymbol(Oh)
     || lfit bush logdens, lpattern(solid) lwidth(medthick)
```

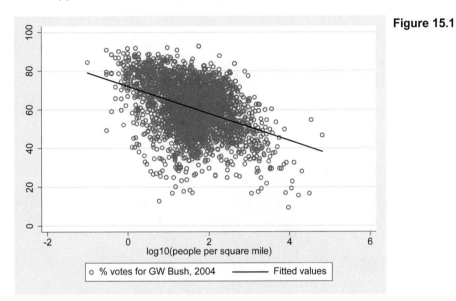

Figure 15.1

An improved version of this voting–density scatterplot appears in Figure 15.2. The logarithmic *x*-axis values have been relabeled (1 becomes "10", 2 becomes "100," and so forth) to make them more reader-friendly. Using *votes* as frequency weights for the scatterplot causes marker symbol size to be proportional to the total number of votes cast, visually distinguishing counties with small or large populations. Otherwise, the analyses in this chapter do not make use of weighting. We focus here on the patterns of county voting, rather than the votes of individuals within those counties.

```
. graph twoway scatter bush logdens [fw=votes], msymbol(Oh)

    || lfit bush logdens, lpattern(solid) lwidth(medthick)

    || , xlabel(-1 "0.1" 0 "1" 1 "10" 2 "100" 3 "1,000"
          4 "10,000", grid) legend(off)
    xtitle("Population per square mile")
    ytitle("Percent vote for GW Bush")
```

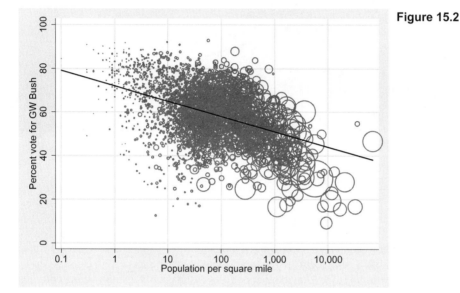

Figure 15.2

As Figure 15.2 confirms, the percent voting for George W. Bush tended to be lower in high-density, urban counties. It tended also to be lower in counties with substantial minority populations, or a greater proportion of adults with college degrees.

```
. regress bush logdens minority colled
```

Source	SS	df	MS
Model	127236.86	3	42412.2868
Residual	359970.423	3050	118.02309
Total	487207.284	3053	159.583126

Number of obs = 3054
F(3, 3050) = 359.36
Prob > F = 0.0000
R-squared = 0.2612
Adj R-squared = 0.2604
Root MSE = 10.864

bush	Coef.	Std. Err.	t	P>|t|	[95% Conf. Interval]
logdens	-5.580708	.2972768	-18.77	0.000	-6.163591 -4.997825
minority	-.2529574	.0124597	-20.30	0.000	-.2773877 -.2285272
colled	-.1677876	.0327761	-5.12	0.000	-.232053 -.1035221
_cons	75.809	.5729391	132.32	0.000	74.68561 76.93238

In mixed-modeling terms, we just estimated a model containing only fixed effects — intercept and coefficients that describe the sample as a whole. The same fixed-effects model can be estimated using **xtmixed**, with similar syntax.

```
. xtmixed bush logdens minority colled
```

```
Mixed-effects REML regression                    Number of obs    =      3054

                                                 Wald chi2(3)     =   1078.07
Log restricted-likelihood = -11623.618           Prob > chi2      =    0.0000
```

bush	Coef.	Std. Err.	z	P>|z|	[95% Conf. Interval]	
logdens	-5.580708	.2972768	-18.77	0.000	-6.16336	-4.998057
minority	-.2529574	.0124597	-20.30	0.000	-.277378	-.2285369
colled	-.1677876	.0327761	-5.12	0.000	-.2320275	-.1035476
_cons	75.809	.5729391	132.32	0.000	74.68606	76.93194

Random-effects Parameters	Estimate	Std. Err.	[95% Conf. Interval]	
sd(Residual)	10.86384	.1390061	10.59478	11.13974

Maximum restricted likelihood (REML) is the default estimation method for **xtmixed**, but could be requested explicitly with the option **reml**. Alternatively, the option **ml** would call for maximum likelihood estimation. See **help xtmixed** for a list of estimation, specification and reporting options.

The geographical pattern of voting seen in red state/blue state maps of this election are not captured by the fixed-effects model above, which assumes that the same intercept and slopes characterize all 3,054 counties. One way to model the tendency towards different voting patterns in different parts of the country (and to reduce the problem of spatially correlated errors) is to allow each of the nine census divisions (New England, Middle Atlantic, Mountain, Pacific, etc.) to have its own random intercept. Instead of the usual (fixed-effects) regression model with a form such as

$$y_i = \beta_0 + \beta_1 x_{1i} + \beta_2 x_{2i} + \beta_3 x_{3i} + \epsilon_i \qquad [15.1]$$

we could include not only a set of β coefficients that describe all the counties, but also a random intercept u_0, which varies from one census division to the next.

$$y_{ij} = \beta_0 + \beta_1 x_{1ij} + \beta_2 x_{2ij} + \beta_3 x_{3ij} + u_{0j} + \epsilon_{ij} \qquad [15.2]$$

Equation [15.2] depicts the value of y for the ith county and the jth census division as a function of x_1, x_2, and x_3 effects that are the same for all divisions. The random intercept u_{0j}, however, allows for the possibility that the mean level of y (e.g., mean percent voting for Bush) is systematically higher or lower among the counties of some divisions. That seems appropriate with regard to U.S. voting, given its obvious geographical patterns. We can estimate a model with random intercepts for each census division by adding a new random-effects part to the **xtmixed** command, as follows:

`. xtmixed `**`bush logdens minority colled`**` || `**`cendiv:`**

```
Performing EM optimization:

Performing gradient-based optimization:

Iteration 0:    log restricted-likelihood = -11392.938
Iteration 1:    log restricted-likelihood = -11392.938

Computing standard errors:
```

```
Mixed-effects REML regression            Number of obs      =        3054
Group variable: cendiv                   Number of groups   =           9

                                         Obs per group: min =          67
                                                        avg =       339.3
                                                        max =         618

                                         Wald chi2(3)       =     1205.68
Log restricted-likelihood = -11392.938   Prob > chi2        =      0.0000
```

bush	Coef.	Std. Err.	z	P>|z|	[95% Conf. Interval]	
logdens	-4.703101	.3506895	-13.41	0.000	-5.390439	-4.015762
minority	-.3658849	.0129338	-28.29	0.000	-.3912347	-.3405351
colled	-.0407444	.0346952	-1.17	0.240	-.1087458	.027257
_cons	72.13473	2.429822	29.69	0.000	67.37237	76.8971

Random-effects Parameters	Estimate	Std. Err.	[95% Conf. Interval]	
cendiv: Identity				
sd(_cons)	7.040655	1.800949	4.264635	11.6237
sd(Residual)	10.00944	.1283304	9.761053	10.26415

LR test vs. linear regression: <u>chibar2(01) =</u>   **461.36** Prob >= chibar2 = **0.0000**

The upper section in the **xtmixed** output table shows the fixed-effects part of our model. Random intercepts do not appear in the output. This model implies nine separate intercepts, one for each census division, but these are not directly estimated. Instead, the lower section of the table gives an estimated standard deviation of the random intercepts (7.04), along with a standard error (1.80) and 95% confidence interval for that standard deviation. Our model is

$$bush_{ij} = 72.13 - 4.70 logdens_{ij} - .37 minority_{ij} - .04 colled_{ij} + u_{0j} + \epsilon_{ij} \qquad [15.3]$$

If the standard deviation of u_0 appears significantly different from zero, we conclude that these intercepts do vary from place to place. That seems to be the case here — the standard deviation is almost four standard errors from zero, and its value is substantial (7.04 percentage points) in the metric of our dependent variable, percent voting for Bush. A likelihood-ratio test reported on the output's final line confirms that this random-intercept model offers significant improvement over a linear regression model with fixed effects only ($p \approx .0000$).

Although **xtmixed** does not directly calculate random effects, we can obtain the best linear unbiased predictions (BLUPS) of random effects through **predict**. The following commands create a new variable named *randint0*, containing the predicted random intercepts, and then graph each census division's intercept in a bar chart (Figure 15.3).

```
. predict randint0, reffects

. graph hbar (mean) randint0, over(cendiv)
      ytitle("Random intercepts by census division")
```

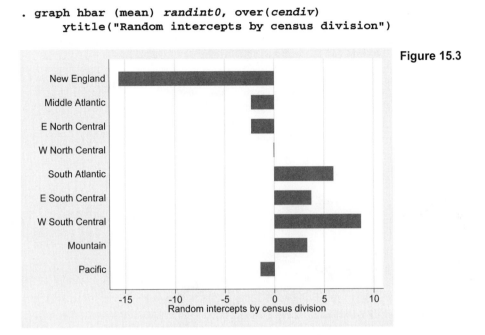

Figure 15.3

Figure 15.3 reveals that, at any given level of *logdens*, *minority*, and *colled*, the percentage of votes going to Bush averaged about 16 points lower among New England counties, or about 8 points higher in the W South Central (counties in Arkansas, Louisiana, Oklahoma, and Texas), compared with the middle-of-the-road W North Central division.

Random Intercepts and Slopes

In Figure 15.2 we saw that, overall, the percentage of Bush votes tended to decline as population density increased. Our random-intercept model in the previous section accepted this generalization, while allowing intercepts to vary across regions. But what if the slope of the votes–density relationship also varies across regions? A quick look at scatterplots for each region (Figure 15.4) gives us reason to suspect that it does.

```
. graph twoway scatter bush logdens, msymbol(Oh)

    || lfit bush logdens, lpattern(solid) lwidth(medthick)

    || , xlabel(-1(1)4, grid) ytitle("Percent vote for GW Bush")
    by(cendiv, legend(off) note(""))
```

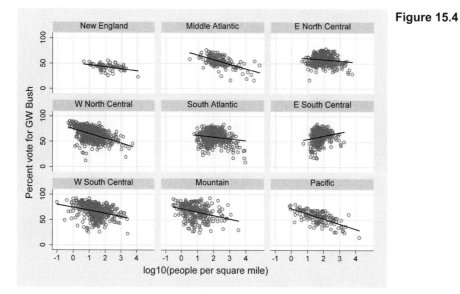

Figure 15.4

Bush votes decline most steeply with rising density in the Pacific and W North Central regions, but there appears to be little relationship in the E North Central, and even a positive effect in the E South Central. The negative fixed-effect coefficient on *logdens* in our previous model was averaging these downwards, flat, and upwards trends together.

A mixed model including random slopes (u_{1j}) on predictor x_1, and random intercepts (u_{0j}) for each of j groups could have the general form

$$y_{ij} = \beta_0 + \beta_1 x_{1ij} + \beta_2 x_{2ij} + \beta_3 x_{3ij} + u_{0j} + u_{1j} x_{1ij} + \epsilon_{ij} \qquad [15.4]$$

To estimate such a model, we include the predictor variable *logdens* to the mixed-effect part of the **xtmixed** command.

```
. xtmixed bush logdens minority colled || cendiv: logdens
```

Performing EM optimization:

Performing gradient-based optimization:

```
Iteration 0:   log restricted-likelihood = -11356.36
Iteration 1:   log restricted-likelihood = -11356.36
```

Computing standard errors:

Mixed-effects REML regression
Group variable: **cendiv**

Number of obs =	3054
Number of groups =	9

Obs per group:	
min =	67
avg =	339.3
max =	618

Log restricted-likelihood = -11356.36

Wald chi2(3) =	816.40
Prob > chi2 =	0.0000

bush	Coef.	Std. Err.	z	P>|z|	[95% Conf. Interval]	
logdens	-3.58512	1.124943	-3.19	0.001	-5.789967	-1.380273
minority	-.3645748	.0130074	-28.03	0.000	-.3900687	-.3390808
colled	-.0857994	.0347314	-2.47	0.013	-.1538718	-.017727
_cons	70.18571	3.164845	22.18	0.000	63.98272	76.38869

Random-effects Parameters	Estimate	Std. Err.	[95% Conf. Interval]	
cendiv: Independent				
sd(logdens)	3.15621	.8773938	1.830378	5.442409
sd(_cons)	9.236917	2.5247	5.405932	15.78278
sd(Residual)	9.851028	.1265092	9.606169	10.10213

LR test vs. linear regression: chi2(2) =   534.52   Prob > chi2 = 0.0000

Note: <u>LR test is conservative</u> and provided only for reference.

As usual, the random effects are not directly estimated. Instead, the **xtmixed** table gives estimates of their standard deviations. The standard deviation for coefficients on log density is 3.16 — more than three standard errors (.88) from zero — suggesting that there exists significant division-to-division variation in the slope coefficients. A more definitive likelihood-ratio test will support this inference. To perform this test, we quietly re-estimate the intercept-only model, store those estimates as *A* (an arbitrary name) then re-estimate the intercept-and-slope model, store those estimates as *B*, and finally perform a likelihood-ratio test for whether *B* fits significantly better than *A*. In this example it does ($p \approx .0000$), so we conclude that adding random slopes brought significant improvement.

```
. quietly xtmixed bush logdens minority colled || cendiv:
. estimates store A
. quietly xtmixed bush logdens minority colled || cendiv: logdens
. estimates store B
. lrtest A B
```

```
Likelihood-ratio test                        LR chibar2(01)  =      73.16
(Assumption: A nested in B)                  Prob > chibar2  =     0.0000
```

Note: LR tests based on REML are valid only when the fixed-effects
      specification is identical for both models.

The likelihood-ratio test output reminds us that tests based on REML models, the **xtmixed**
default, are valid only when the fixed-effects parts of our two models are identical. In this case,
they are. If that were not true, we could estimate both models using maximum likelihood
instead (the **ml** option), and then perform **lrtest**.

The previous model assumes that random intercept and slopes are uncorrelated, equivalent to
adding a **cov(independent)** option specifying the covariance structure. Other possibilities
include **cov(unstructured)**, which would allow for a distinct, nonzero covariance between the
random effects.

```
. xtmixed bush logdens minority colled
       || cendiv: logdens, cov(unstructured)

Performing EM optimization:

Performing gradient-based optimization:

Iteration 0:   log restricted-likelihood = -11354.196
Iteration 1:   log restricted-likelihood = -11354.196

Computing standard errors:

Mixed-effects REML regression          Number of obs      =        3054
Group variable: cendiv                 Number of groups   =           9

                                       Obs per group: min =          67
                                                      avg =       339.3
                                                      max =         618

                                       Wald chi2(3)       =      810.51
Log restricted-likelihood = -11354.196 Prob > chi2        =      0.0000
```

bush	Coef.	Std. Err.	z	P>|z|	[95% Conf. Interval]	
logdens	-3.435127	1.179726	-2.91	0.004	-5.747348	-1.122907
minority	-.3642832	.0130293	-27.96	0.000	-.3898201	-.3387462
colled	-.0904012	.0347532	-2.60	0.009	-.1585161	-.0222862
_cons	69.92388	3.391184	20.62	0.000	63.27728	76.57048

Random-effects Parameters	Estimate	Std. Err.	[95% Conf. Interval]	
cendiv: Unstructured				
sd(logdens)	3.326393	.920379	1.933992	5.721269
sd(_cons)	9.92502	2.712817	5.808601	16.95865
corr(logdens,_cons)	-.6769334	.2073502	-.9175927	-.0731567
sd(Residual)	9.849335	.1264463	9.604597	10.10031

```
LR test vs. linear regression:      chi2(3) =    538.84   Prob > chi2 = 0.0000
```

Note: LR test is conservative and provided only for reference.

The estimated correlation between the random slope on *logdens* and the random intercept is –.68, more than three standard errors from zero. A likelihood-ratio test agrees that allowing for this correlation results in significant (*P* = .0375) improvement over our previous model.

```
. estimates store C
. lrtest B C
```

Likelihood-ratio test	LR chi2(1) =	4.33
(Assumption: B nested in C)	Prob > chi2 =	0.0375

Note: LR tests based on REML are valid only when the fixed-effects specification is identical for both models.

The current model is

$$bush_{ij} = 69.92 - 3.44\ logdens_{ij} - .36\ minority_{ij} - .09\ colled_{ij}$$
$$+ u_{0j} + u_{1j} logdens_{ij} + \epsilon_{ij} \quad\quad [15.5]$$

So what are the slopes relating votes to density for each census division? Again, we can obtain values for the random effects (here named *randint1* and *randslo1*) through **predict**. Our dataset by now contains several new variables.

```
. predict randslo1 randint1, reffects
. describe
```

Contains data from C:\data\election_2004i.dta
```
  obs:         3,054                          US counties -- 2004 election
                                              (Robinson 2005)
  vars:           16                          8 Mar 2008 17:47
  size:      229,050 (99.6% of memory free)
```

variable name	storage type	display format	value label	variable label
fips	long	%9.0g		FIPS code
state	str2	%9s		State name
region	byte	%9.0g	region	Region (4)
cendiv	byte	%15.0g	division	Census division (9)
county	str24	%24s		County name
votes	float	%9.0g		Total # of votes cast, 2004
bush	float	%9.0g		% votes for GW Bush, 2004
logdens	float	%9.0g		log10(people per square mile)
minority	float	%9.0g		% population minority
colled	float	%9.0g		% adults >25 w/4+ years college
randint0	float	%9.0g		BLUP r.e. for cendiv: _cons
_est_A	byte	%8.0g		esample() from estimates store
_est_B	byte	%8.0g		esample() from estimates store
_est_C	byte	%8.0g		esample() from estimates store
randslo1	float	%9.0g		BLUP r.e. for cendiv: logdens
randint1	float	%9.0g		BLUP r.e. for cendiv: _cons

```
Sorted by:
    Note:  dataset has changed since last saved
```

The random slope coefficients range from –5.27 for counties in the W North Central division, to +4.59 in the E South Central.

```
. table cendiv, contents(mean randslo1 mean randint1)
```

Census division (9)	mean(randslo1)	mean(randint1)
New England	1.97963	-19.48989
Middle Atlantic	.0704282	-2.432464
E North Central	3.712087	-9.102283
W North Central	-5.274884	7.740558
South Atlantic	.4770184	5.311593
E South Central	4.594014	-4.056114
W South Central	-.6417558	10.59901
Mountain	-1.833838	6.785816
Pacific	-3.0827	4.643779

To clarify the relationship between votes and population density we could rearrange equation [15.5], combining the fixed and random slopes on *logdens*,

$$bush_{ij} = 69.92 + (u_{1j} - 3.44)\, logdens_{ij} - .36 minority_{ij} - .09 colled_{ij} + u_{0j} + \epsilon_{ij} \qquad [15.6]$$

In other words, the slope for each census division equals the fixed-effect slope for the whole sample, plus the random-effect slope for that division. Among Pacific counties, for instance, the combined slope is $-3.44 - 3.08 = -6.52$. The nine combined slopes are calculated and graphed in Figure 15.5.

```
. gen slope1 = randslo1 + _b[logdens]
. graph hbar (mean) slope1, over(cendiv)
       ytitle("Change in % Bush vote, with each tenfold increase
       in density")
```

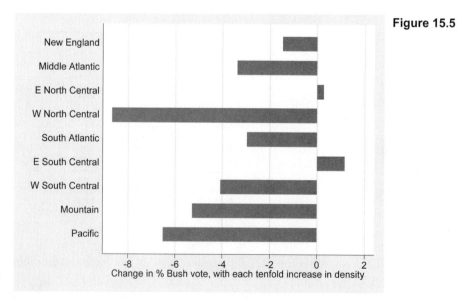

Figure 15.5

Figure 15.5 shows how the rural-urban gradient in voting behavior worked differently in different places. In counties of the W North Central, Pacific and Mountain regions, the percent voting for Bush declined most steeply as population density increased. In the E North Central and E South Central it went the other way — Bush votes increased slightly as density increased. The combined slopes in Figure 15.5 generally resemble those in the separate scatterplots of Figure 15.4, but do not match them exactly because our combined slopes (equation [15.6] or Figure 15.5) also adjust for the effects of minority and college-educated populations. In the next section we consider whether those effects, too, might have random components.

Multiple Random Slopes

To specify random coefficients on *logdens*, *minority*, and *colled* we can simply add these variable names to the random-effects part of an **xtmixed** command. For later comparison tests, we save the estimation results with name *full*. Some of the iteration output has been omitted in the following output.

```
. xtmixed bush logdens minority colled
     || cendiv: logdens minority colled
```

Mixed-effects REML regression   Number of obs   =   3054
Group variable: **cendiv**   Number of groups =   9

                                                 Obs per group: min =   67
                                                                avg =   339.3
                                                                max =   618

                                                 Wald chi2(3)   =   48.11
Log restricted-likelihood = **-11245.885**   Prob > chi2   =   0.0000

bush	Coef.	Std. Err.	z	P>|z|	[95% Conf. Interval]	
logdens	-3.117957	1.354377	-2.30	0.021	-5.772487	-.4634264
minority	-.3774872	.0594014	-6.35	0.000	-.4939119	-.2610625
colled	-.1429522	.1786875	-0.80	0.424	-.4931733	.2072689
_cons	71.23176	3.632311	19.61	0.000	64.11256	78.35096

Random-effects Parameters	Estimate	Std. Err.	[95% Conf. Interval]	
cendiv: Independent				
sd(logdens)	3.811876	1.039395	2.233774	6.504866
sd(minority)	.163672	.0489699	.0910539	.294205
sd(colled)	.5216929	.136224	.3127158	.870322
sd(_cons)	10.61635	2.853342	6.26901	17.97842
sd(Residual)	9.421733	.1212993	9.186965	9.6625

LR test vs. linear regression: chi2(4) =   755.46   Prob > chi2 = 0.0000

```
. estimates store full
```

Taking the *full* model as our baseline, likelihood-ratio tests establish that the random coefficients on *logdens*, *minority*, and *colled* each have statistically significant variation, so these should be kept in the model. For example, to evaluate the random coefficients on *colled* we quietly estimate a new model without them (*nocolled*), then compare that model with *full*. The *nocolled* model fits significantly worse ($p \approx .0000$) than the *full* model seen earlier.

```
. quietly xtmixed bush logdens minority colled
      || cendiv: logdens minority
. estimates store nocolled
. lrtest nocolled full
```

| Likelihood-ratio test | LR chibar2(01) | = | 188.18 |
| (Assumption: <u>nocolled</u> nested in <u>full</u>) | Prob > chibar2 | = | 0.0000 |

Similar steps with two further models (*nologdens* and *nominority*) and likelihood-ratio tests show that the *full* model also fits significantly better than models without either a random coefficient on *logdens* or one on *minority*.

```
. quietly xtmixed bush logdens minority colled
      || cendiv: minority colled
. estimates store nologdens
. lrtest nologdens full
```

| Likelihood-ratio test | LR chibar2(01) | = | 103.81 |
| (Assumption: <u>nologdens</u> nested in <u>full</u>) | Prob > chibar2 | = | 0.0000 |

```
. quietly xtmixed bush logdens minority colled
      || cendiv: logdens colled
. estimates store nominority
. lrtest nominority full
```

| Likelihood-ratio test | LR chibar2(01) | = | 39.04 |
| (Assumption: <u>nominority</u> nested in <u>full</u>) | Prob > chibar2 | = | 0.0000 |

We could investigate the details of all these random effects, or the combined effects they produce, through calculations along the lines of those shown earlier for Figure 15.5.

Mixed-modeling research often focuses on the fixed effects, with random effects included to represent heterogeneity in the data, but not of substantive interest. For example, our analysis thus far has demonstrated that population density, percent minority and percent college educated predict county voting patterns nationally, even after adjusting for regional differences in mean votes and for regional effects of density and college grads. On the other hand, random effects might themselves be quantities of interest. To look more closely at how the relationship between voting and percent college graduates (or percent minority, or log density) varies across census divisions, we can predict the random effects and from these calculate total effects. These steps are illustrated for the total effect of *colled* in our full model below, and graphed in Figure 15.6.

```
. quietly xtmixed bush logdens minority colled
      || cendiv: logdens minority colled
. predict relogdens reminority recolled re_cons, reffects
. describe relogdens-re_cons
```

variable name	storage type	display format	value label	variable label
relogdens	float	%9.0g		BLUP r.e. for cendiv: logdens
reminority	float	%9.0g		BLUP r.e. for cendiv: minority
recolled	float	%9.0g		BLUP r.e. for cendiv: colled
re_cons	float	%9.0g		BLUP r.e. for cendiv: _cons

```
. generate tecolled = recolled + _b[colled]
. label variable tecolled "random + fixed effect of colled"
. table cendiv, contents(mean recolled mean tecolled)
```

Census division (9)	mean(recolled)	mean(tecolled)
New England	-.2547574	-.3977096
Middle Atlantic	-.1089157	-.2518679
E North Central	-.0926856	-.2356378
W North Central	.3526636	.2097114
South Atlantic	.0607289	-.0822233
E South Central	.4656368	.3226846
W South Central	.8802846	.7373324
Mountain	-.7271097	-.8700619
Pacific	-.5758455	-.7187977

```
. graph hbar (mean) tecolled, over(cendiv)
       ytitle("Change in % Bush vote, per 1% increase in
       college graduates")
```

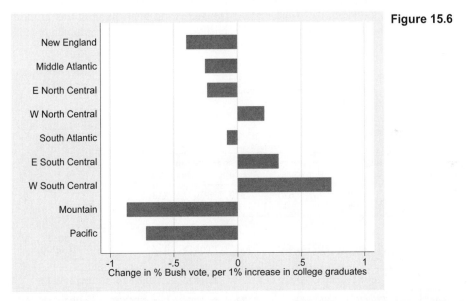

Figure 15.6

Figure 15.6 visualizes the reason our model was significantly improved ($p \approx .0000$) by including a random slope for *colled*. The total effects of *colled* on Bush votes range from substantially negative (percent Bush votes lower among counties with more college graduates) in the Mountain (−.87) and Pacific (−.72) census divisions, through negligible in the South Atlantic or W North Central, to substantially positive (+.74) in the W South Central where Bush votes were substantially higher in counties with more graduates — controlling for population density, percent minority, and other regional effects. If we estimate only a fixed effect for *colled*, our model effectively would average these negative, near zero, and positive random effects of *colled* into one weakly negative fixed coefficient, specifically the −.17 coefficient in the two fixed-effects regressions that started off this chapter.

The examples seen so far were handled successfully by **xtmixed**, but that is not always the case. Mixed-model estimation can fail to converge for a variety of different reasons, resulting in repeatedly "nonconcave" or "backed-up" iterations, or error messages about Hessian or standard error calculations. The *Longitudinal/Panel Data Reference Manual* discusses how to diagnose and work around convergence problems. A frequent cause seems to be models with near-zero variance components, such as random coefficients that really do not vary much, or have low covariances. In such cases the offending components do not appear useful, and could be dropped.

Nested Levels

Mixed-effects models can include more than one nested level. The counties of our voting data, for example, are nested not only within census divisions, but also within states that are nested within census divisions. Might there exist random effects not only at the level of census divisions, but also at the smaller level of states? The **xtmixed** command allows for such hierarchical models. Additional random-effects parts are added to the command, with successively smaller (nested) units to the right. The following analysis specifies random intercepts and slopes on all three predictors for each census division, and also random intercepts and slopes on percent college graduates, *colled*, for each state.

```
. xtmixed bush logdens minority colled
    || cendiv: logdens minority colled
    || state: colled
```

Mixed-effects REML regression   Number of obs   =   3054

Group Variable	No. of Groups	Observations per Group		
		Minimum	Average	Maximum
cendiv	9	67	339.3	618
state	49	1	62.3	254

Log restricted-likelihood = -10774.461

Wald chi2(3)   =   60.72
Prob > chi2   =   0.0000

bush	Coef.	Std. Err.	z	P>|z|	[95% Conf. Interval]	
logdens	-2.729832	.9856788	-2.77	0.006	-4.661727	-.7979371
minority	-.4038836	.0577132	-7.00	0.000	-.5169994	-.2907678
colled	-.1586315	.1344411	-1.18	0.238	-.4221313	.1048682
_cons	71.30659	3.208566	22.22	0.000	65.01791	77.59526

Random-effects Parameters	Estimate	Std. Err.	[95% Conf. Interval]	
cendiv: Independent				
sd(logdens)	2.67561	.8340451	1.452396	4.929021
sd(minority)	.1600201	.0481924	.0886793	.288753
sd(colled)	.3825822	.1058628	.2224294	.6580475
sd(_cons)	8.928818	2.684593	4.952979	16.09613
state: Independent				
sd(colled)	.1392784	.0371954	.0825209	.2350736
sd(_cons)	5.932112	.7492981	4.631186	7.598475
sd(Residual)	7.884882	.1027931	7.685963	8.088949

```
LR test vs. linear regression:        chi2(6) =  1698.31    Prob > chi2 = 0.0000
```

Note: <u>LR test is conservative</u> and provided only for reference.

At first glance, all of the random effects at both *cendiv* and *state* level in this output appear significant, judging from their standard errors and confidence intervals. The standard deviation of state-level random coefficients on *colled* (.14) is smaller than the standard deviation of corresponding census division-level coefficients (.38), but both are substantial relative to the fixed-effect coefficient on *colled* (–.16). Our confidence interval for the state-level coefficient ranges from .08 to .24. A likelihood-ratio test indicates that this model (here named *state*) with state-level random intercepts and slopes fits much better than our earlier model (*full*), which had only census division-level random intercepts and slopes.

```
. estimates store state
. lrtest full state
```

```
Likelihood-ratio test                    LR chi2(2)   =    942.85
(Assumption: full nested in state)       Prob > chi2  =    0.0000
```

Note: <u>LR test is conservative</u>
Note: LR tests based on REML are valid only when the fixed-effects
      specification is identical for both models.

As before, we can predict the random effects, then use these to calculate and graph the total effects. For *colled*, we now have random effects for 49 different states. Box plots work well to show their distribution (Figure 15.7), which follows the general pattern of census division effects seen earlier in Figure 15.6, but now also within-division variation. Indiana (E North Central) and Oklahoma (W South Central) plot as outliers, unusual in their respective divisions.

```
. predict re*, reffects
. describe re1-re6
```

variable name	storage type	display format	value label	variable label
re1	float	%9.0g		BLUP r.e. for cendiv: logdens
re2	float	%9.0g		BLUP r.e. for cendiv: minority
re3	float	%9.0g		BLUP r.e. for cendiv: colled
re4	float	%9.0g		BLUP r.e. for cendiv: _cons
re5	float	%9.0g		BLUP r.e. for state: colled
re6	float	%9.0g		BLUP r.e. for state: _cons

```
. gen tecolled2 = re3 + re5 + _b[colled]
. label variable tecolled2
     "cendiv + state + fixed effect of % college grads"
. graph hbox tecolled2, over(cendiv) yline(-.16)
     marker(1, mlabel(state))
```

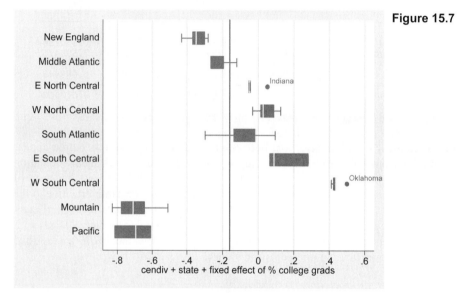

Figure 15.7

For other **xtmixed** postestimation tools, type **help xtmixed_postestimation**. The *Longitudinal/Panel Data Reference Manual*, and also Rabe-Hesketh and Skrondal (2008), present further applications of mixed modeling such as blocked-diagonal covariance structures and crossed-effects models.

Cross-Sectional Time Series

This section applies **xtmixed** to a different kind of multilevel data: cross-sectional time series. Dataset *rural_Alaska.dta* contains population and other information describing 12 boroughs or census areas of the Alaska bush (Northwest Arctic Borough, Dillingham Census Area, and so forth). For each of these 12 areas we have multiple years of data, ranging from 1969 to 2003 but with many missing values, comprising a set of 12 incomplete time series.

```
Contains data from C:\data\rural_Alaska.dta
  obs:          420                    Rural Alaska boroughs and census
                                         areas, 1969-2003
  vars:           8                    9 Mar 2008 18:53
  size:      19,320 (99.9% of memory free)

              storage  display   value
variable name   type   format    label    variable label

areaname       str19   %19s                Area name
year           int     %9.0g               Year
fips           int     %12.0g              FIPS code
```

```
inc          float  %9.0g           per cap personal income
transper     float  %9.0g           Transfers as % personal income
popt         float  %9.0g           Population in 1000s
year0        byte   %9.0g           years since 1968
year2        int    %9.0g           year0 squared
```

Sorted by: **areaname   year**

During the first half of the time period covered by these data, population grew substantially in many parts of rural Alaska. In more recent years, however, the rate of growth leveled off, and populations in some areas declined. The trends are relevant to discussions about sustainable economic development for these places, and also for their cultural importance to the Alaska Native populations who live there.

Because population trends have not simply gone upwards, they cannot realistically be modeled as a linear function of *year*. The mixed model below represents population trends as a curvilinear function, by regressing population in thousands (*popt*) on years since 1968 (*year0*) and also on *year0* squared. We allow for fixed (β) and random (u_j) intercepts and slopes on both terms.

$$population_{ij} = \beta_0 + \beta_1 year0_{ij} + \beta_2 year0^2_{ij} + u_{0j} + u_{1j} year0_{ij} + u_{2j} year0^2_{ij} + \epsilon_{ij} \qquad [15.7]$$

. **xtmixed** *popt year0 year2* || **areaname:** *year0 year2*

```
Mixed-effects REML regression          Number of obs     =        376
Group variable: areaname               Number of groups  =         12

                                       Obs per group: min =         24
                                                      avg =       31.3
                                                      max =         35

                                       Wald chi2(2)      =      67.15
Log restricted-likelihood = -304.95366 Prob > chi2       =     0.0000
```

popt	Coef.	Std. Err.	z	P>|z|	[95% Conf. Interval]	
year0	.1578059	.0196835	8.02	0.000	.119227	.1963848
year2	-.0025241	.0007368	-3.43	0.001	-.0039683	-.0010799
_cons	3.886339	.8408574	4.62	0.000	2.238289	5.53439

Random-effects Parameters	Estimate	Std. Err.	[95% Conf. Interval]	
areaname: Independent				
sd(year0)	.0549767	.0165487	.0304758	.0991748
sd(year2)	.0023566	.0005591	.0014803	.0037517
sd(_cons)	2.887339	.6290316	1.883893	4.425266
sd(Residual)	.4323029	.0165595	.4010354	.4660083

LR test vs. linear regression: chi2(3) = 1344.84   Prob > chi2 = 0.0000

All the random effects in this output show significant variation from place to place. The fixed-effect coefficient on *year0* is positive (.1578), and the coefficient on its squared value *year2* is negative (–.0025), indicative of a general trend towards slowing growth. Graphing both predicted population (with a median-spline curve) and actual population against calendar year (*year*) helps to visualize the details of area-to-area variation (Figure 15.8). In some areas the population grew steadily, while in others it first grew, then declined. The model does a decent job of smoothing past some visible measurement issues in the data.

```
. predict yhat, fitted

. graph twoway scatter popt year, msymbol(Oh)
      ||   mspline yhat year, lpattern(solid) lwidth(medthick)
      ||   , by(areaname, note("") legend(off))
    ylabel(0(5)20, angle(horizontal)) xtitle("")
    ytitle("Population in 1,000s") xlabel(1970(10)2000, grid)
```

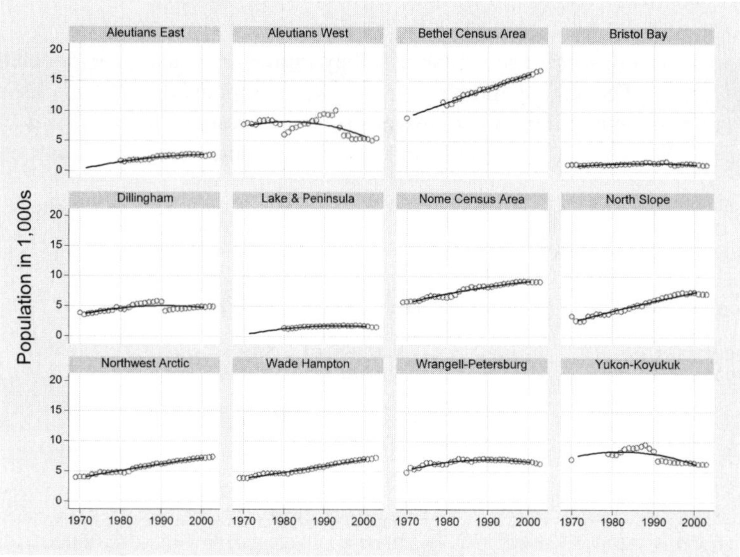

Figure 15.8

Mixed-Effects Logit Regression

Since 1972, the General Social Survey (Davis et al. 2005) has tracked U.S. public opinion through a series of annual or biannual polls, and made the data available for teaching and research. The 2006 poll, which asked respondents how they had voted in 2004, provides individual-level data in keeping with this chapter's election theme. Dataset *GSS_SwS1* contains a subset of GSS 2006 variables and observations selected for this chapter. Consult the website of the Inter-University Consortium for Political and Social Research for detailed information and more representative GSS datasets,
 http://www.icpsr.umich.edu/cocoon/ICPSR/SERIES/00028.xml

```
. use GSS_SwS1.dta, clear
. describe
```

```
Contains data from C:\data\GSS_SwS1.dta
  obs:         1,595                          General Social Survey 2006
  vars:           17                          14 Jun 2008 13:45
  size:       60,610 (99.9% of memory free)
```

variable name	storage type	display format	value label	variable label
id	int	%9.0g		Case number (Statistics w/Stata)
finalwt	float	%9.0g		weight variable -- wtssall
cendiv	byte	%12.0g	cendiv	Census division (9)
age	byte	%8.0g	age	Age in years
educ	byte	%8.0g	educ	Highest year of schooling completed
gender	byte	%9.0g	gender	Respondent gender
income	float	%9.0g	income	Family income in constant dollars
loginc	float	%9.0g		log10(family income, +1)
logsize	float	%9.0g		log10(size of place in thousands, +1)
married	byte	%9.0g	married	Married or unmarried?
minority	byte	%11.0g	minority	Minority status
politics	byte	%20.0g	politics	Think of self as liberal or conservative?
bush	byte	%11.0g	bush	Voted for GW Bush in 2004?
grass	byte	%9.0g	grass	Should marijuana be made legal?
gunlaw	byte	%8.0g	gunlaw	Favor or oppose gun permits?
postlife	byte	%8.0g	postlife	Believe in life after death?
enviro	byte	%11.0g	enviro	Govt spending on environment?

```
Sorted by: id
```

The GSS question about 2004 voting (*bush*) will be our focus in this section. Coded 1 if respondents said they voted for George W. Bush, and 0 if they voted for John Kerry or Ralph Nader, this dummy variable is suitable for logit modeling. (Respondents who said they did not vote, or could not recall, are set aside for purposes of this chapter.) The unweighted GSS results on this question come reasonably close to the actual vote outcome — 51.35% for Bush in the survey, compared with 50.7% in the U.S. election, reflecting well on the GSS efforts to secure a representative random sample.

```
. tab bush
```

Voted for GW Bush in 2004	Freq.	Percent	Cum.
Kerry/Nader	776	48.65	48.65
Bush	819	51.35	100.00
Total	1,595	100.00	

Previously in this chapter we saw that the percentage of Bush votes across U.S. counties decreased with population density, percent minority, and percent college educated, although there are significant regional variations. We should not assume that relationships existing at an aggregated level (such as counties) will necessarily also be found at the individual level, but these GSS data provide an opportunity to investigate that possibility. The GSS does not identify respondents' counties, but it does report the population of the place where they live, which gives an alternative indicator of urban-ness. Due to the positive skewness of place size, we work with its base-10 logarithm, *logsize*. A {0, 1} variable indicating minority status (*minority*), and a variable for respondent's years of education (*educ*), provide other individual-level predictors roughly analogous to our county-level predictors. The GSS also notes census divisions (*cendiv*), referring to the same nine geographical units that proved important for our county-level modeling.

The **xtmelogit** command, with syntax similar to **xtmixed**, fits mixed-effects logit models. We start with a basic model predicting individual Bush votes from three fixed effects — logarithm of place size, minority status, and education — and random intercepts for each census division.

```
. xtmelogit bush logsize minority educ || cendiv:

Mixed-effects logistic regression          Number of obs    =      1595
Group variable: cendiv                     Number of groups =         9

                                           Obs per group: min =        62
                                                          avg =     177.2
                                                          max =       336

Integration points =    7                  Wald chi2(3)     =    156.51
Log likelihood = -1001.4226                Prob > chi2      =    0.0000
```

bush	Coef.	Std. Err.	z	P>|z|	[95% Conf.	Interval]
logsize	-.2456232	.0723751	-3.39	0.001	-.3874758	-.1037707
minority	-1.988091	.1693794	-11.74	0.000	-2.320069	-1.656114
educ	-.0435621	.020059	-2.17	0.030	-.082877	-.0042472
_cons	1.331863	.3044779	4.37	0.000	.7350978	1.928629

Random-effects Parameters	Estimate	Std. Err.	[95% Conf.	Interval]
cendiv: Identity				
sd(_cons)	.2631985	.0937197	.1309747	.5289069

```
LR test vs. logistic regression: chibar2(01) =     9.78 Prob>=chibar2 = 0.0009
```

All three predictors have significant effects, with the negative signs we might expect based on our county-level findings. Inclusion of a random intercept for each census division improves over a fixed-effects logit model ($P = .0009$). Should we also specify a random slope on predictor *logsize*, and an unstructured covariance matrix?

```
. xtmelogit bush logsize minority educ
       || cendiv: logsize, covariance(unstructured)
```

```
Mixed-effects logistic regression          Number of obs    =     1595
Group variable: cendiv                      Number of groups =        9

                                            Obs per group: min =      62
                                                           avg =    177.2
                                                           max =      336

Integration points =   7                    Wald chi2(3)     =   136.94
Log likelihood = -993.18808                 Prob > chi2      =   0.0000
```

bush	Coef.	Std. Err.	z	P>|z|	[95% Conf. Interval]	
logsize	-.3067214	.1351802	-2.27	0.023	-.5716697	-.0417731
minority	-1.929399	.1705008	-11.32	0.000	-2.263575	-1.595224
educ	-.041411	.0201692	-2.05	0.040	-.0809419	-.0018801
_cons	1.361903	.3014101	4.52	0.000	.7711505	1.952656

Random-effects Parameters	Estimate	Std. Err.	[95% Conf. Interval]	
cendiv: Unstructured				
sd(logsize)	.324985	.1153992	.1620353	.6518042
sd(_cons)	.2166181	.1880604	.0395103	1.187623
corr(logsize,_cons)	-.8708507	.2677071	-.9982074	.683149

```
LR test vs. logistic regression:     chi2(3) =     26.25   Prob > chi2 = 0.0000
```

```
. estimates store A
```

The random coefficient on *logsize* has substantial variation (standard deviation .32, standard error .12), but the random intercept's variation now appears less distinct (standard deviation .22, standard error .19). The correlation between these two random effects appears nonsignificant as well.

```
. xtmelogit bush logsize minority educ
      || cendiv: logsize, nocons
```

```
Mixed-effects logistic regression          Number of obs    =      1595
Group variable: cendiv                      Number of groups =         9

                                            Obs per group: min =        62
                                                           avg =     177.2
                                                           max =       336

Integration points =   7                    Wald chi2(3)     =    150.81
Log likelihood = -994.02453                 Prob > chi2      =    0.0000
```

bush	Coef.	Std. Err.	z	P>|z|	[95% Conf. Interval]	
logsize	-.2988801	.1097208	-2.72	0.006	-.5139289	-.0838312
minority	-1.969727	.1666701	-11.82	0.000	-2.296395	-1.64306
educ	-.0414891	.0201459	-2.06	0.039	-.0809745	-.0020038
_cons	1.361989	.2909969	4.68	0.000	.791646	1.932333

Random-effects Parameters	Estimate	Std. Err.	[95% Conf. Interval]	
cendiv: Identity				
sd(logsize)	.2340086	.071626	.128438	.4263536

LR test vs. logistic regression: chibar2(01) =   24.58 Prob>=chibar2 = 0.0000

```
. estimates store B
. lrtest B A
```

```
Likelihood-ratio test                        LR chi2(2)  =      1.67
(Assumption: B nested in A)                  Prob > chi2 =    0.4332
```

Note: LR test is conservative

The random-slope-only model *B* fits better than a random-intercept-only model, but not significantly worse (*P* = .4332) than model *A* which included random slope, intercept, and covariance. Regional variations in these individual-level survey data appear best modeled as variations in the effect of urban-ness (*logsize*) on the odds of voting for Bush. In some census divisions this effect was strong and negative, whereas in others it was weaker or even positive. Adjusting for these regional differences, however, the individual-level fixed effects of place size, minority status, and education follow patterns consistent with earlier county-level findings.

16

Introduction to Programming

As mentioned earlier, we can create a simple type of program by writing any sequence of Stata commands in a text (ASCII) file. Stata's Do-file Editor (click on Window > Do-file Editor or the icon ▢) provides a convenient way to do this. After saving the do-file, we enter Stata and type a command with the form **do** *filename* that tells Stata to read *filename.do* and execute whatever commands it contains. More sophisticated programs are possible as well, making use of Stata's built-in programming language. Many of the commands used in previous chapters actually involve programs written in Stata. These programs might have originated either from StataCorp or from users who wanted something beyond Stata's built-in features to accomplish a particular task.

Stata programs can access all the existing features of Stata, call other programs that call other programs in turn, and use model-fitting aids including matrix algebra and maximum likelihood estimation. Whether our purposes are broad, such as adding new statistical techniques, or narrowly specialized, such as managing a particular database, our ability to write programs in Stata greatly extends what we can do.

Programming is a deep topic in Stata. This brief chapter introduces just a few of the basic concepts and tools, with some examples of how they can be used to facilitate common data-analysis tasks. If you are interested in learning more, Stata's expertly-taught online NetCourses (www.stata.com/netcourse) are a good place to start. The main reference sources are the *Programming Reference Manual* and the two-volume *Mata Matrix Programming Manual*. Details about maximum-likelihood estimation and programming are presented in *Maximum Likelihood Estimation with Stata* (Gould, Pitblado, and Sribney 2006).

Basic Concepts and Tools

Some elementary concepts and tools, combined with the Stata capabilities described in earlier chapters, suffice to get started.

Do-files

Do-files are ASCII (text) files, created by Stata's Do-file Editor, a word processor, or any other text editor. They are typically saved with a .do extension. The file can contain any sequence of legitimate Stata commands. In Stata, typing the following command causes Stata to read *filename.do* and execute the commands it contains:

```
. do filename
```

Each command in *filename.do*, including the last, must end with a hard return, unless we have reset the delimiter through a **#delimit** command:

```
#delimit ;
```

This sets a semicolon as the end-of-line delimiter, so that Stata does not consider a line finished until it encounters a semicolon. Setting the semicolon as delimiter permits a single command to extend over more than one physical line. Later, we can reset "carriage return" as the usual end-of-line delimiter with another **#delimit** command:

```
#delimit cr
```

A typographical note: Many commands illustrated in this chapter are most likely to be used inside a do-file or ado-file, instead of being typed as a stand-alone command in the Command window. I have written such within-program commands without showing a preceding "." prompt, as with the two **#delimit** examples above (but not with the **do filename** command, which would have been typed in the Command window as usual).

Ado-files

Ado (automatic do) files are ASCII files containing sequences of Stata commands, much like do-files. The difference is that we need not type **do *filename*** in order to run an ado-file. Suppose we type the command

```
. clear
```

As with any command, Stata reads this and checks whether an intrinsic command by this name exists. If a **clear** command does not exist as part of the base Stata executable (and, in fact, it does not), then Stata next searches in its usual "ado" directories, trying to find a file named clear.ado. If Stata finds such a file (as it should), it then executes whatever commands the file contains.

Ado-files have the extension .ado. User-written programs commonly go in a directory named C:\ado\personal, whereas the hundreds of official Stata ado-files get installed in C:\Program Files\Stata\ado. Type **sysdir** to see a list of the directories Stata currently uses. Type **help sysdir** or **help adopath** for advice on changing them.

The **which** command reveals whether a given command really is an intrinsic, hardcoded Stata command or one defined by an ado-file; and if it is an ado-file, where that resides. For example, **anova** is a built-in command, but the **ttest** command is defined by an ado-file named ttest.ado, updated in December 2004.

```
. which anova
built-in command:   anova
```

```
. which ttest
C:\Program Files\Stata10\ado\base\t\ttest.ado
*! version 4.1.1  30dec2004
```

This distinction makes no difference to most users, because **anova** and **ttest** work with similar ease when called. Studying examples and borrowing code from Stata's thousands of official ado-files can be helpful as you get started writing your own. The **which** output above gave the location for file ttest.ado. To see its actual code, type

```
. viewsource ttest.ado
```

The ado-files defining many Stata estimation commands have grown noticeably more complicated-looking over the years, as they accommodate new capabilities such as **svy:** prefixes. This has not happened with ttest.ado, however.

Programs

Both do-files and ado-files might be viewed as types of programs, but Stata uses the word "program" in a narrower sense, to mean a sequence of commands stored in memory and executed by typing a particular program name. Do-files, ado-files, or commands typed interactively can define such programs. The definition begins with a statement that names the program. For example, to create a program named *count5*, we start with

```
program count5
```

Next should be the lines that actually define the program. Finally, we give an **end** command, followed by a hard return:

```
end
```

Once Stata has read the program definition commands, it retains that definition of the program in memory and will run it any time we type the program's name as a command:

```
. count5
```

Programs effectively make new commands available within Stata, so most users do not need to know whether a given command comes from Stata itself or from an ado-file-defined program.

As we start to write a new program, we often create preliminary versions that are incomplete or just unsuccessful. The **program drop** command provides essential help here, allowing us to clear programs from memory so that we can define a new version. For example, to clear program *count5* from memory, type

```
. program drop count5
```

To clear all programs (but not the data) from memory, type

```
. program drop _all
```

Local Macros

Macros are names (up to 31 characters) that can stand for strings, program-defined numerical results, or user-defined values. A local macro exists only within the program that defines it, and cannot be referred to in another program. To create a local macro named *iterate*, standing for the number 0, type

```
local iterate 0
```

To refer to the contents of a local macro (0 in this example), place the macro name within left and right single quotes. For example,

```
display `iterate'
```
0

Thus, to increase the value of *iterate* by one, we write

```
local iterate = `iterate' + 1
display `iterate'
```
1

Instead of a number, the macro's contents could be a string or list of words, such as

```
local islands Iceland Greenland
```

To see the string contents, place double quotes around the single-quoted macro name:

```
display "`islands'"
```
Iceland Greenland

We can concatenate further words or numbers, adding to the macro's contents. For example,

```
local islands `islands' Newfoundland Nantucket
display "`islands'"
```
Iceland Greenland Newfoundland Nantucket

Type **help extended fcn** for information about Stata's "extended macro functions," which extract information from the contents of macros. For instance, we could obtain a count of words in the macro, and store this count as a new macro named *howmany*:

```
local howmany: word count `islands'
display `howmany'
```
4

Many other extended macro functions exist, with applications to programming.

Global Macros

Global macros are similar to local macros, but once defined, they remain in memory and can be used by other programs for the duration of your current Stata session. To refer to a global macro's contents, we preface the macro name with a dollar sign (instead of enclosing the name in left and right single quotes as done with local macros):

```
global distance = 73
display $distance * 2
```
146

Unless we specifically want to keep macro contents for re-use later in our session, it is better (less confusing, faster to execute, and potentially less hazardous) if we use local rather than global macros in writing programs. To drop a macro from memory, issue a **macro drop** command.

```
macro drop distance
```

We could also drop all macros from memory.

```
macro drop _all
```

Scalars

Scalars can be individual numbers or strings, referenced by a name much as local macros are. To retrieve the contents, however, we do not need to enclose the scalar name in quotes. For example,

```
scalar onethird = 1/3
display onethird
```
.33333333
```
display onethird*6
```
2

Scalars are most useful in storing numerical results from calculations, at full numerical precision. Many Stata analytical procedures retain results such as degrees of freedom, test statistics, log likelihoods, and so forth as scalars — as can be seen by typing **return list** or **ereturn list** after the analysis. The scalars, local macros, matrices, and functions automatically stored by Stata programs supply building blocks that could be used within new programs.

Version

Stata's capabilities and features have changed over the years. Consequently, programs written for an older version of Stata might not run directly under the current version. The **version** command works around this problem so that old programs remain usable. Once we tell Stata for what version the program was written, Stata makes the necessary adjustments and the old program can run under a new version of Stata. For example, if we begin our program with the following statement, Stata interprets all the program's commands as it would have in Stata 9:

```
version 9
```

Typed by itself, the command **version** simply reports the version to which the interpreter is currently set.

Comments

Stata does not attempt to execute any line that begins with an asterisk. Such lines can therefore be used to insert comments and explanations into a program, or interactively during a Stata session. For example,

```
* This entire line is a comment.
```

Alternatively, we can include a comment within an executable line. The simplest way to do so is to place the comment after a double slash, // (with at least one space before the double slash). For example,

```
summarize income education     // this part is the comment
```

A triple slash (also preceded by at least one space) indicates that what follows, to the end of the line, is a comment; but then the following physical line should be executed as a continuation of the first. For example,

```
summarize income education     /// this part is the comment
occupation age
```

will be executed as if we had typed

```
summarize income education occupation age
```

With or without comments, a triple slash tells Stata to read the next line as a continuation of the present line. For example, the following two lines would be read as one **table** command, even though they are separated by a hard return.

```
table gender kids school if contam==1, contents(mean lived ///
median lived count lived)
```

The triple slash thus provides an alternative to the **#delimit ;** approach described earlier, for writing program commands that are more than one physical line long.

It is also possible to include comments in the middle of a command line, bracketed by /* and */ . For example,

```
summarize income /* this is the comment */ education occupation
```

If one line ends with /* and the next begins with */ then Stata skips over the line break and reads both lines as a single command — another line-lengthening trick sometimes found in programs, although /// is now favored.

Looping

There are a number of ways to create program loops. One simple method employs the **forvalues** command. For example, the following program counts from 1 to 5.

```
* Program that counts from one to five
program count5
    version 10.0
    forvalues i = 1/5 {
        display `i'
        }
end
```

By typing these commands, we define program *count5*. Alternatively, we could use the Do-file Editor to save the same series of commands as an ASCII file named *count5.do*. Then, typing the following causes Stata to read the file:

```
. do count5
```

Either way, by defining program *count5* we make this available as a new command:

```
. count5
1
2
3
4
5
```

The command

```
forvalues i = 1/5 {
```

assigns to local macro *i* the consecutive integers from 1 through 5. The command

```
display `i'
```

shows the contents of this macro. The name *i* is arbitrary. A slightly different notation would allow us to count from 0 to 100 by fives (0, 5, 10, ... , 100):

```
forvalues j = 0(5)100 {
```

The steps between values need not be integers, so long as the endpoints are. To count from 4 to 5 by increments of .01 (4.00, 4.01, 4.02, ... , 5.00), write

```
forvalues k = 4(.01)5 {
```

Any lines containing valid Stata commands, between the opening and closing curly brackets { }, will be executed repeatedly for each of the values specified. Note that apart from optional comments, nothing on that line follows the opening bracket, and the closing bracket requires a line of its own.

The **foreach** command takes a different approach. Instead of specifying a set of consecutive numeric values, we give a list of items for which iteration occurs. These items could be variables, files, strings, or numeric values. Type **help foreach** to see the syntax of this command.

forvalues and **foreach** create loops that repeat for a pre-specified number of times. If we want looping to continue until some other condition is met, the **while** command is useful. A section of program with the following general form will repeatedly execute the commands within curly brackets, so long as *expression* evaluates to "true":

```
while expression {
    command A
    command B
    . . . .
}
command Z
```

As in previous examples, the closing bracket } should be on its own separate line, not at the end of a command line.

When *expression* evaluates to "false," the looping stops and Stata goes on to execute *command Z*. Parallel to our previous example, here is a simple program that uses a **while** loop to display onscreen the iteration numbers from 1 through 6:

```
* Program that counts from one to six
  program count6
  version 10.0
  local iterate = 1
  while `iterate' <= 6 {
      display `iterate'
      local iterate = `iterate' + 1
  }
end
```

A more substantial loop appears in the *multicat.ado* program described later in this chapter. The *Programming Reference Manual* contains more about programming loops.

If . . . else

The **if** and **else** commands tell a program to do one thing if an expression is true, and something else otherwise. They are set up as follows:

```
if expression {
    command A
    command B
    . . . .
}
else {
    command Z
}
```

For example, the following program segment checks whether the content of local macro *span* is an odd number, and informs the user of the result.

```
if int(`span'/2) != (`span' - 1)/2 {
    display "span is NOT an odd number"
}
else {
    display "span IS an odd number"
}
```

Arguments

Programs define new commands. In some instances (as with the earlier example, **count5**), we intend our command to do exactly the same thing each time it is used. Often, however, we need a command that is modified by arguments such as variable names or options. There are two ways we can tell Stata how to read and understand a command line that includes arguments. The simplest of these is the **args** command.

The following do-file (*listres1.do*) defines a program that performs a two-variable regression, and then lists the observations with the largest absolute residuals. listres1 exhibits several bad habits, such as dropping variables and leaving new ones in memory, which could have unwanted side effects. It serves to illustrate the use of temporary variables, however.

```
* Perform simple regression and list observations with #
*   largest absolute residuals.
*   syntax:  listres1 Yvariable Xvariable # IDvariable
capture drop program listres1
program listres1, sortpreserve
    version 10.0
    args Yvar Xvar number id
    quietly regress `Yvar' `Xvar'
    capture drop Yhat_
    capture drop Resid_
    capture drop Absres_
    quietly predict Yhat_
    quietly predict Resid_, resid
    quietly gen Absres_ = abs(Resid_)
    gsort -Absres_
    drop Absres_
    list `id' `Yvar' Yhat_ Resid_ in 1/`number'
end
```

The line **args Yvar Xvar number id** tells Stata that the command **listres1** should be followed by four arguments. These arguments could be numbers, variable names, or other strings separated by spaces. The first argument becomes the contents of a local macro named *Yvar*, the second a local macro named *Xvar*, and so forth. The program then uses the contents of these macros in other commands, such as the regression:

```
quietly regress `Yvar' `Xvar'
```

The program calculates absolute residuals (*Absres*), and then uses the **gsort** command (with minus sign before the variable name) to sort data in high-to-low order, with missing values last:

```
gsort -Absres_
```

The option **sortpreserve** on the command line makes this program "sort-preserving," insuring that the order of the observations is the same after the program runs as it was before.

Dataset *nations.dta*, seen previously in Chapter 8, contains variables indicating life expectancy (*life*), per capita daily calories (*food*), and country name (*country*) for 109 countries. We can open this file, and use it to demonstrate our new program. A **do** command runs do-file *listres1.do*, thereby defining the program and new command **listres1**:

```
. do listres1.do
```

Next, we use the newly-defined **listres1** command, followed by its four arguments. The first argument specifies the *y* variable, the second *x*, the third how many observations to list, and the fourth gives the case identifier. In this example, our command asks for a list of observations that have the five largest absolute residuals.

```
. listres1 life food 5 country
```

	country	life	Yhat_	Resid_
1.	Libya	60	76.6901	-16.69011
2.	Bhutan	44	60.49577	-16.49577
3.	Panama	72	58.13118	13.86882
4.	Malawi	45	58.58232	-13.58232
5.	Ecuador	66	52.45305	13.54695

Life expectancies are lower than predicted in Libya, Bhutan, and Malawi. Conversely, life expectancies in Panama and Ecuador are higher than predicted, based on food supplies.

Syntax

The **syntax** command provides a more complicated but also more powerful way to read a command line. The following do-file named *listres2.do* is similar to our previous example, but it uses **syntax** instead of **args**.

```
* Perform simple or multiple regression and list
* observations with # largest absolute residuals.
*  listres2 yvar xvarlist [if] [in], number(#) [id(varname)]
capture drop program listres2
program listres2, sortpreserve
version 10.0
syntax varlist(min=1) [if] [in], Number(integer) [Id(varlist)]
   marksample touse
   quietly regress `varlist' if `touse'
   capture drop Yhat_
   capture drop Resid_
   capture drop Absres_
   quietly predict Yhat_ if `touse'
   quietly predict Resid_ if `touse', resid
   quietly gen Absres_ = abs(Resid_)
   gsort -Absres_
   drop Absres_
   list `id' `1' Yhat_ Resid_ in 1/`number'
end
```

listres2 has the same purpose as the earlier **listres1**: it performs regression, then lists observations with the largest absolute residuals. This newer version contains several enhancements, made possible by the **syntax** command. It is not restricted to two-variable regression, as was **listres1**. **listres2** will work with any number of predictor variables, including none (in which case, predicted values equal the mean of *y*, and residuals are deviations from the mean). **listres2** permits optional **if** and **in** qualifiers. A variable identifying the observations is optional with **listres2**, instead of being required as it was with **listres1**. For example, we could regress life expectancy on *food* and *energy*, while restricting our analysis to only those countries where per capita GNP is above 500 dollars:

```
. do listres2.do
. listres2 life food energy if gnpcap > 500, n(6) i(country)
```

	country	life	Yhat_	Resid_
1.	YemenPDR	46	61.34964	-15.34964
2.	YemenAR	45	59.85839	-14.85839
3.	Libya	60	73.62516	-13.62516
4.	S_Africa	55	67.9146	-12.9146
5.	HongKong	76	64.64022	11.35978
6.	Panama	72	61.77788	10.22212

The **syntax** line in this example illustrates some general features of the command:

```
syntax varlist(min=1) [if] [in], Number(integer) [Id(varlist)]
```

The variable list for a **listres2** command is required to contain at least one variable name (**varlist(min=1)**). Square brackets denote optional arguments — in this example, the **if** and **in** qualifiers, and also the **id( )** option. Capitalization of initial letters for the options indicates the minimum abbreviation that can be used. Because the **syntax** line in our example specified **Number(integer) Id(varlist)**, an actual command could be written:

```
. listres2 life food, number(6) id(country)
```

or, equivalently,

```
. listres2 life food, n(6) i(country)
```

The contents of local macro *number* must be an integer, and *id* one or more variable names.

This example also illustrates the **marksample** command, which marks the subsample (as qualified by **if** and **in**) to be used in subsequent analyses.

The syntax of **syntax** is outlined in the *Programming Manual*. Experimentation and studying other programs help in gaining fluency with this command.

Example Program: Moving Autocorrelation

The preceding sections presented basic ideas and example short programs. In this section, we apply those ideas to a slightly longer program that defines a new statistical procedure. The procedure obtains moving autocorrelations through a time series, as proposed for ocean-atmosphere data by Topliss (2002). The following do-file, *gossip.do*, defines a program that makes available a new command called **gossip**. Comments, in lines that begin with * or in phrases set off by //, explain what the program is doing. Indentation of lines has no effect on the program's execution, but makes it easier for the programmer to read.

```
capture program drop gossip  // FOR WRITING/DEBUGGING; DELETE LATER
program gossip
version 10.0
* Syntax requires user to specify two variables (Yvar and TIMEvar), and
* the span of the moving window.  Optionally, the user can ask to generate
* a new variable holding autocorrelations, to draw a graph, or both.
syntax varlist(min=1 max=2 numeric), SPan(integer) [GENerate(namelist) GRaph]
if mod(`span',2) != 1 {
    display as error "Span must be an odd integer"
}
else {
* The first variable in `varlist' becomes Yvar, the second TIMEvar.
    tokenize `varlist'
        local Yvar `1'
        local TIMEvar `2'
    tempvar NEWVAR
    quietly gen `NEWVAR' = .
    local miss = 0
* spanlo and spanhi are local macros holding the observation number at the
* low and high ends of a particular window.  spanmid holds the observation
* number at the center of this window.
    local spanlo = 0
    local spanhi = `span'
    local spanmid = int(`span'/2)
    while `spanlo' <= _N -`span' {
        local spanhi = `span' + `spanlo'
        local spanlo = `spanlo' + 1
        local spanmid = `spanmid' + 1
```

```
* The next lines check whether missing values exist within the window.
* If they do exist, then no autocorrelation is calculated and we
* move on to the next window.  Users are informed that this occurred.
        quietly count if !missing(`Yvar') in `spanlo'/`spanhi'
        if r(N) != `span' {
            local miss = 1
        }
* The value of NEWVAR in observation `spanmid' is set equal to the first
* row, first column (1,1) element of the row vector of autocorrelations
* r(AC) saved by corrgram.
        else {
            quietly corrgram `Yvar' in `spanlo'/`spanhi', lag(1)
            quietly replace `NEWVAR' = el(r(AC),1,1) in `spanmid'
        }
    }
    if "`graph'" != "" {
* The following graph command illustrates the use of comments to cause
* Stata to skip over line breaks, so it reads the next two lines as if
* they were one.
        graph twoway spike `NEWVAR' `TIMEvar', yline(0) ///
        ytitle("First-order autocorrelations of `Yvar' (span `span')")
    }
    if `miss' == 1 {
        display as error "Caution:  missing values exist"
    }
    if "`generate'" != "" {
        rename `NEWVAR' `generate'
        label variable `generate' ///
        "First-order autocorrelations of `Yvar' (span `span')"
    }
}
end
```

As the comments note, **gossip** requires time series (**tsset**) data. From an existing time series variable, **gossip** calculates a second time series consisting of lag-1 autocorrelation coefficients within a moving window of observations — for example, a moving 9-year span. Dataset *nao.dta* contains North Atlantic climate time series that can be used for illustration:

```
Contains data from c:\data\nao.dta
  obs:           159                       North Atlantic Oscillation &
                                           mean air temperature at
                                           Stykkisholmur, Iceland
  vars:            5                       2 Jun 2008 09:38
  size:        4,134 (99.9% of memory free)
```

variable name	storage type	display format	value label	variable label
year	int	%ty		Year
wNAO	float	%9.0g		Winter NAO
wNAO4	float	%9.0g		Winter NAO smoothed
temp	float	%9.0g		Mean air temperature (C)
temp4	float	%9.0g		Mean air temperature smoothed

```
Sorted by:  year
```

The variable *temp* records annual mean air temperatures at Stykkishólmur in west Iceland from 1841 to 1999. *temp4* contains smoothed values of *temp* (see Chapter 13). Figure 16.1 graphs these two time series. To visually distinguish between raw (*temp*) and smoothed (*temp4*) variables, we connect the former with very thin lines, **clwidth(vthin)**, and the latter with thick lines, **clwidth(thick)**. Type **help linewidthstyle** for a list of other line-width choices.

```
. graph twoway line temp year, clpattern(solid) clwidth(vthin)
       || line temp4 year, clpattern(solid) clwidth(thick)
       || , ytitle("Temperature, degrees C") legend(off)
```

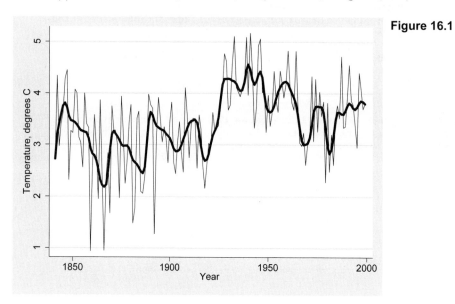

Figure 16.1

To calculate and graph a series of autocorrelations of *temp*, within a moving window of 9 years, we type the following commands. They produce the graph shown in Figure 16.2.

```
. do gossip.do
. gossip temp year, span(9) generate(autotemp) graph
```

Figure 16.2

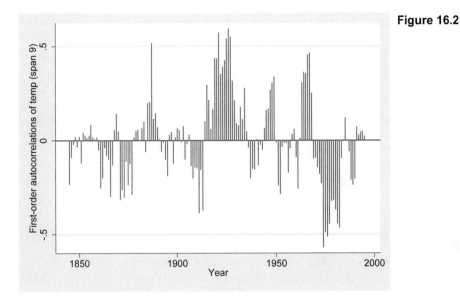

In addition to drawing Figure 16.2, **gossip** created a new variable named *autotemp*:

```
. describe autotemp
```

variable name	storage type	display format	value label	variable label
autotemp	float	%9.0g		First-order autocorrelations of temp (span 9)

```
. list year temp autotemp in 1/10
```

	year	temp	autotemp
1.	1841	2.73	.
2.	1842	4.34	.
3.	1843	2.97	.
4.	1844	3.41	.
5.	1845	3.62	-.2324837
6.	1846	4.28	-.0883512
7.	1847	4.45	-.0194607
8.	1848	2.32	.0175247
9.	1849	3.27	-.03303
10.	1850	3.23	.0181154

autotemp values are missing for the first four years (1841 to 1844). In 1845, the *autotemp* value (–.2324837) equals the lag-1 autocorrelation of *temp* over the 9-year span from 1841 to 1849. This is the same coefficient we would obtain by typing

```
. corrgram temp in 1/9, lag(1)
```

					-1 0 1	-1 0 1
LAG	AC	PAC	Q	Prob>Q	[Autocorrelation]	[Partial Autocor]
1	-0.2325	-0.2398	.66885	0.4135	─┤	─┤

In 1846, *autotemp* (–.0883512) equals the lag-1 autocorrelation of *temp* over the 9 years from 1842 to 1850, and so on through the data. *autotemp* values are missing for the last four years in the data (1996 to 1999), as they are for the first four.

The pronounced Arctic warming of the 1920s, visible in the temperatures of Figure 16.1, manifests in Figure 16.2 as a period of consistently positive autocorrelations. A briefer period of positive autocorrelations in the 1960s coincides with a cooling climate. Topliss (2002) suggests interpretation of such autocorrelations as indicators of changing feedbacks in ocean-atmosphere systems.

The do-file *gossip.do* was written incrementally, starting with input components such as the syntax statement and span macros, running the do-file to check how these work, and then adding other components. Not all of the trial runs produced satisfactory results. Typing the command

```
. set trace on
```

causes Stata to display programs line-by-line as they execute, so we can see exactly where an error occurs. Later, we can turn this feature off by typing

```
. set trace off
```

gossip.do contains a first line, **capture program drop *gossip***, that discards the program from memory before defining it again. This is helpful during the writing and debugging stage, when a previous version of our program might have been incomplete or incorrect. Such lines should be deleted once the program is mature, however. The next section describes further steps toward making **gossip** available as a regular Stata command.

Ado-File

Once we believe our do-file defines a program that we will want to use again, we can create an ado-file to make it available like any other Stata command. For the previous example, *gossip.do*, the change involves two steps:

1. With the Do-file Editor, delete the initial "DELETE LATER" line that had been inserted to streamline the program writing and debugging phase. We can also delete the comment lines. Doing so makes the program more compact and easier to read.

2. Save our modified file, renaming it to have an .ado extension (in this example, *gossip.ado*), in a new directory. The recommended location is in ado\personal; you might need to create this directory and subdirectory if they do not already exist. Other locations are possible, but

review the *User's Manual* section on "Where does Stata look for ado-files?" before proceeding.

Once this is done, we can use **gossip** as a regular command within Stata. A listing of *gossip.ado* follows.

```
*!  version 2.0  3jul2008
*!  L. Hamilton, Statistics with Stata (2009)
program gossip
version 10.0
syntax varlist(min=1 max=2 numeric), SPan(integer) [GENerate(namelist) GRaph]
if mod(`span',2) != 1 {
    display as error "Span must be an odd integer"
}
else {
    tokenize `varlist'
        local Yvar `1'
        local TIMEvar `2'
    tempvar NEWVAR
    quietly gen `NEWVAR' = .
    local miss = 0
    local spanlo = 0
    local spanhi = `span'
    local spanmid = int(`span'/2)
    while `spanlo' <= _N -`span' {
        local spanhi = `span' + `spanlo'
        local spanlo = `spanlo' + 1
        local spanmid = `spanmid' + 1
        quietly count if !missing(`Yvar') in `spanlo'/`spanhi'
        if r(N) != `span' {
            local miss = 1
        }
        else {
            quietly corrgram `Yvar' in `spanlo'/`spanhi', lag(1)
            quietly replace `NEWVAR' = el(r(AC),1,1) in `spanmid'
        }
    }
    if "`graph'" != "" {
        graph twoway spike `NEWVAR' `TIMEvar', yline(0) ///
        ytitle("First-order autocorrelations of `Yvar' (span `span')")
    }
    if `miss' == 1 {
        display as error "Caution:  missing values exist"
    }
    if "`generate'" != "" {
        rename `NEWVAR' `generate'
        label variable `generate' ///
        "First-order autocorrelations of `Yvar' (span `span')"
    }
}
end
```

The program could be refined further to make it more flexible, elegant, and user-friendly. Note the inclusion of comments stating the source and "version 2.0" in the first two lines, which both begin *! The comment refers to version 2.0 of *gossip.ado*, not Stata (an earlier version of *gossip.ado* appeared in a previous edition of this book). The Stata version suitable for this program is specified as 10.0 by the **version** command a few lines later. Although the ***!** comments do not affect how the program runs, they are visible to a **which** command:

```
. which gossip
c:\ado\personal\gossip.ado
*!  version 2.0  3jul2008
*!  L. Hamilton, Statistics with Stata (2009)
```

Once *gossip.ado* has been saved in the ado\personal directory, the command **gossip** could be used at any time. If we are following the steps in this chapter, which previously defined a preliminary version of **gossip**, then before running the new ado-file version we should drop the old definition from memory by typing

```
. program drop gossip
```

We are now prepared to run the final ado-file version. To see a graph of span-15 autocorrelations of variable *wNAO* from dataset *nao.dta*, for example, we would simply open *nao.dta* and type

```
. gossip wNAO year, span(15) graph
```

Help File

Help files are an integral aspect of using Stata. For a user-written program such as *gossip.ado*, they become even more important because no documentation exists in the printed manuals. We can write a help file for *gossip.ado* by using Stata's Do-file Editor to create a text file named *gossip.sthlp*. This help file should be saved in the same ado-file directory (for example, C:\ado\personal) as *gossip.ado*.

Any text file, saved in one of Stata's recognized ado-file directories with a name of the form *filename.sthlp*, will be displayed onscreen by Stata when we type **help** *filename*. For example, we might write the following in the Do-file Editor, and save it in directory C:\ado\personal as file *gossip.sthlp*. Typing **help gossip** at any time would then cause Stata to display the text.

```
gossip -- Moving first-order autocorrelations

gossip yvar timevar, span(#) [generate(newvar) graph]

Description

calculates first-order autocorrelations of time series
yvar, within a moving window of span #.  For example, if we
specify span(7) gen(new), then the first
through 3rd values of new are missing.  The 4th value of new
equals the lag-1 autocorrelation of yvar across observations 1
through 7.  The 5th value of new equals the lag-1 autocorrelation
of yvar across observations 2 through 8, and so forth.  The last
3 values of new are missing.  See Topliss (2002) for a rationale
and applications of this statistic to atmosphere-ocean data.
Statistics with Stata (2009) discusses the gossip program itself.
```

```
gossip requires tsset data.  timevar is the time
variable to be used for graphing.
```

```
Options
```

```
span(#) specifies the width of the window for
calculating autocorrelations.  This option is required; # should be
 an odd integer.
```

```
gen(newvar) creates a new variable holding the
autocorrelation coefficients.
```

```
graph requests a spike plot of lag-1 autocorrelations vs. timevar.
```

```
Examples
. gossip water month, span(13) graph
. gossip water month, span(9) gen(autowater)
. gossip water month, span(17) gen(autowater) graph
```

```
References
```

```
Hamilton, Lawrence C.  2009.
Statistics with Stata.  Belmont, CA:  Cengage.
```

```
Topliss, Brenda J.  2002. "Ocean-atmosphere feedback:  Using the
non-stationarity in the climate system."  Geophysical Research
Letters 29(8):1196.
```

Nicer help files containing links, text formatting, dialog boxes, and other features can be designed using Stata Markup and Control Language (SMCL). All official Stata help files, as well as log files and onscreen results, employ SMCL. A recommended standard outline for help files appears in the *User's Guide*.

The following is an SMCL version of the help file for **gossip**, roughly following the *User's Guide* outline. Once this file has been saved in ado\personal with the file name *gossip.sthlp*, typing **help gossip** will produce a readable and official-looking display.

```
{smcl}
{* *! version 2.0 3jul2008}{...}
{cmd:help gossip}
{hline}

{title:Title}
{phang}
{bf:gossip -- Moving first-order autocorrelations}

{title:Syntax}
{p 8 17 2}
{cmd:gossip} {it:yvar timevar} {cmd:,} {cmdab:sp:an}{cmd:(}
{it:#}{cmd:)} [ {cmdab:gen:erate}{cmd:(}{it:newvar}{cmd:)}
{cmdab:gr:aph} ]
```

{title:Description}

{pstd}
{cmd:gossip} calculates first-order autocorrelations of time series
{it:yvar}, within a moving window of span {it:#}. For example, if we
specify {cmd:span(}7{cmd:)} {cmd:gen(}{it:new}{cmd:)}, then the first
through 3rd values of {it:new} are missing. The 4th value of {it:new}
equals the lag-1 autocorrelation of {it:yvar} across observations 1
through 7. The 5th value of {it:new} equals the lag-1 autocorrelation
of {it:yvar} across observations 2 through 8, and so forth. The last
3 values of {it:new} are missing. See Topliss (2002) for a rationale
and applications of this statistic to atmosphere-ocean data.
{browse "http://www.stata.com/bookstore/sws.html":Statistics with Stata}
  (2009) discusses the {cmd:gossip} program itself.

{pstd}{cmd:gossip} requires {cmd:tsset} data. {it:timevar} is the time
variable to be used for graphing.

{title:Options}

{phang}
{cmd:span(}{it:#}{cmd:)} specifies the width of the window for
calculating autocorrelations. This option is required; {it:#} should be
 an odd integer.

{phang}
{cmd:gen(}{it:newvar}{cmd:)} creates a new variable holding the
autocorrelation coefficients.

{phang}
{cmd:graph} requests a spike plot of lag-1 autocorrelations vs. {it:timevar}.

{title:Examples}

{phang}
{cmd:. gossip water month, span(13) graph}

{phang}
{cmd:. gossip water month, span(9) gen(autowater)}

{phang}
{cmd:. gossip water month, span(17) gen(autowater) graph}

{title:References}

{pstd}
Hamilton, Lawrence C. 2009.
{browse "http://www.stata.com/bookstore/sws.html":Statistics with Stata}.
Belmont, CA: Cengage.

{pstd}
Topliss, Brenda J. 2002. "Ocean-atmosphere feedback: Using the
non-stationarity in the climate system." Geophysical Research
Letters 29(8):1196.

The help file begins with **{smcl}**, which tells Stata to process the file as SMCL. Curly brackets { } enclose SMCL codes, many of which have the form **{command:text}** or **{command arguments:text}**. The following examples illustrate how these codes are interpreted.

{cmd:help gossip}   Display the text "help gossip" as a command. That is, show "gossip" with whatever colors and font attributes are presently defined as appropriate for a command.

{hline}   Draw a horizontal line.

{hi:gossip}   Highlight the text "gossip".

{title:Syntax}   Display the text "Syntax" as a title.

{p 8 17 12}   Format the following text as a paragraph, with the first line indented 8 characters, subsequent lines indented 17, and the right margin brought in 12 characters.

{it:yvar}   Display the text "yvar" in italics.

{cmdab:sp:an}   Display "span" as a command, with the letters "sp" marked as the minimum abbreviation.

{phang}   Hanging indent, equivalent tp **{p 4 8 2}**

{browse "http://www.stata.com/bookstore/sws.html":Statistics...}
Link the text "Statistics with Stata" to the web address (URL) http://www.stata.com/bookstore/sws.html. Clicking on "Statistics with Stata" should then launch your browser and connect it to this URL.

The *Programming Manual* supplies details about using these and many other SMCL commands.

Example Program: Plot Survey Variables

Survey research produces datasets containing many categorical variables — sometimes 100 or more. As a first step in exploring such data, or in preparing a preliminary report, the analyst might simply construct tables showing percentage distributions for each variable. A command such as **tab1** *educ-enviro* will accomplish this for all variables in the dataset from *educ* to *enviro*. Stata provides no similarly easy way to draw and save bar charts for a variable list, however. As a second programming example, this section shows a makeshift program that was written to meet that particular need.

Chapter 15 introduced dataset *GSS_SwS1.dta*, a small subset of the General Social Survey for 2006. These data are used again here for illustration.

```
Contains data from c:\data\gss_sws1.dta
   obs:          1,595                          General Social Survey 2006
   vars:            17                          14 Jun 2008 13:45
   size:        60,610 (99.9% of memory free)
```

variable name	storage type	display format	value label	variable label
id	int	%9.0g		Case number (Statistics w/Stata)
finalwt	float	%9.0g		weight variable -- wtssall
cendiv	byte	%12.0g	cendiv	Census division (9)
age	byte	%8.0g	age	Age in years
educ	byte	%8.0g	educ	Highest year of schooling completed
gender	byte	%9.0g	gender	Respondent gender
income	float	%9.0g	income	Family income in constant dollars
loginc	float	%9.0g		log10(family income, +1)
logsize	float	%9.0g		log10(size of place in thousands, +1)
married	byte	%9.0g	married	Married or unmarried?
minority	byte	%11.0g	minority	Minority status
politics	byte	%20.0g	politics	Think of self as liberal or conservative?
bush	byte	%11.0g	bush	Voted for GW Bush in 2004?
grass	byte	%9.0g	grass	Should marijuana be made legal?
gunlaw	byte	%8.0g	gunlaw	Favor or oppose gun permits?
postlife	byte	%8.0g	postlife	Believe in life after death?
enviro	byte	%11.0g	enviro	Govt spending on environment?

In Chapter 15, we looked at predictors of voting in the 2004 elections. For this chapter, we turn instead to the last four items in the dataset: respondent opinions concerning the legalization of marijuana, gun permit laws, life after death, and government spending on the environment. Responses for the marijuana question *grass*, for example (and for this subset of the GSS), broke down as follows.

```
. tab grass, miss
```

Should marijuana be made legal?	Freq.	Percent	Cum.
not legal	399	25.02	25.02
legal	253	15.86	40.88
dk	58	3.64	44.51
NAP	885	55.49	100.00
Total	1,595	100.00	

Although the GSS interviewed more than 4,500 people in 2006, not all of them were asked every question. Among the 1,595 respondents in our subset of the GSS, 885 (55%) were not asked the marijuana question and hence coded as NAP. About 16% of the respondents thought marijuana should be made legal, 25% disagreed, and 4% said they did not know.

Program **multicat**, defined by the ado-file on the next page, builds upon another user-written program named **catplot**, discussed in Chapter 14. **catplot** can draw a variety of graphs showing the distribution of a categorical variable. **multicat** is more specialized, producing only horizontal bar charts for the percentage in each category; but this is a particularly useful format for presenting survey data. **multicat** adds the ability to handle a list of many variables, which

catplot and other Stata graphing commands cannot do. Thus, we could ask **multicat** to draw bar charts of all the variables in our dataset, saving each graph individually along the way. As written, the program saves graphs both in Stata (gph) format and in one other format (emf, eps, or pdf depending on the operating system), with file names based on the variable names. You could change any of these specifications by editing *multicat.ado*, tailoring the program to your own analytical needs.

```
*!  version 1.1  11jul2008
*!  L. Hamilton, Statistics with Stata (2009)
*   Requires catplot.ado installed.
program define multicat
version 10.0
    syntax varlist [if] [in] [aweight fweight iweight]  ///
    [, MISSing BY(varname) OVER(varname) ]
    if "`over'" != ""    {
    display as error "over() option not allowed with multicat;"
    display as error "use by() option or try catplot command instead."
    exit 198
    }
    marksample touse, strok novarlist
    if "`weight'" != "" local Weighted_ = "Weighted"
    if "`c(os)'"=="Windows" {
        local filetype "emf"
        }
    else if "`c(os)'"=="Unix" {
        local filetype "eps"
        }
    else if "`c(os)'"=="MacOSX" {
        local filetype = "pdf"
        }
    else {
        display as error "unknown operating system: `c(os)'"
        exit 799
        }
    capture {
        if "`by'" != ""  {
            foreach var of varlist `varlist' {
                local Vlab_: variable label `var'
                catplot hbar `var' [`weight' `exp'] if `touse',  ///
                  blabel(bar, format(%3.0f))  ///
                  percent(`by') ytitle("`Weighted_' Percent")  ///
                  `missing' by(`by', title("`Vlab_'"))
                graph save -`by'-`var'.gph, replace
                graph export -`by'-`var'.`filetype', replace
                }
            }
        else   {
            foreach var of varlist `varlist' {
                quietly tab `var' if `touse', `missing'
                local Nofobs_ = r(N)
                local Vlab_: variable label `var'
                catplot hbar `var' [`weight' `exp'] if `touse',  ///
                  blabel(bar, format(%3.0f))  ///
                  percent ytitle("`Weighted_' Percent, N = `Nofobs_'")  ///
                  title("`Vlab_'") `missing' `options'
                graph save Graph -`var'.gph, replace
                graph export -`var'.`filetype', replace
                }
```

```
        }
     }
   error _rc
end
```

The heart of **multicat** is its **syntax** statement, and then a **foreach** loop that repeatedly calls **catplot** for each variable in the variable list. Local macros pass information to the **catplot** command that actually draws the graphs. The command allows analytical weights (aweights) which have an effect here similar to that of probability weights (pweights) in a **svy: tab** command. It allows **in** or **if** qualifiers. Optionally we could include **missing** values, and use **by( )** but not **over( )**.

Once *multicat.ado* has been saved (for example, in C:\ado\personal), the command **multicat** becomes usable as if it were an ordinary (albeit, unfinished) feature of Stata. Figure 16.3 shows responses regarding legalization of marijuana. Both percentages and the number of observations appear in the plot.

```
. multicat grass, missing
```

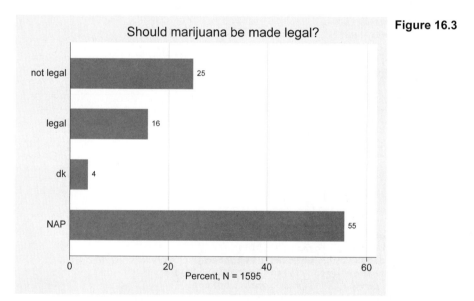

Figure 16.3

Survey data commonly are analyzed using probability weights, as discussed in Chapter 14. **multicat** (based on **catplot**) does not understand probability weights, but analytical weights (aweights) have the same effect here. Using survey weights, and focusing just on those respondents who were asked and expressed some opinion, we get a better view of opinions regarding marijuana: about 38% favored legalization, and 62% did not.

```
. svy: tab grass, percent
```

```
(running tabulate on estimation sample)

Number of strata   =        1        Number of obs    =        652
Number of PSUs     =      652        Population size  =   650.8525
                                     Design df        =        651
```

```
Should
marijuana
be made
legal?       percentages

not lega          62.16
    legal         37.84

   Total            100
```

Key: percentages = **cell percentages**

Figure 16.4 shows a **multicat** visualization of this table. Note that percentages in Figure 16.4 match the weighted values above, and the graphic again gives the relevant sample size. A set of many graphs such as this one, dropped into a document or slide show, could be read and annotated by the analyst for a quick presentation of results.

. **multicat** *grass* **[aw=***finalwt***]**

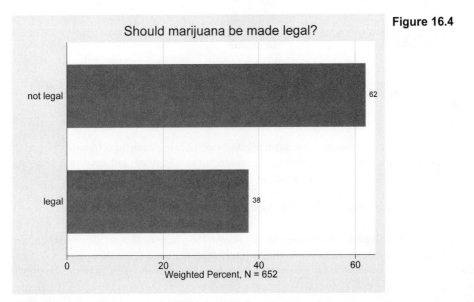

Figure 16.4

If we need only a few graphs, **catplot** or other existing **graph** commands provide much more robust flexibility than **multicat**. When we need many graphs, on the other hand, **multicat**'s variable-list capability becomes important. For example, we could draw separate graphs showing the weighted distributions of *grass*, *gunlaw*, *postlife*, and *enviro* in these data simply by typing the following command (results not shown).

. **multicat** *grass-enviro* **[aw=***finalwt***]**

With a larger dataset, it would be just as easy to draw a hundred such graphs, or to redraw them all as new data come in.

Survey researchers often want to compare several groups.   **multicat** allows graphical comparisons with the **by( )** option. For example, Figure 16.5 reveals that 45% of males, but only 31% of females, favor legalization of marijuana.

. **multicat** *grass* **[aw=***finalwt***], by(***gender***)**

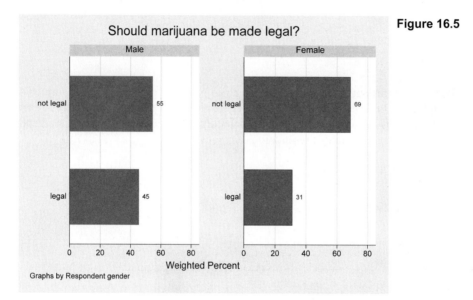

Figure 16.5

With a single command, we could draw similar graphs comparing male and female responses across many different variables. The following command would draw and save four separate graphs, for the variables *grass* through *enviro*.

. **multicat** *grass-enviro* **[aw=***finalwt***], by(***gender***)**

Figure 16.6 used a **graph combine** command (not shown) to place these four individual graphs into one image. We see that females were more likely than males to favor gun permits (83 vs. 71%), and to report a belief in life after death (89 vs. 80%). Women and men held similar views regarding government spending for environmental protection: 69% of women and 67% of men thought the government was spending too little. A **multicat** command could just as easily draw 100 graphs comparing survey responses by men and women, or 100 more comparing responses by education level, age group, political party, or any other categories of interest. Most analysts will never need this particular trick, but when specialized needs arise in a project, makeshift programs of this sort can become very handy.

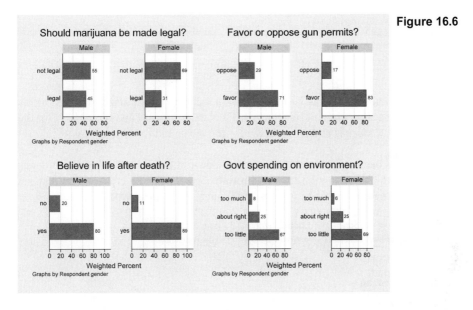

Figure 16.6

Monte Carlo Simulation

Monte Carlo simulations generate and analyze many samples of artificial data, allowing researchers to investigate the long-run behavior of their statistical techniques. The **simulate** command makes designing a simulation straightforward so that it only requires a small amount of additional programming. This section gives two examples.

To begin a simulation, we need to define a program that generates one sample of random data, analyzes it, and stores the results of interest in memory. Below, we see a file defining an r-class program (one capable of storing r( ) results) named **meanmedian**. This program randomly generates 100 values of variable *x* from a standard normal distribution. It next generates 100 values of variable *w* from a "contaminated normal" distribution: N(0,1) with probability .95, and N(0,10) with probability .05. Contaminated normal distributions have often been used in robustness studies to simulate variables that contain occasional wild errors. For both variables, **meanmedian** obtains means and medians.

```
* Creates a sample containing n=100 observations of variables x and w.
* x~N(0,1)                                   x is standard normal
* w~N(0,1) with p=.95, w~N(0,10) with p=.05   w is contaminated normal
* Calculates the mean and median of x and w.
* Stored results:   r(xmean)   r(xmedian)   r(wmean)   r(wmedian)
program meanmedian, rclass
    version 10.0
    drop _all
    set obs 100
    generate x = invnormal(uniform())
    summarize x, detail
    return scalar xmean = r(mean)
    return scalar xmedian = r(p50)
    generate w = invnormal(uniform())
```

```
    replace w = 10*w if uniform() < .05
    summarize w, detail
    return scalar wmean = r(mean)
    return scalar wmedian = r(p50)
end
```

Because we defined **meanmedian** as an r-class command, like **summarize**, it can store its results in r( ) scalars. **meanmedian** creates four such scalars: *r(xmean)* and *r(xmedian)* for the mean and median of *x*; *r(wmean)* and *r(wmedian)* for the mean and median of *w*.

Once **meanmedian** has been defined, whether through a do-file, ado-file, or typing commands interactively, we can call this program with a **simulate** command. To create a new dataset containing means and medians of *x* and *w* from 5,000 random samples, type

```
. simulate xmean = r(xmean) xmedian = r(xmedian) wmean = r(wmean)
     wmedian = r(wmedian), reps(5000): meanmedian

     command:  meanmedian
       xmean:  r(xmean)
     xmedian:  r(xmedian)
       wmean:  r(wmean)
     wmedian:  r(wmedian)

Simulations (5000)
```

This command creates new variables *xmean*, *xmedian*, *wmean*, and *wmedian*, based on the r( ) results from each iteration of **meanmedian**.

```
. describe
```

```
Contains data
  obs:          5,000                       simulate: central
  vars:             4                       17 Jun 2008 20:27
  size:       120,000 (99.8% of memory free)
```

variable name	storage type	display format	value label	variable label
xmean	float	%9.0g		r(xmean)
xmedian	float	%9.0g		r(xmedian)
wmean	float	%9.0g		r(wmean)
wmedian	float	%9.0g		r(wmedian)

```
Sorted by:
```

```
. summarize
```

Variable	Obs	Mean	Std. Dev.	Min	Max
xmean	5000	-.0001127	.1009648	-.3607487	.3920952
xmedian	5000	.0003337	.1255395	-.4114866	.5152998
wmean	5000	-.0013991	.2449764	-1.067707	1.176946
wmedian	5000	-.0005912	.1298139	-.4642514	.4808787

The means of these means and medians, across 5,000 samples, are all close to 0 — consistent with our expectation that the sample mean and median should both provide unbiased estimates of the true population means (0) for *x* and *w*. Also as theory predicts, the mean exhibits less sample-to-sample variation than the median when applied to a normally distributed variable (*x*). The standard deviation of *xmedian* is .126, noticeably larger than the standard deviation of *xmean* (.101). When applied to the non-normal, outlier-prone variable *w*, on the other hand, the opposite holds true: the standard deviation of *wmedian* is much lower than the standard deviation of *wmean* (.130 vs. .245). This Monte Carlo experiment demonstrates that the median remains a relatively stable measure of center despite wild outliers in the contaminated distribution, whereas the mean breaks down and varies much more from sample to sample. Figure 16.7 draws the comparison graphically, with box plots (and, incidentally, demonstrates how to control the shapes of box plot outlier-marker symbols).

```
. graph box xmean xmedian wmean wmedian, yline(0) legend(col(4))
     marker(1, msymbol(+)) marker(2, msymbol(Th))
     marker(3, msymbol(Oh)) marker(4, msymbol(Sh))
```

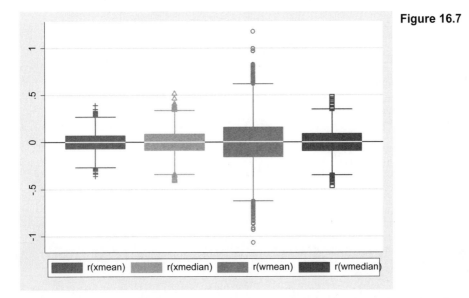

Figure 16.7

Our final example extends the inquiry to robust methods, bringing together several themes from this book. Program **regsim** generates 100 observations of *x* (standard normal) and two *y* variables. *y1* is a linear function of *x* plus standard normal errors. *y2* is also a linear function of *x*, but adding contaminated normal errors. These variables permit us to explore how various regression methods behave in the presence of normal and nonnormal, heavy-tailed error distributions. Four methods are employed: ordinary least squares (**regress**), robust regression (**rreg**), quantile regression (**qreg**), and quantile regression with bootstrapped standard errors (**bsqreg**, with 500 repetitions). Robust and quantile regression, introduced in Chapter 9, should prove more resistant to the effects of outliers. **regsim** applies each method to the regression of *y1* on *x* and then to the regression of *y2* on *x*. For this exercise, the program is defined by an ado-file, *regsim.ado*, saved in the ado\personal directory.

```
program regsim, rclass
* Performs one iteration of a Monte Carlo simulation comparing
* OLS regression (regress) with robust (rreg) and quantile
* (qreg and bsqreg) regression.  Generates one n = 100 sample
* with x ~ N(0,1) and y variables defined by the models:
*
*    MODEL 1:        y1 = 2x + e1        e1 ~ N(0,1)
*
*    MODEL 2:        y2 = 2x + e2        e2 ~ N(0,1) with p = .95
*                                        e2 ~ N(0,10) with p = .05
*
* Bootstrap standard errors for qreg involve 500 repetitions.
*
    version 10.0
    if "`1'" == "?" {
        global S_1 "b1 b1r se1r b1q se1q se1qb ///
           b2 b2r se2r b2q se2q se2qb"
        exit
    }
    drop _all
    set obs 100
    generate x = invnormal(uniform())
    generate e = invnormal(uniform())
    generate y1 = 2*x + e
    reg y1 x
        return scalar B1 = _b[x]
    rreg y1 x, iterate(25)
        return scalar B1R = _b[x]
        return scalar SE1R = _se[x]
    qreg y1 x
        return scalar B1Q = _b[x]
        return scalar SE1Q = _se[x]
    bsqreg y1 x, reps(500)
        return scalar SE1QB = _se[x]
    replace e = 10 * e if uniform() < .05
    generate y2 = 2*x + e
    reg y2 x
        return scalar B2 = _b[x]
    rreg y2 x, iterate(25)
        return scalar B2R = _b[x]
        return scalar SE2R = _se[x]
    qreg y2 x
        return scalar B2Q = _b[x]
        return scalar SE2Q = _se[x]
    bsqreg y2 x, reps(500)
        return scalar SE2QB = _se[x]
end
```

The r-class program stores coefficient or standard error estimates from eight regression analyses. These results are given names such as

       r(B1)   coefficient from OLS regression of *y1* on *x*
       r(B1R)   coefficient from robust regression of *y1* on *x*
       r(SE1R)   standard error of robust coefficient from model 1

and so forth. All the robust and quantile regressions involve multiple iterations: typically five to ten iterations for **rreg**, about five for **qreg**, and several thousand for **bsqreg** with its 500 bootstrap re-estimations of about five iterations each, *per sample.* Thus, a single execution of *regsim* demands more than 2,000 regressions. The following command calls for five repetitions, requiring more than 10,000 regressions.

```
. simulate b1 = r(B1)   b1r = r(B1R)   se1r = r(SE1R)
     b1q = r(B1Q)   se1q = r(SE1Q)   se1qb = r(SE1QB)   b2 = r(B2)
     b2r = r(B2R)   se2r = r(SE2R)   b2q = r(B2Q)   se2q = r(SE2Q)
     se2qb = r(SE2QB), reps(5):  regsim
```

You might want to run a small simulation like this as a trial to get a sense of the time required on your computer. For research purposes, however, we would need a much larger experiment. Dataset *regsim.dta* contains results from an overnight experiment involving 5,000 repetitions of **regsim** — more than 10 million regressions. The regression coefficients and standard error estimates produced by this experiment are summarized below.

```
. describe
```

```
Contains data from c:\data\regsim.dta
  obs:           5,000                         Monte Carlo estimates of b in
                                                 5000 samples of n=100
  vars:             12                         4 Jun 2008 11:06
  size:        280,000  (99.6% of memory free)
```

variable name	storage type	display format	value label	variable label
b1	float	%9.0g		OLS b (normal errors)
b1r	float	%9.0g		Robust b (normal errors)
se1r	float	%9.0g		Robust SE[b] (normal errors)
b1q	float	%9.0g		Quantile b (normal errors)
se1q	float	%9.0g		Quantile SE[b] (normal errors)
se1qb	float	%9.0g		Quantile bootstrap SE[b] (normal errors)
b2	float	%9.0g		OLS b (contaminated errors)
b2r	float	%9.0g		Robust b (contaminated errors)
se2r	float	%9.0g		Robust SE[b] (contaminated errors)
b2q	float	%9.0g		Quantile b (contaminated errors)
se2q	float	%9.0g		Quantile SE[b] (contaminated errors)
se2qb	float	%9.0g		Quantile bootstrap SE[b] (contaminated errors)

```
Sorted by:
```

```
. summarize
```

Variable	Obs	Mean	Std. Dev.	Min	Max
b1	5000	2.000828	.102018	1.631245	2.404814
b1r	5000	2.000989	.1052277	1.603106	2.391946
se1r	5000	.1041399	.0109429	.0693786	.1515421
b1q	5000	2.001135	.1309186	1.471802	2.536621
se1q	5000	.1262578	.0281738	.0532731	.2371508
se1qb	5000	.1362755	.032673	.0510808	.29979
b2	5000	2.006001	.2484688	.9001114	3.050552
b2r	5000	2.000399	.1092553	1.633241	2.411423
se2r	5000	.1081348	.0119274	.0743103	.1560973
b2q	5000	2.000701	.137111	1.471802	2.536621
se2q	5000	.1328431	.0299644	.0542015	.2594844
se2qb	5000	.1436366	.0346679	.0589409	.3006417

Figure 16.8 draws the distributions of coefficients as box plots. To make the plot more readable we use the **legend(symxsize(2) colgap(4))** options, which set the width of symbols and the gaps between columns within the legend at less than their default size. **help legend option** and **help relativesize** supply further information about these options.

```
. graph box b1 b1r b1q b2 b2r b2q,
      ytitle("Estimates of slope (b=2)") yline(2)
      legend(row(1) symxsize(2) colgap(4)
         label(1 "OLS 1") label(2 "robust 1") label(3 "quantile 1")
         label(4 "OLS 2") label(5 "robust 2") label(6 "quantile 2"))
```

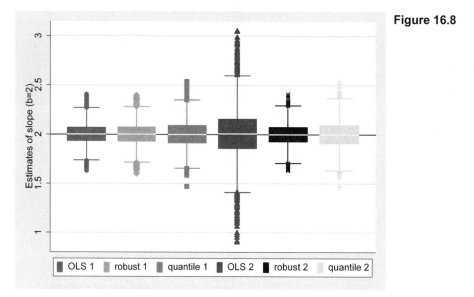

Figure 16.8

All three regression methods (OLS, robust, and quantile) produced mean coefficient estimates for both models that are not significantly different from the true value, $\beta = 2$. This can be confirmed through t tests such as

```
. ttest b2r = 2
```

One-sample t test

Variable	Obs	Mean	Std. Err.	Std. Dev.	[95% Conf. Interval]
b2r	5000	2.000399	.0015451	.1092553	1.99737 2.003428

mean = mean(b2r)		t = 0.2585
Ho: mean = 2		degrees of freedom = 4999

Ha: mean < 2	Ha: mean != 2	Ha: mean > 2
Pr(T < t) = 0.6020	Pr(|T| > |t|) = 0.7960	Pr(T > t) = 0.3980

All the regression methods thus yield unbiased estimates of β, but they differ in their sample-to-sample variation or efficiency. Applied to the normal-errors model 1, OLS proves the most efficient, as the famous Gauss–Markov theorem would lead us to expect. The observed standard deviation of OLS coefficients is .1020, compared with .1052 for robust regression and .1309 for quantile regression. Relative efficiency, expressing the OLS coefficient's observed variance as a percentage of another estimator's observed variance, provides a standard way to compare such statistics:

```
. quietly summarize b1
. scalar Varb1 = r(Var)
. quietly summarize b1r
. display 100*(Varb1/r(Var))
93.992612
. quietly summarize b1q
. display 100*(Varb1/r(Var))
60.722696
```

The calculations above use the r(Var) variance result from **summarize**. We first obtain the variance of the OLS estimates *b1*, and make this value scalar *Varb1*. Next the variances of the robust estimates *b1r*, and the quantile estimates *b1q*, are obtained and each is compared with *Varb1*. This reveals that robust regression was about 94% as efficient as OLS when applied to the normal-errors model — close to the large-sample efficiency of 95% that this robust method theoretically should have (Hamilton 1992a). Quantile regression, in contrast, achieves a relative efficiency of only 61% with the normal-errors model.

Similar calculations for the contaminated-errors model tell a different story. OLS was the best (most efficient) estimator with normal errors, but with contaminated errors it becomes the worst:

```
. quietly summarize b2
. scalar Varb2 = r(Var)
. quietly summarize b2r
. display 100*(Varb2/r(Var))
517.20057
. quietly summarize b2q
. display 100*(Varb2/r(Var))
328.3971
```

Outliers in the contaminated-errors model cause OLS coefficient estimates to vary wildly from sample to sample, as can be seen in the fourth box plot of Figure 16.8. The variance of these OLS coefficients is more than five times greater than the variance of the corresponding robust coefficients, and more than three times greater than that of quantile coefficients. Put another way, both robust and quantile regression prove to be much more stable than OLS in the presence of outliers, yielding correspondingly lower standard errors and narrower confidence intervals. Robust regression outperforms quantile regression with both the normal-errors and the contaminated-errors models.

Figure 16.9 illustrates the comparison between OLS and robust regression with a scatterplot showing 5,000 pairs of regression coefficients. The OLS coefficients (vertical axis) vary much more widely around the true value, 2.0, than **rreg** coefficients (horizontal axis) do.

```
. graph twoway scatter b2 b2r, msymbol(p) ylabel(1(.5)3, grid)
    yline(2) xlabel(1(.5)3, grid) xline(2)
```

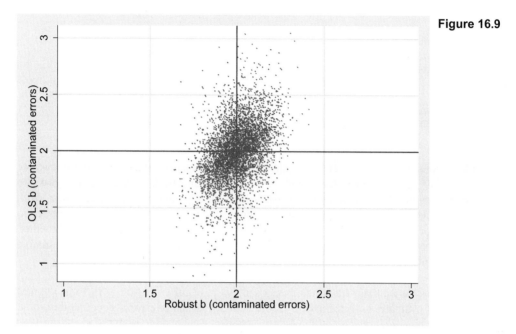

Figure 16.9

The experiment also provides information about the estimated standard errors under each method and model. Mean estimated standard errors differ from the observed standard deviations of coefficients. Discrepancies for the robust standard errors are small — less than 1%. For the theoretically-derived quantile standard errors the discrepancies appear a bit larger, between 3% and 4%. The least satisfactory estimates appear to be the bootstrapped quantile standard errors obtained by **bsqreg**. Means of the bootstrap standard errors exceed the observed standard deviation of $b1q$ and $b2q$ by 4% to 5%. Bootstrapping apparently over-estimated the sample-to-sample variation.

Monte Carlo simulation has become a key method in modern statistical research, and it plays a growing role in statistical teaching as well. These examples demonstrate some easy ways to get started.

Matrix Programming with Mata

With version 9, Stata added a matrix programming language called Mata. This rich topic, described in the two-volume *Mata Matrix Programming* manual, lies beyond the introductory scope of *Statistics with Stata*. It seems fitting, however, to conclude the book with a brief look at Mata. Its programming tools open paths for Stata's future development.

Rather than undertaking the large task of explaining Mata's concepts and features, we will proceed inductively and jump right to an example: writing a program that performs ordinary least squares (OLS) regression. The basic regression model is

$$\mathbf{y} = \mathbf{Xb} + \mathbf{u}$$

where $\mathbf{y}$ is an ($n{\times}1$) column vector of dependent-variable values, $\mathbf{X}$ an ($n{\times}k$) matrix containing values of (usually) $k{-}1$ predictor variables and a column of 1's, and $\mathbf{u}$ an ($n{\times}1$) vector of errors. $\mathbf{b}$ is a ($k{\times}1$) vector of regression coefficients, estimated as

$$\mathbf{b} = (\mathbf{X'X})^{-1}\mathbf{X'y}$$

This matrix calculation, familiar to generations of statistics students, provides a good entry point for seeing Mata at work.

Dataset *reactor.dta* contains information about the decommissioning costs of five nuclear power plants that were shut down over 1968–1982. This example has the pedagogical advantage that its small matrices could be written easily on blackboard or paper, if desired (e.g., Hamilton 1992a:340). In any event, it invites the question of how decommissioning costs might be related to reactor capacity and years of operation.

```
Contains data from C:\data\reactor.dta
  obs:             5                          Reactor decommissioning costs
                                                (from Brown et al. 1986)
  vars:            6                          15 Jun 2008 21:59
  size:          150 (99.9% of memory free)
```

variable name	storage type	display format	value label	variable label
site	str14	%14s		Reactor site
decom	byte	%8.0g		Decommissioning cost, millions
capacity	int	%8.0g		Generating capacity, megawatts
years	byte	%9.0g		Years in operation
start	int	%8.0g		Year operations started
close	int	%8.0g		Year operations closed

```
Sorted by:  start
```

Performing OLS regression with Stata is very easy, of course. We find that decommissioning costs among these five reactors increased by about .176 million dollars ($175,874) with each megawatt of generating capacity, and by about 3.9 million dollars with each year of operation. The two predictors explain almost 99% of the variance in decommissioning costs ($R^2_a = .9895$).

```
. regress decom capacity years
```

Source	SS	df	MS
Model	4666.16571	2	2333.08286
Residual	24.6342883	2	12.3171442
Total	4690.8	4	1172.7

```
Number of obs =        5
F(  2,     2) =  189.42
Prob > F      =  0.0053
R-squared     =  0.9947
Adj R-squared =  0.9895
Root MSE      =  3.5096
```

decom	Coef.	Std. Err.	t	P>|t|	[95% Conf. Interval]	
capacity	.1758739	.0247774	7.10	0.019	.0692653	.2824825
years	3.899314	.2643087	14.75	0.005	2.762085	5.036543
_cons	-11.39963	4.330311	-2.63	0.119	-30.03146	7.23219

The ado-file below defines program **ols0** using Mata commands. It simply calculates the vector of regression coefficients **b**. Mata commands start with **mata:** in this example. (Several other ways to use these commands interactively or in programs are described in the manuals.) The first two **mata:** commands define vector **y** and matrix **X** as "views" of the data in memory, specified by whatever left-hand-side (lhs) and right-hand-side (rhs) variables appeared in the **ols0** command line. A constant, 1, forms the last column of matrix **X**. **ols0** permits **in** or **if** qualifiers, or missing values. The estimating equation

$$\mathbf{b} = (\mathbf{X'X})^{-1}\mathbf{X'y}$$

is written in Mata as

```
mata: b = invsym(X'X)*X'y
```

The fourth **mata:** command displays the resulting contents of **b**.

```
*!  17jun2008
*!  L. Hamilton, Statistics with Stata (2009)
program ols0
    version 10
    syntax varlist(min=1 numeric) [in] [if]
    marksample touse
    gen cons_ = 1
    tokenize `varlist'
    local lhs "`1'"
    mac shift
    local rhs "`*'"
    mata: st_view(y=., ., "`lhs'", "`touse'")
    mata: st_view(X=., ., (tokens("`rhs'"), "cons_"), "`touse'")
    mata: b = invsym(X'X)*X'y
    mata: b
    drop cons_
end
```

Applied to the reactor decommissioning data, **ols0** obtains regression coefficients identical to those found earlier by **regress**.

```
. ols0 decom capacity years
```

	1
1	.1758738974
2	3.899313867
3	-11.39963279

Using Mata versions of the standard equations, program **ols1** (next page) adds the calculation of standard errors, *t* statistics, and *t* test probabilities. Again, the calculations lead to the same results we saw earlier with **regress**. Commas in the final **mata** statement of **ols1** are operators, meaning "join the columns of the following matrices."

```
*!  17jun2008
*!  L. Hamilton, Statistics with Stata (2009)
program ols1
    version 10
    syntax varlist(min=1 numeric) [in] [if]
    marksample touse
    gen cons_ = 1
    tokenize `varlist'
    local lhs "`1'"
    mac shift
    local rhs "`*'"
    mata: st_view(y=., ., "`lhs'", "`touse'")
    mata: st_view(X=., ., (tokens("`rhs'"), "cons_"), "`touse'")
    mata: b = invsym(X'X)*X'y
    mata: e = y - X*b
    mata: n = rows(X)
    mata: k = cols(X)
    mata: s2 = (e'e)/(n-k)
    mata: V = s2*invsym(X'X)
    mata: se = sqrt(diagonal(V))
    mata: (b, se, b:/se, 2*ttail(n-k, abs(b:/se)))
    drop cons_
end
```

```
. ols1 decom capacity years
```

	1	2	3	4
1	.1758738974	.0247774037	7.098156835	.0192756353
2	3.899313867	.26430873	14.75287581	.0045631637
3	-11.39963279	4.330310729	-2.632520735	.1190686843

We could expand this program to store results, and post them in a nicely-formatted output table similar to that of **regress**. Program **ols2** (next page) accomplishes something different, in order to demonstrate how Mata joins matrices together. It combines the numerical results seen above into a string matrix that also contains column headings and a list of independent-variable names. This happens through several additional **mata** commands. One defines row vector **vnames_** containing a list of variable names. The commas in this expression join three sets of columns: (1) the word "Yvar:" followed by the left-hand-side variable's name; (2) the names of all right-hand-side variables; and (3) the word "_cons".

```
    mata: vnames_ = "Yvar: `lhs'", tokens("`rhs'"), "_cons"
```

The next long **mata** command uses within-line comment delimiters, /* and */, so that Mata reads past the end of two physical lines and sees this as all one command:

```
    mata: vnames_', ("Coef." \ strofreal(b)),      /*
    */ ("Std. Err." \ strofreal(se)),          /*
    */ ("t" \ strofreal(t)), ("P>|t|" \ strofreal(Prt))
```

The command displays a matrix in which the first column is the transpose of **vnames_** (that is, a column of variable names). The column of variable names is joined, using a comma, to a second column vector created with the word "Coefs" as its first row; remaining rows are filled by the coefficients in **b** converted from real numbers to strings. The backslash operator "\"

joins rows to a matrix, just as "," joins columns. The real-to-string conversion of **b** values is necessary to make the matrix types compatible. Similar operations in **ols2** form labeled columns of standard errors, *t* statistics, and probabilities.

```
*!  17jun2008
*!  L. Hamilton, Statistics with Stata (2009)
program ols2
    version 10
    syntax varlist(min=1 numeric) [in] [if]
    marksample touse
    gen cons_ = 1
    tokenize `varlist'
    local lhs "`1'"
    mac shift
    local rhs "`*'"
    mata: st_view(y=., ., "`lhs'", "`touse'")
    mata: st_view(X=., ., (tokens("`rhs'"), "cons_"), "`touse'")
    mata: b = invsym(X'X)*X'y
    mata: e = y - X*b
    mata: n = rows(X)
    mata: k = cols(X)
    mata: s2 = (e'e)/(n-k)
    mata: V = s2*invsym(X'X)
    mata: se = sqrt(diagonal(V))
    mata: t = b:/se
    mata: Prt = 2*ttail(n-k, abs(b:/se))
    mata: vnames_ = "Yvar: `lhs'", tokens("`rhs'"), "_cons"
    mata: vnames_', ("Coef." \ strofreal(b)),   /*
        */ ("Std. Err." \ strofreal(se)),      /*
        */ ("t" \ strofreal(t)), ("P>|t|" \ strofreal(Prt))
    drop cons_
end
```

```
. ols2 decom capacity years
```

	1	2	3	4	5
1	Yvar: decom	Coef.	Std. Err.	t	P>|t|
2	capacity	.1758739	.0247774	7.098157	.0192756
3	years	3.899314	.2643087	14.75288	.0045632
4	_cons	-11.39963	4.330311	-2.632521	.1190687

These Mata exercises, like other examples in this chapter, give only a glimpse of Stata programming. The *Stata Journal* publishes more inspired applications, and each update of Stata involves new or improved ado-files. Online NetCourses provide a guided route to fluency in writing your own programs.

References

Barron's Educational Series. 1992. *Barron's Compact Guide to Colleges*, 8th ed. New York: Barron's Educational Series.

Beatty, J. Kelly, Brian O'Leary and Andrew Chaikin (eds.). 1981. *The New Solar System*. Cambridge, MA: Sky.

Belsley, D. A., E. Kuh and R. E. Welsch. 1980. *Regression Diagnostics: Identifying Influential Data and Sources of Collinearity*. New York: John Wiley & Sons.

Box, G. E. P., G. M. Jenkins and G. C. Reinsel. 1994. *Time Series Analysis: Forecasting and Control*. 3rd ed. Englewood Cliffs, NJ: Prentice–Hall.

Brown, Lester R., William U. Chandler, Christopher Flavin, Cynthia Pollock, Sandra Postel, Linda Starke and Edward C. Wolf. 1986. *State of the World 1986*. New York: W. W. Norton.

Buch, E. 2000. *Oceanographic Investigations off West Greenland 1999*. Copenhagen: Danish Meteorological Institute.

CDC (Centers for Disease Control). 2003. Website: http://www.cdc.gov

Chambers, John M., William S. Cleveland, Beat Kleiner and Paul A. Tukey (eds.). 1983. *Graphical Methods for Data Analysis*. Belmont, CA: Wadsworth.

Chatfield, C. 2004. *The Analysis of Time Series: An Introduction*, 6th edition. Boca Raton, FL: CRC.

Chatterjee, S., A. S. Hadi and B. Price. 2000. *Regression Analysis by Example*, 3rd edition. New York: John Wiley & Sons.

Cleveland, William S. 1994. *The Elements of Graphing Data*. Monterey, CA: Wadsworth.

Cleves, Mario, William Gould, Roberto Gutierrez and Yulia Marchenko. 2008. *An Introduction to Survival Analysis Using Stata*, revised edition. College Station, TX: Stata Press.

Cook, R. Dennis and Sanford Weisberg. 1982. *Residuals and Influence in Regression*. New York: Chapman & Hall.

Cook, R. Dennis and Sanford Weisberg. 1994. *An Introduction to Regression Graphics*. New York: John Wiley & Sons.

Council on Environmental Quality. 1988. *Environmental Quality 1987–1988*. Washington, DC: Council on Environmental Quality.

Davis, James A. Tom W. Smith, and Peter V. Marsden. 2005. General Social Surveys, 1972–2004 Cumulative File [computer data file]. Chicago: National Opinion Research Center [producer]. Ann Arbor, MI: Inter-University Consortium for Political and Social Research [distributor].

DFO (Canadian Department of Fisheries and Oceans). 2003. Website: http://www.meds-sdmm.dfo-mpo.gc.ca/alphapro/zmp/climate/IceCoverage_e.shtml

Diggle, P. J. 1990. *Time Series: A Biostatistical Introduction*. Oxford: Oxford University Press.

Enders, W. 2004. *Applied Econometric Time Series*, 2nd edition. New York: John Wiley & Sons.

Everitt, Brian S., Savine Landau and Morven Leese. 2001. *Cluster Analysis*, 4th edition. London: Arnold.

Federal, Provincial, and Territorial Advisory Commission on Population Health. 1996. *Report on the Health of Canadians*. Ottawa: Health Canada Communications.

Fox, John. 1991. *Regression Diagnostics*. Newbury Park, CA: Sage Publications.

Fox, John and J. Scott Long. 1990. *Modern Methods of Data Analysis*. Beverly Hills: Sage Publications.

Frigge, Michael, David C. Hoaglin and Boris Iglewicz. 1989. "Some implementations of the boxplot." *The American Statistician* 43(1):50–54.

Gould, William, Jeffrey Pitblado and William Sribney. 2006. *Maximum Likelihood Estimation with Stata*, 3rd edition. College Station, TX: Stata Press.

Haedrich, Richard L. and Lawrence C. Hamilton. 2000. "The fall and future of Newfoundland's cod fishery." *Society and Natural Resources* 13:359–372.

Hamilton, Dave C. 2003. "The Effects of Alcohol on Perceived Attractiveness." Senior Thesis. Claremont, CA: Claremont McKenna College.

Hamilton, James D. 1994. *Time Series Analysis*. Princeton, NJ: Princeton University Press.

Hamilton, Lawrence C. 1985a. "Concern about toxic wastes: Three demographic predictors." *Sociological Perspectives* 28(4):463–486.

Hamilton, Lawrence C. 1985b. "Who cares about water pollution? Opinions in a small-town crisis." *Sociological Inquiry* 55(2):170–181.

Hamilton, Lawrence C. 1992a. *Regression with Graphics: A Second Course in Applied Statistics*. Pacific Grove, CA: Brooks/Cole.

Hamilton, Lawrence C. 1992b. "Quartiles, outliers and normality: Some Monte Carlo results." Pp. 92–95 in Joseph Hilbe (ed.) *Stata Technical Bulletin Reprints, Volume 1*. College Station, TX: Stata Press.

Hamilton, Lawrence C. and Carole L. Seyfrit. 1993. "Town-village contrasts in Alaskan youth aspirations." *Arctic* 46(3):255–263.

Hamilton, Lawrence C., Benjamin C. Brown and Rasmus Ole Rasmussen. 2003. "Local dimensions of climatic change: West Greenland's cod-to-shrimp transition." *Arctic* 56(3):271–282.

Hamilton, Lawrence C., Richard L. Haedrich and Cynthia M. Duncan. 2004. "Above and below the water: Social/ecological transformation in northwest Newfoundland." *Population and Environment* 25(3):195–215.

Hamilton, Lawrence C. 2006a. "Rural voting in the 2004 election." Fact Sheet No. 2, The Carsey Institute, University of New Hampshire. http://www.carseyinstitute.unh.edu/documents/RuralVote_final.pdf (accessed 2/2/2008).

Hamilton, Lawrence C. 2006b. "Migration and Population in the Rural South." The Carsey Institute, University of New Hampshire. http://www.carseyinstitute.unh.edu/snapshot_south_migration.html (accessed 2/2/2008).

Hamilton, Lawrence C. 2007. "Climate, fishery and society interactions: Observations from the North Atlantic." *Deep Sea Research II* 54:2958–2969.

Hamilton, Lawrence C. 2008. "Who cares about polar regions? Results from a survey of U.S. public opinion." *Arctic, Antarctic, and Alpine Research*.

Hamilton, Lawrence C., Benjamin C. Brown and Barry D. Keim. 2007. "Ski areas, weather and climate: Time series models for New England case studies." *International Journal of Climatology* 27:2113–2124.

Hamilton, Lawrence C. and Rasmus Ole Rasmussen. 2008. "The demographic transition in Greenland." Paper presented at the Sixth International Congress of Arctic Social Sciences. Nuuk, Greenland, August 22–26.

Hardin, James and Joseph Hilbe. 2007. *Generalized Linear Models and Extensions*, 2nd edition. College Station, TX: Stata Press.

Hoaglin, David C., Frederick Mosteller and John W. Tukey (eds.). 1983. *Understanding Robust and Exploratory Data Analysis*. New York: John Wiley & Sons.

Hoaglin, David C., Frederick Mosteller and John W. Tukey (eds.). 1985. *Exploring Data Tables, Trends and Shape*. New York: John Wiley & Sons.

Hosmer, David W., Jr., Stanley Lemeshow and Susanne May. 2008. *Applied Survival Analysis: Regression Modeling of Time to Event Data*, 2nd edition. New York: John Wiley & Sons.

Hosmer, David W., Jr. and Stanley Lemeshow. 2000. *Applied Logistic Regression*, 2nd edition. New York: John Wiley & Sons.

Jentoft, Svein and Trond Kristoffersen. 1989. "Fishermen's co-management: The case of the Lofoten fishery." *Human Organization* 48(4):355–365.

Johnson, Anne M., Jane Wadsworth, Kaye Wellings, Sally Bradshaw and Julia Field. 1992. "Sexual lifestyles and HIV risk." *Nature* 360(3 December):410–412.

Korn, Edward L. and Barry I. Graubard. 1999. *Analysis of Health Surveys*. New York: Wiley.

League of Conservation Voters. 1990. *The 1990 National Environmental Scorecard*. Washington, DC: League of Conservation Voters.

Lee, Elisa T. 1992. *Statistical Methods for Survival Data Analysis*, 2nd edition. New York: John Wiley & Sons.

Lee, Eun Sul and Ronald N. Forthofer. 2006. *Analyzing Complex Survey Data*, second edition. Thousand Oaks, CA: Sage.

Levy, Paul S. and Stanley Lemeshow. 1999. *Sampling of Populations: Methods and Applications*, 3rd Edition. New York: Wiley.

Li, Guoying. 1985. "Robust regression." Pp. 281–343 in D. C. Hoaglin, F. Mosteller and J. W. Tukey (eds.) *Exploring Data Tables, Trends and Shape*. New York: John Wiley & Sons.

Long, J. Scott. 1997. *Regression Models for Categorical and Limited Dependent Variables*. Thousand Oaks, CA: Sage Publications.

Long, J. Scott and Jeremy Freese. 2006. *Regression Models for Categorical Dependent Variables Using Stata*, 2nd edition. College Station, TX: Stata Press.

Luke, Douglas A. 2004. *Multilevel Modeling*. Thousand Oaks, CA: Sage.

MacKenzie, Donald. 1990. *Inventing Accuracy: A Historical Sociology of Nuclear Missile Guidance*. Cambridge, MA: MIT.

Mallows, C. L. 1986. "Augmented partial residuals." *Technometrics* 28:313–319.

Mayewski, P. A., G. Holdsworth, M. J. Spencer, S. Whitlow, M. Twickler, M. C. Morrison, K. K. Ferland and L. D. Meeker. 1993. "Ice-core sulfate from three northern hemisphere sites: Source and temperature forcing implications." *Atmospheric Environment* 27A(17/18):2915–2919.

Mayewski, P. A., L. D. Meeker, S. Whitlow, M. S. Twickler, M. C. Morrison, P. Bloomfield, G. C. Bond, R. B. Alley, A. J. Gow, P. M. Grootes, D. A. Meese, M. Ram, K. C. Taylor and W. Wumkes. 1994. "Changes in atmospheric circulation and ocean ice cover over the North Atlantic during the last 41,000 years." *Science* 263:1747–1751.

McCullagh, D. W. Jr. and J. A. Nelder. 1989. *Generalized Linear Models*, 2nd edition. London: Chapman & Hall.

McCulloch, Charles E. 2005. "Repeated measures ANOVA, R.I.P.?" *Chance* 18(3):29–33.

McCulloch, Charles E. and S.R. Searle. 2001. *Generalized, Linear, and Mixed Models*. New York: Wiley.

Mitchell, Michael N. 2008. *A Visual Guide to Stata Graphics*, 2nd edition. College Station, TX: Stata Press.

Moore, David. 2008. *The Opinion Makers: An Insider Reveals the Truth about Opinion Polls*. Boston: Beacon Press.

Nash, James and Lawrence Schwartz. 1987. "Computers and the writing process." *Collegiate Microcomputer* 5(1):45–48.

National Center for Education Statistics. 1992. *Digest of Education Statistics 1992*. Washington, DC: U.S. Government Printing Office.

National Center for Education Statistics. 1993. *Digest of Education Statistics 1993*. Washington, DC: U.S. Government Printing Office.

Rabe–Hesketh, Sophia and Brian Everitt. 2000. *A Handbook of Statistical Analysis Using Stata*, 2nd edition. Boca Raton, FL: Chapman & Hall.

Rabe-Hesketh, Sophia and Anders Skrondal. 2008. *Multilevel and Longitudinal Modeling Using Stata*, 2nd edition. College Station, TX: Stata Press.

Report of the Presidential Commission on the Space Shuttle Challenger Accident. 1986. Washington, DC.

Robinson, Anthony. 2005. "Geovisualization of the 2004 presidential election." Available at http://www.personal.psu.edu/users/a/c/acr181/election.html (accessed 3/8/2008).

Rosner, Bernard. 1995. *Fundamentals of Biostatistics*, 4th edition. Belmont, CA: Duxbury Press.

Selvin, Steve. 1995. *Practical Biostatistical Methods*. Belmont, CA: Duxbury Press.

Selvin, Steve. 1996. *Statistical Analysis of Epidemiologic Data*, 2nd edition. New York: Oxford University.

Seyfrit, Carole L. 1993. *Hibernia's Generation: Social Impacts of Oil Development on Adolescents in Newfoundland*. St. John's: Institute of Social and Economic Research, Memorial University of Newfoundland.

Shumway, R. H. 1988. *Applied Statistical Time Series Analysis*. Upper Saddle River, NJ: Prentice–Hall.

Skrondal, A. and Sophia Rabe-Hesketh. 2004. *Generalized Latent Variable Modeling: Multilevel, Longitudinal, and Structural Equation Models*. Boca Raton, FL: Chapman & Hall/CRC.

StataCorp. 2007. *Getting Started with Stata for Macintosh*. College Station, TX: Stata Press.

StataCorp. 2007. *Getting Started with Stata for Unix*. College Station, TX: Stata Press.

StataCorp. 2007. *Getting Started with Stata for Windows*. College Station, TX: Stata Press.

StataCorp. 2007. *Mata Reference Manual*. College Station, TX: Stata Press.

StataCorp. 2007. *Stata Base Reference Manual* (3 volumes). College Station, TX: Stata Press.

StataCorp. 2007. *Stata Data Management Reference Manual*. College Station, TX: Stata Press.

StataCorp. 2007. *Stata Graphics Reference Manual*. College Station, TX: Stata Press.

StataCorp. 2007. *Stata Programming Reference Manual*. College Station, TX: Stata Press.

StataCorp. 2007. *Stata Longitudinal/Panel Data Reference Manual*. College Station, TX: Stata Press.

StataCorp. 2007. *Stata Multivariate Statistics Reference Manual*. College Station, TX: Stata Press.

StataCorp. 2007. *Stata Quick Reference and Index*. College Station, TX: Stata Press.

StataCorp. 2007. *Stata Survey Data Reference Manual*. College Station, TX: Stata Press.

StataCorp. 2007. *Stata Survival Analysis and Epidemiological Tables Reference Manual*. College Station, TX: Stata Press.

StataCorp. 2007. *Stata Time-Series Reference Manual*. College Station, TX: Stata Press.

StataCorp. 2007. *Stata User's Guide*. College Station, TX: Stata Press.

Topliss, Brenda J. 2002. "Ocean-atmosphere feedback: Using the non-stationarity in the climate system." *Geophysical Research Letters* 29(8):1196.

Tufte, Edward R. 1990. *Envisioning Information*. Cheshire CT: Graphics Press.

Tufte, Edward R. 1997. *Visual Explanations: Images and Quantities, Evidence and Narrative*. Cheshire CT: Graphics Press.

Tufte, Edward R. 2001. *The Visual Display of Quantitative Information*, 2nd edition Cheshire CT: Graphics Press.

Tukey, John W. 1977. *Exploratory Data Analysis*. Reading, MA: Addison–Wesley.

Velleman, Paul F. 1982. "Applied Nonlinear Smoothing," pp.141–177 in Samuel Leinhardt (ed.) *Sociological Methodology 1982*. San Francisco: Jossey-Bass.

Velleman, Paul F. and David C. Hoaglin. 1981. *Applications, Basics and Computing of Exploratory Data Analysis*. Boston: Wadsworth.

Verbeke, G. and G. Molenberghs. 2000. *Linear Mixed Models for Longitudinal Data*. New York: Springer.

Voss, Paul R., Scott McNiven, Roger B. Hammer, Kenneth M. Johnson, and Glenn V. Fuguitt. 2005. "County-specific Net Migration by Five-year Age Groups, Hispanic Origin, Race, and Sex, 1990–2000." Ann Arbor, MI: Inter-university Consortium for Political and Social Research, 2005-05-23.

Ward, Sally and Susan Ault. 1990. "AIDS knowledge, fear, and safe sex practices on campus." *Sociology and Social Research* 74(3):158–161.

Werner, Al. 1990. "Lichen growth rates for the northwest coast of Spitsbergen, Svalbard." *Arctic and Alpine Research* 22(2):129–140.

Wood, B.D. and A. Vedlitz. 2007. "Issue definition, information processing, and the politics of global warming." *American Journal of Political Science* 51(3):552–568.

World Bank. 1987. *World Development Report 1987*. New York: Oxford University.

World Resources Institute. 1993. *The 1993 Information Please Environmental Almanac*. Boston: Houghton Mifflin.

Index

insert
    graph into document, 6
    table into document, 4
insheet (read spreadsheet data), 46–47
instrumental variables (2SLS), 175
interaction effect
    ANOVA, 170–171
    regression, 196–199, 406–411
interquartile range (IQR), 36, 59, 98, 138–141, 150
iteratively reweighted least squares (IRLS), 256
iweight (importance weights), 61–62

J

jackknife
    residuals, 181
    standard errors, 332–335, 393, 411–412

K

Kaplan–Meier survivor function, 307, 313–316
Kendall's tau, 136, 146, 190
kernel density, 72, 92
Kruskal–Wallis test, 155–156, 165–166
kurtosis, 135, 138, 141–142

L

L-estimator, 257
label data, 19
label define, 28, 31
label values, 28, 31
label variable, 17, 19
ladder of powers, 142–145
lag (time series), 361–363, 372–374
lead (time series), 372–374
legend in graph, 85–86, 88–89, 90–91, 100–101
letter-value display, 140–141
leverage, 181, 211, 217–219, 224
leverage-vs.-squared-residuals plot, 211, 217–218
likelihood-ratio chi-squared. *See* chi-squared
line plot. *See* **graph twoway line**
link function (GLM), 308, 332–335
log file, 2, 7, 68, 70

logical operator, 22
logistic growth model, 230, 247
logistic regression, 279–303, 392, 405–408, 436–438
looping, 445–446
lowess smoothing, 92, 95–96, 229–230, 233–236
lroc (logistic ROC), 279
lrtest (likelihood-ratio test), 287–288, 294, 298–299
lsens (logistic sensitivity graph), 279

M

M-estimator, 257
macro, 442–443
Mann–Whitney U test, 155, 157, 162–163
marker label in graph, 217–218, 223, 291–292
marker symbol in graph, 72, 79–81
marksample, 449–450
mata (matrix programming), 472–476
matched-pairs test, 157, 160
mean, 4, 137–139
    robust mean, 254, 268, 270
median, 58, 73, 98, 135, 137–141, 269
median regression. *See* quantile regression
memory, 69–70
merge, 15, 49–52
missing value, 14–16, 18, 23–24, 29–32
mixed effects, 413–438
Monte Carlo, 465, 472
moving average, 361–369
multicollinearity, 191, 210, 224–228
multilevel modeling (mixed effects), 413–438
multinomial logistic regression, 277, 279, 295–303
multiple-comparison test
    correlation matrix, 187–188
    one-way ANOVA, 165

N

negative binomial regression, 308, 332–333, 392, 413
negative exponential growth model, 247
nolabel option, 30, 37–39
nonlinear regression, 229–230, 245–251
nonlinear smoothing, 363, 368–369